# Lecture Notes in Computer Science  14557

Founding Editors

Gerhard Goos
Juris Hartmanis

The series Lecture Notes in Computer Science (LNCS), including its subseries Lecture Notes in Artificial Intelligence (LNAI) and Lecture Notes in Bioinformatics (LNBI), has established itself as a medium for the publication of new developments in computer science and information technology research, teaching, and education.

LNCS enjoys close cooperation with the computer science R & D community, the series counts many renowned academics among its volume editors and paper authors, and collaborates with prestigious societies. Its mission is to serve this international community by providing an invaluable service, mainly focused on the publication of conference and workshop proceedings and postproceedings. LNCS commenced publication in 1973.

Stevan Rudinac · Alan Hanjalic · Cynthia Liem ·
Marcel Worring · Björn Þór Jónsson · Bei Liu ·
Yoko Yamakata
Editors

# MultiMedia Modeling

30th International Conference, MMM 2024
Amsterdam, The Netherlands, January 29 – February 2, 2024
Proceedings, Part IV

 Springer

*Editors*
Stevan Rudinac 🆔
University of Amsterdam
Amsterdam, The Netherlands

Alan Hanjalic 🆔
Delft University of Technology
Delft, The Netherlands

Cynthia Liem 🆔
Delft University of Technology
Delft, The Netherlands

Marcel Worring 🆔
University of Amsterdam
Amsterdam, The Netherlands

Björn Þór Jónsson 🆔
Reykjavik University
Reykjavik, Iceland

Bei Liu 🆔
Microsoft Research Lab – Asia
Beijing, China

Yoko Yamakata 🆔
The University of Tokyo
Tokyo, Japan

ISSN 0302-9743       ISSN 1611-3349 (electronic)
Lecture Notes in Computer Science
ISBN 978-3-031-53301-3       ISBN 978-3-031-53302-0 (eBook)
https://doi.org/10.1007/978-3-031-53302-0

This Springer imprint is published by the registered company Springer Nature Switzerland AG
The registered company address is: Gewerbestrasse 11, 6330 Cham, Switzerland

Paper in this product is recyclable.

# Preface

These four proceedings volumes contain the papers presented at MMM 2024, the International Conference on Multimedia Modeling. This 30th anniversary edition of the conference was held in Amsterdam, The Netherlands, from 29 January to 2 February 2024. The event showcased recent research developments in a broad spectrum of topics related to multimedia modelling, particularly: audio, image, video processing, coding and compression, multimodal analysis for retrieval applications, and multimedia fusion methods.

We received 297 regular, special session, Brave New Ideas, demonstration and Video Browser Showdown paper submissions. Out of 238 submitted regular papers, 27 were selected for oral and 86 for poster presentation through a double-blind review process in which, on average, each paper was judged by at least three program committee members and reviewers. In addition, the conference featured 23 special session papers, 2 Brave New Ideas and 8 demonstrations. The following four special sessions were part of the MMM 2024 program:

- FMM: Special Session on Foundation Models for Multimedia
- MDRE: Special Session on Multimedia Datasets for Repeatable Experimentation
- ICDAR: Special Session on Intelligent Cross-Data Analysis and Retrieval
- XR-MACCI: Special Session on eXtended Reality and Multimedia - Advancing Content Creation and Interaction

The program further included four inspiring keynote talks by Anna Vilanova from the Eindhoven University of Technology, Cees Snoek from the University of Amsterdam, Fleur Zeldenrust from the Radboud University and Ioannis Kompatsiaris from CERTH-ITI.

In addition, the annual MediaEval workshop was organised in conjunction with the conference. The attractive and high-quality program was completed by the Video Browser Showdown, an annual live video retrieval competition, in which 13 teams participated.

We would like to thank the members of the organizing committee, special session and VBS organisers, steering and technical program committee members, reviewers, keynote speakers and authors for making MMM 2024 a success.

December 2023

Stevan Rudinac
Alan Hanjalic
Cynthia Liem
Marcel Worring
Björn Þór Jónsson
Bei Liu
Yoko Yamakata

# Organization

## General Chairs

| | |
|---|---|
| Stevan Rudinac | University of Amsterdam, The Netherlands |
| Alan Hanjalic | Delft University of Technology, The Netherlands |
| Cynthia Liem | Delft University of Technology, The Netherlands |
| Marcel Worring | University of Amsterdam, The Netherlands |

## Technical Program Chairs

| | |
|---|---|
| Björn Þór Jónsson | Reykjavik University, Iceland |
| Bei Liu | Microsoft Research, China |
| Yoko Yamakata | University of Tokyo, Japan |

## Community Direction Chairs

| | |
|---|---|
| Lucia Vadicamo | ISTI-CNR, Italy |
| Ichiro Ide | Nagoya University, Japan |
| Vasileios Mezaris | Information Technologies Institute, Greece |

## Demo Chairs

| | |
|---|---|
| Liting Zhou | Dublin City University, Ireland |
| Binh Nguyen | University of Science, Vietnam National University Ho Chi Minh City, Vietnam |

## Web Chairs

| | |
|---|---|
| Nanne van Noord | University of Amsterdam, The Netherlands |
| Yen-Chia Hsu | University of Amsterdam, The Netherlands |

## Video Browser Showdown Organization Committee

| | |
|---|---|
| Klaus Schoeffmann | Klagenfurt University, Austria |
| Werner Bailer | Joanneum Research, Austria |
| Jakub Lokoc | Charles University in Prague, Czech Republic |
| Cathal Gurrin | Dublin City University, Ireland |
| Luca Rossetto | University of Zurich, Switzerland |

## MediaEval Liaison

| | |
|---|---|
| Martha Larson | Radboud University, The Netherlands |

## MMM Conference Liaison

| | |
|---|---|
| Cathal Gurrin | Dublin City University, Ireland |

## Local Arrangements

| | |
|---|---|
| Emily Gale | University of Amsterdam, The Netherlands |

## Steering Committee

| | |
|---|---|
| Phoebe Chen | La Trobe University, Australia |
| Tat-Seng Chua | National University of Singapore, Singapore |
| Kiyoharu Aizawa | University of Tokyo, Japan |
| Cathal Gurrin | Dublin City University, Ireland |
| Benoit Huet | Eurecom, France |
| Klaus Schoeffmann | Klagenfurt University, Austria |
| Richang Hong | Hefei University of Technology, China |
| Björn Þór Jónsson | Reykjavik University, Iceland |
| Guo-Jun Qi | University of Central Florida, USA |
| Wen-Huang Cheng | National Chiao Tung University, Taiwan |
| Peng Cui | Tsinghua University, China |
| Duc-Tien Dang-Nguyen | University of Bergen, Norway |

# Special Session Organizers

### FMM: Special Session on Foundation Models for Multimedia

| | |
|---|---|
| Xirong Li | Renmin University of China, China |
| Zhineng Chen | Fudan University, China |
| Xing Xu | University of Electronic Science and Technology of China, China |
| Symeon (Akis) Papadopoulos | Centre for Research and Technology Hellas, Greece |
| Jing Liu | Chinese Academy of Sciences, China |

### MDRE: Special Session on Multimedia Datasets for Repeatable Experimentation

| | |
|---|---|
| Klaus Schöffmann | Klagenfurt University, Austria |
| Björn Þór Jónsson | Reykjavik University, Iceland |
| Cathal Gurrin | Dublin City University, Ireland |
| Duc-Tien Dang-Nguyen | University of Bergen, Norway |
| Liting Zhou | Dublin City University, Ireland |

### ICDAR: Special Session on Intelligent Cross-Data Analysis and Retrieval

| | |
|---|---|
| Minh-Son Dao | National Institute of Information and Communications Technology, Japan |
| Michael Alexander Riegler | Simula Metropolitan Center for Digital Engineering, Norway |
| Duc-Tien Dang-Nguyen | University of Bergen, Norway |
| Binh Nguyen | University of Science, Vietnam National University Ho Chi Minh City, Vietnam |

### XR-MACCI: Special Session on eXtended Reality and Multimedia - Advancing Content Creation and Interaction

| | |
|---|---|
| Claudio Gennaro | Information Science and Technologies Institute, National Research Council, Italy |
| Sotiris Diplaris | Information Technologies Institute, Centre for Research and Technology Hellas, Greece |
| Stefanos Vrochidis | Information Technologies Institute, Centre for Research and Technology Hellas, Greece |
| Heiko Schuldt | University of Basel, Switzerland |
| Werner Bailer | Joanneum Research, Austria |

## Program Committee

| | |
|---|---|
| Alan Smeaton | Dublin City University, Ireland |
| Anh-Khoa Tran | National Institute of Information and Communications Technology, Japan |
| Chih-Wei Lin | Fujian Agriculture and Forestry University, China |
| Chutisant Kerdvibulvech | National Institute of Development Administration, Thailand |
| Cong-Thang Truong | Aizu University, Japan |
| Fan Zhang | Macau University of Science and Technology/Communication University of Zhejiang, China |
| Hilmil Pradana | Sepuluh Nopember Institute of Technology, Indonesia |
| Huy Quang Ung | KDDI Research, Inc., Japan |
| Jakub Lokoc | Charles University, Czech Republic |
| Jiyi Li | University of Yamanashi, Japan |
| Koichi Shinoda | Tokyo Institute of Technology, Japan |
| Konstantinos Ioannidis | Centre for Research & Technology Hellas/Information Technologies Institute, Greece |
| Kyoung-Sook Kim | National Institute of Advanced Industrial Science and Technology, Japan |
| Ladislav Peska | Charles University, Czech Republic |
| Li Yu | Huazhong University of Science and Technology, China |
| Linlin Shen | Shenzhen University, China |
| Luca Rossetto | University of Zurich, Switzerland |
| Maarten Michiel Sukel | University of Amsterdam, The Netherlands |
| Martin Winter | Joanneum Research, Austria |
| Naoko Nitta | Mukogawa Women's University, Japan |
| Naye Ji | Communication University of Zhejiang, China |
| Nhat-Minh Pham-Quang | Aimesoft JSC, Vietnam |
| Pierre-Etienne Martin | Max Planck Institute for Evolutionary Anthropology, Germany |
| Shaodong Li | Guangxi University, China |
| Sheng Li | National Institute of Information and Communications Technology, Japan |
| Stefanie Onsori-Wechtitsch | Joanneum Research, Austria |
| Takayuki Nakatsuka | National Institute of Advanced Industrial Science and Technology, Japan |
| Tao Peng | UT Southwestern Medical Center, USA |

| | |
|---|---|
| Thitirat Siriborvornratanakul | National Institute of Development Administration, Thailand |
| Vajira Thambawita | SimulaMet, Norway |
| Wei-Ta Chu | National Cheng Kung University, Taiwan |
| Wenbin Gan | National Institute of Information and Communications Technology, Japan |
| Xiangling Ding | Hunan University of Science and Technology, China |
| Xiao Luo | University of California, Los Angeles, USA |
| Xiaoshan Yang | Institute of Automation, Chinese Academy of Sciences, China |
| Xiaozhou Ye | AsiaInfo, China |
| Xu Wang | Shanghai Institute of Microsystem and Information Technology, China |
| Yasutomo Kawanishi | RIKEN, Japan |
| Yijia Zhang | Dalian Maritime University, China |
| Yuan Lin | Kristiania University College, Norway |
| Zhenhua Yu | Ningxia University, China |
| Weifeng Liu | China University of Petroleum, China |

## Additional Reviewers

Alberto Valese
Alexander Shvets
Ali Abdari
Bei Liu
Ben Liang
Benno Weck
Bo Wang
Bowen Wang
Carlo Bretti
Carlos Cancino-Chacón
Chen-Hsiu Huang
Chengjie Bai
Chenlin Zhao
Chenyang Lyu
Chi-Yu Chen
Chinmaya Laxmikant Kaundanya
Christos Koutlis
Chunyin Sheng
Dennis Hoppe
Dexu Yao
Die Yu

Dimitris Karageorgiou
Dong Zhang
Duy Dong Le
Evlampios Apostolidis
Fahong Wang
Fang Yang
Fanran Sun
Fazhi He
Feng Chen
Fengfa Li
Florian Spiess
Fuyang Yu
Gang Yang
Gopi Krishna Erabati
Graham Healy
Guangjie Yang
Guangrui Liu
Guangyu Gao
Guanming Liu
Guohua Lv
Guowei Wang

Gylfi Þór Guðmundsson

Hai Yang Zhang

Hannes Fassold

Hao Li

Hao-Yuan Ma

Haochen He

Haotian Wu

Haoyang Ma

Haozheng Zhang

Herng-Hua Chang

Honglei Zhang

Honglei Zheng

Hu Lu

Hua Chen

Hua Li Du

Huang Lipeng

Huanyu Mei

Huishan Yang

Ilias Koulalis

Ioannis Paraskevopoulos

Ioannis Sarridis

Javier Huertas-Tato

Jiacheng Zhang

Jiahuan Wang

Jianbo Xiong

Jiancheng Huang

Jiang Deng

Jiaqi Qiu

Jiashuang Zhou

Jiaxin Bai

Jiaxin Li

Jiayu Bao

Jie Lei

Jing Zhang

Jingjing Xie

Jixuan Hong

Jun Li

Jun Sang

Jun Wu

Jun-Cheng Chen

Juntao Huang

Junzhou Chen

Kai Wang

Kai-Uwe Barthel

Kang Yi

Kangkang Feng

Katashi Nagao

Kedi Qiu

Kha-Luan Pham

Khawla Ben Salah

Konstantin Schall

Konstantinos Apostolidis

Konstantinos Triaridis

Kun Zhang

Lantao Wang

Lei Wang

Li Yan

Liang Zhu

Ling Shengrong

Ling Xiao

Linyi Qian

Linzi Xing

Liting Zhou

Liu Junpeng

Liyun Xu

Loris Sauter

Lu Zhang

Luca Ciampi

Luca Rossetto

Luotao Zhang

Ly-Duyen Tran

Mario Taschwer

Marta Micheli

Masatoshi Hamanaka

Meiling Ning

Meng Jie Zhang

Meng Lin

Mengying Xu

Minh-Van Nguyen

Muyuan Liu

Naomi Ubina

Naushad Alam

Nicola Messina

Nima Yazdani

Omar Shahbaz Khan

Panagiotis Kasnesis

Pantid Chantangphol

Peide Zhu

Pingping Cai

Qian Cao

Qian Qiao
Qiang Chen
Qiulin Li
Qiuxian Li
Quoc-Huy Trinh
Rahel Arnold
Ralph Gasser
Ricardo Rios M. Do Carmo
Rim Afdhal
Ruichen Li
Ruilin Yao
Sahar Nasirihaghighi
Sanyi Zhang
Shahram Ghandeharizadeh
Shan Cao
Shaomin Xie
Shengbin Meng
Shengjia Zhang
Shihichi Ka
Shilong Yu
Shize Wang
Shuai Wang
Shuaiwei Wang
Shukai Liu
Shuo Wang
Shuxiang Song
Sizheng Guo
Song-Lu Chen
Songkang Dai
Songwei Pei
Stefanos Iordanis Papadopoulos
Stuart James
Su Chang Quan
Sze An Peter Tan
Takafumi Nakanishi
Tanya Koohpayeh Araghi
Tao Zhang
Theodor Clemens Wulff
Thu Nguyen
Tianxiang Zhao
Tianyou Chang
Tiaobo Ji
Ting Liu
Ting Peng
Tongwei Ma

Trung-Nghia Le
Ujjwal Sharma
Van-Tien Nguyen
Van-Tu Ninh
Vasilis Sitokonstantinou
Viet-Tham Huynh
Wang Sicheng
Wang Zhou
Wei Liu
Weilong Zhang
Wenjie Deng
Wenjie Wu
Wenjie Xing
Wenjun Gan
Wenlong Lu
Wenzhu Yang
Xi Xiao
Xiang Li
Xiangzheng Li
Xiaochen Yuan
Xiaohai Zhang
Xiaohui Liang
Xiaoming Mao
Xiaopei Hu
Xiaopeng Hu
Xiaoting Li
Xiaotong Bu
Xin Chen
Xin Dong
Xin Zhi
Xinyu Li
Xiran Zhang
Xitie Zhang
Xu Chen
Xuan-Nam Cao
Xueyang Qin
Xutong Cui
Xuyang Luo
Yan Gao
Yan Ke
Yanyan Jiao
Yao Zhang
Yaoqin Luo
Yehong Pan
Yi Jiang

Yi Rong
Yi Zhang
Yihang Zhou
Yinqing Cheng
Yinzhou Zhang
Yiru Zhang
Yizhi Luo
Yonghao Wan
Yongkang Ding
Yongliang Xu
Yosuke Tsuchiya
Youkai Wang
Yu Boan
Yuan Zhou
Yuanjian He
Yuanyuan Liu
Yuanyuan Xu
Yufeng Chen
Yuhang Yang
Yulong Wang

Yunzhou Jiang
Yuqi Li
Yuxuan Zhang
Zebin Li
Zhangziyi Zhou
Zhanjie Jin
Zhao Liu
Zhe Kong
Zhen Wang
Zheng Zhong
Zhengye Shen
Zhenlei Cui
Zhibin Zhang
Zhongjie Hu
Zhongliang Wang
Zijian Lin
Zimi Lv
Zituo Li
Zixuan Hong

# Contents – Part IV

**Demonstrations**

**Video Browser Showdown**

# FMM: Special Session on Foundation Models for Multimedia

# Removing Stray-Light for Wild-Field Fundus Image Fusion Based on Large Generative Models

Jun Wu[1(✉)], Mingxin He[1], Yang Liu[1], Jingjie Lin[1], Zeyu Huang[1], and Dayong Ding[2]

[1] School of Electronics and Information, Northwestern Polytechnical University, Xi'an 710072, China
junwu@nwpu.edu.cn, {hemingxin,2021202082,jingjielin}@mail.nwpu.edu.cn
[2] Vistel AI Lab, Visionary Intelligence Ltd., Beijing 100080, China
dayong.ding@vistel.cn

**Abstract.** In low-cost wide-field fundus cameras, the built-in lighting sources are prone to generate stray-light nearby, leading to low-quality image regions. To visualize retinal structures clearer, when fusing two images with complementary patterns of different lighting sources, the fused image might still have stray-light phenomenon near the hard fusing boundaries, i.e., typically in the diagonal directions. In this paper, an image enhancement algorithm based on generative adversarial network is proposed to eliminate the stray-light in wide-field fundus fusing images. First, a haze density estimation module is introduced to guide the model to pay attention to more serious stray-light regions. Second, a detail recovery module is introduced to reduce the loss of details caused by stray-light. Finally, a domain discriminator with unsupervised domain adaptation is employed to achieve better performance generalization on clinical data. Experiments show that our method obtains the best results on both public synthesized traditional fundus image dataset EyePACS-K and private wide-field fundus images dataset Retivue. Compared to the SOTA, the PSNR and Structural Similarity on average upon two above datasets are increased by 1.789 dB and 0.021 respectively.

**Keywords:** Wide Field Fundus Camera · Image Enhancement · Deep Learning · Generative Adversarial Network

## 1 Introduction

In typical wide field fundus cameras, there will be strong stray-light (fog-alike) phenomenon close to the built-in light sources, which makes it difficult to visualize high-quality retinal imaging in a single shoot. A low-cost solution is to make two lighting beam groups (left-right vs. top-bottom) to alternately light on, so as to shoot two complementary images, where the high-quality regions far from lighting are fused together [19]. It has satisfactory results when the stray-light

4        J. Wu et al.

is not serious, as group (a) of Fig. 1. Otherwise, the fused image will have significant stray-light on fusing boundary, as group (b) of Fig. 1. The stray-light is prone to cause problems such as contrast reduction and color deviation. So, it is necessary to eliminate stray-light defects as much as possible.

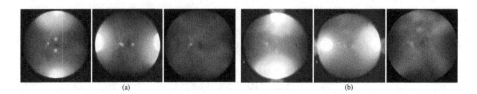

**Fig. 1.** Wide-field fundus images pairs and fusion images. From left to right, are input image A, input image B and the fused image.

Typical fundus image enhancement methods can be divided into two main branches: traditional image processing methods and deep-learning methods.

**Traditional Methods.** Restricted contrast adaptive histogram equalization has been widely used in retinal images to enhance contrast. CLAHE [15] is proposed for image segmentation and histogram equalization in each region separately, so that local information can be used to enhance the image to solve the problem in global viewpoint. The method of limiting contrast also solves the problem of background noise enhancement. On the other hand, based on the image formation model, Xiong et al. [20] applied the image generation model to estimate the background illuminance and transmission map by extracting the background and foreground. Then, the fuzzy fundus images are divided into high-intensity regions and low-intensity regions for further processing respectively. He et al. [7] estimated the transmittance of the scene by finding the minimum value in the dark channel of image, and then realized haze removal. In addition, the Retinex based method is also used to enhance retinal images. The low-frequency component of the fundus image contains local brightness, and the high-frequency component contains structural information such as blood vessels, lesions, optic cup and optic disc. Based on Retinex theory, Cao et al. [3] adopted Gaussian filtering to eliminate the influence of low-frequency signals, and then utilize grayscale adjustment to further enhance image contrast and adjust colors.

**Deep Learning Methods.** Cheng et al. [4] proposed a new semi-supervised contrast learning method based on importance guidance, I-SECRET, for fundus image enhancement. The framework consists of unsupervised component, supervised component and importance estimation component. The unsupervised part used unpaired high-quality and low-quality images by contrast constraints, while the supervised part takes advantage of paired images by reducing the quality of pre-selected high-quality images. Importance estimation provides a pixel-level importance map to guide unsupervised and supervised learning.

Shen et al. [16] modeled the common defects of fundus images, which can randomly degenerate high-quality images and obtain the corresponding degraded images of high-quality images. On the basis of regression data set, it is proposed that Cofe-Net network inhibits global degradation factors while preserving retinal anatomical structure and pathological features. But these methods are often multi-step, and the enhancement depends on their degradation algorithm. Yang et al. [21] proposed an end-to-end fundus image enhancement method. Periodic consistency constraints and unpaired training images are applied at the feature and image levels. In addition, a high frequency extractor is proposed to reduce artifacts effectively. Lee et al. [10] proposed a deep learning model designed to automatically enhance low-quality retinal fundus images with the goal of amplifying image quality and neutralizing multifaceted image degradation. However, a large number of paired high-quality and low-quality fundus images of the same patient are difficult to obtain. Yang et al. [22] proposed a self-augmenting image fog removal framework that removes fog by breaking the transmission graph into density and depth parts.

The generative adversarial network [6] was also applied to image enhancement. Pix2Pix [9] is used to establish a mapping relationship between the input image and the corresponding label image, and realize the image-to-image conversion task. CycleGAN [25] is an unsupervised image translation model that learns mapping from one domain to another, enabling images to be transformed between different domains. You et al. [24] proposed a retinal image enhancement method called CycleCBAM. The algorithm uses convolutional attention module to improve CycleGAN [25], and the enhanced results are beneficial to the classification of diabetic retinopathy. Li et al. [11] proposed a label-free end-to-end network to learn the clinical cataract image enhancement model ArcNet to apply cataract fundus image enhancement in clinical scenarios.

In contrast to common retinal image defects, we observed that stray-light in wide-field fundus images is influenced by camera internal lighting sources and is not uniformly distributed. Instead, it is particularly pronounced at fusing boundaries. Based on these characteristics, we develop a end-to-end framework for fundus image enhancement based on GANs to eliminate stray-light and improve image quality. Our main contributions are summarized as follows:

(1) A stray-light degradation model for fundus images is constructed to generated low-quality images with stray-light defects from high-quality images. Based on these pairs of images, we can apply model training to obtain an image enhancement model.
(2) A fundus image enhancement algorithm based on generative adversarial network is designed to remove stray-light, where a haze density estimation module is integrated to effectively perceive the stray-light distribution, and a detail repair module is introduced to restore and repair the image details.
(3) Perception loss and dark channel loss are employed to guide the training process to further improve the model performance.

## 2   Our Proposed Method: RSL-GAN

Our objective is to generate high-quality images by enhancing low-quality images with stray-light defects. The proposed approach consists of a stray-light image degradation model, a high-frequency extractor, a generator and two discriminators. Its workflow is shown in Fig. 2. The source domain data stream (in blue) is formed by simulating low-quality images from high-quality data, and the target domain data stream (in red) is formed by real-world low-quality wild-filed fundus images with stray-light. The source and target domains share the same generator. Since the paired data is synthetic and the to-be-enhanced images are real-world clinical images, a domain discriminator is introduced to realize domain self-adaptation through adversarial learning, so as to promote the model from the synthetic source image domain to the real-world target domain.

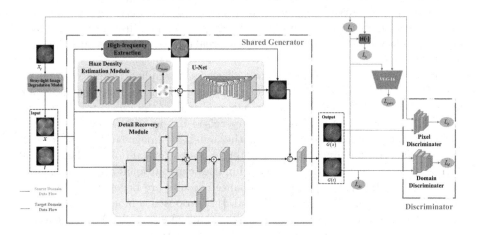

**Fig. 2.** Workflow of our proposed framework: RSL-GAN. (Color figure online)

### 2.1 Stray-Light Image Degradation Model

First, we design a stray-light image degradation model according to the distribution characteristics of stray-light on wide-field fundus fused images, which is used to simulate the stray-light image and generate low-high quality pairs.

Suppose $x = (i, j)$ represents a pixels of an input image $I$. $I(x)$ represents the real-world low-quality retinal image that needs to be recovered or enhanced.

In the natural scenes, the imaging model of an atomization images [7] is

$$I(x) = J(x)t(x) + A(1 - t(x)),  \tag{1}$$

where $J(x)$ represents the natural scene image with fog, $t(x)$ represents the system transmission rate, and $A$ is the global atmospheric light.

For the fundus images, the corresponding optical imaging model [13] is

$$I(x) = \alpha \cdot L \cdot r(x) \cdot t(x) + L \cdot (1 - t(x)),  \tag{2}$$

where $L$ is the fixed illumination of the camera (similar to $A$ in Eq. (1)), $r(x)$ is the effect of all the distortions associated with the optics of the eye and the fundus camera, as well as the slight curvature of the retinal plane in the retinal system, and $\alpha$ represents the attenuation coefficient of retinal illuminance.

For an ideal high-quality fundus image $s$, with the formula $s = L \cdot r$, model $1 - t$ with $J$, $J$ centered on $(a, b)$ is $J = \sqrt{(i - a)^2 + (j - b)^2}$, Eq. (2) will be

$$I(i, j) = \alpha \cdot s(i, j) + \beta \cdot J \cdot g_L(r_L, \sigma_L) \cdot (L - s(i, j)), \tag{3}$$

where $L$ takes the maximum value of pixels in high-quality image $s(i, j)$.

By changing the panel $J$ and limiting the position of generated stray-light to the diagonal, the stray-light-alike fusing boundary can be achieved. Define

$$mask(i, j) = \begin{cases} 255, & if \; -i + H - w \leq j \leq -i + H + w \quad or \quad i - w \leq j \leq i + w \\ 0, & otherwise. \end{cases}$$
$$\tag{4}$$

Then, the stray-light degradation model can be expressed as

$$I(i, j) = \alpha \cdot s(i, j) + \beta \cdot J \cdot mask(i, j) \cdot g_L(r_L, \sigma_L) \cdot (L - s(i, j)). \tag{5}$$

Accordingly, a high-quality image $s$ is degraded to a corresponding low-quality image $I$, which constitutes the low-high quality paired data in the source domain for training a image enhancement model, as illustrated in Fig. 3.

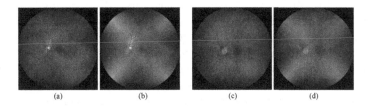

(a)            (b)            (c)            (d)

**Fig. 3.** Synthetic degraded fundus images. (a) High-quality image, (b) Degraded image of (a) with stray-light defects, (c) High-quality image, (d) Degraded image of (c).

## 2.2   High-Frequency Extractor

Incorrect enhancing components in fundus images may seriously distort pathological features. Therefore, we employ high-frequency extractor to preserve the structural features of the retina and guide the restoration process.

According to Retinex theory [23] and frequency separation method, image illumination and image details correspond to low-frequency and high-frequency component respectively. The low-frequency component can be extracted by Gaussian filter, and the high-frequency component of the image can be expressed as

$$H = I - G(I), \tag{6}$$

where $H$ represents the extracted high-frequency signal, $I$ represents the image to be extracted, and $G$ represents the Gaussian filter. Figure 4 shows the extraction results of high-frequency component of two fundus images, which can retain the retinal physiological structural information well.

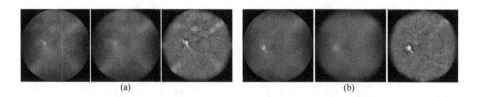

**Fig. 4.** Examples of high-frequency extraction. Images in groups (a) and (b), from left to right, are original image, low frequency and high frequency component.

## 2.3 Shared Generator

We take the U-net as the baseline of the generator, and further improve it by haze density estimation module and detail recovery module. The haze density estimation module effectively handles the problem of non-uniform distribution of stray-light in fundus images, while the detail restoration module enhances the fine details, thus improving overall image quality.

**Haze Density Estimation Module.** Inspired by [14], we utilize DehazeNet [2] to generate a transmission (density) map to guide the network to focus on the severity of stray-light at different locations. First, the convolution and Maxout is applied to extract fog-related features. The convolutional layer is carried out to extract features, followed by the nonlinear mapping and feature dimension reduction using pixel-level Maxout in $k$ channels of the input feature maps. Next, three parallel convolution blocks of different scales extract multi-scale features. Deep-level features help to capture global semantic information, while shallow-level features capture more detailed local details. Multi-scale convolution enables the network to utilize both global and local information to gain a deeper understanding of image content. Then, the features are maximized by pooling operations. By implementing local maximum and extremum operations, the robustness of feature space variation and the local consistency are enhanced. Finally, the range of transmittance is limited and local linearity is guaranteed by the BReLU function of double-sided rectifiers.

**Detail Recovery Module.** We introduce the detail repair module to generate a detail feature map, which aims to recover the inherent image details contained in stray-light images. We fuse these detail features into the coarse defogging image to obtain the defogging image with rich image details.

The module contains global branches and local branches, which are used to capture global features and local details of the image respectively. In order to extract more context information with a larger receptive field, the global branch consists of three smooth dilated convolution of different sizes, whose receptive fields are $1 \times 1$, $3 \times 3$, $5 \times 5$. By merging the features of three smooth dilated convolution, and then using $1 \times 1$ convolution to reduce dimensions, the channel information is integrated. To get more details, a convolution layer is applied as the local branch, and the local features are merged into the global branch by addition operations to better complement the local details. Finally, another convolutional layer is employed for further channel adjustment and information fusion. It can comprehensively consider the global and local information, get richer features, and obtain the final output feature maps.

## 2.4   Pixel and Domain Discriminators

The discriminators include a pixel discriminator and a domain discriminator.

*Pixel discriminator*: is used to evaluate whether the generated image is similar to the real high-quality image. During the training process, the generator constantly tries to generate more realistic enhanced images, while the pixel discriminator constantly tries to improve the accuracy of recognition.

*Domain discriminator*: Due to the gap between the source and target domains, a model trained on the source domain is likely to suffer a significant performance degradation on the target domain. Therefore, an unsupervised domain adaptive is carried out to reduce the domain gap between source and target domain, so that the model learned from source domain can perform well in target domain. Here, both pixel and domain discriminator use PatchGAN [9], which cut the input image into $n$ patches, then distinguish the authenticity of each patch, and finally output the average values.

## 2.5   Loss Function Design

$L_1$ **Loss:** In order to make the enhancement result close to the high-quality fundus image, the $L_1$ loss is calculated as $L_1 = ||x_t - G(x)||_1$, where $x_t$ is the ground truth, $x$ is the degraded image, and $G(x)$ is the enhancement result.

**Structure Consistency Loss:** The structural consistency loss $L_s$ utilizes $H()$ to extract the high-frequency components from both the real and reconstructed fundus images, as $L_s = ||H(x_t) - H(G(x))||_1$. It encourages the generator to focus more on preserving the structural information of fundus image, ensuring that the essential structural features of the real images are retained.

**Transmission Loss:** The haze density estimation module uses $L_2$ loss to constrain the training process, as $L_{trans} = ||t_{trans} - T||_2$. where $L_{trans}$ denotes the loss of the haze density estimation module, $t_{tran}$ is the haze density map, and $T$ represents the ground truth of the transmission map. This loss function enables the module to learn the distribution of stray-light.

**Pixel and Domain Adversarial Loss:** The counter loss of the pixel discriminator $D_p$ is

$$L_p = \mathbb{E}[logD_p(x_t)] + \mathbb{E}[log(1 - D_p(G(x)))].  \quad (7)$$

The counter loss of domain discriminator $D_d$ is

$$L_d = \mathbb{E}[logD_d(G(x))] + \mathbb{E}[log(1 - D_d(G(t)))].  \quad (8)$$

where $t$ is the true stray-light image. In the training phase, in order to deceive the discriminator, the generator will make the degraded stray-light image have the same distribution as the enhancement results of the real stray-light image.

**Dark Channel Loss:** He et al. [7] demonstrated that the dark channel prior can be used for haze removal. The dark channel can be represented as $J^{dark}(x) = \min_{y \in \Omega(x)} \left( \max_{c \in \{b,g,r\}} J^c(y) \right)$. where $x$ and $y$ are pixel coordinates in the image $J$, $J^{dark}$ is the dark channel, $J^c$ represents the color channel, and $\Omega(x)$ denotes a patch centered at $x$. Since in the dark channel of haze-free images, the gray scale value of all pixels is approximately zero, the dark channel loss function can be utilized to regularize the enhanced image. This regularization allows the enhanced image to exhibit the characteristics of a high-quality image.

The dark channel loss function can be defined as $L_{dc} = ||J^{dark}(G(t))||_1$. It can be utilized to regularize the enhanced image.

**Perceptual Loss:** The above mentioned loss function primarily compares images at the pixel level and is insufficient to recover all textures and fine details from degraded retinal images. To address this issue, it is defined as $L_{perc} = ||\phi(x_t) - \phi(G(x))||_2^2$, which incorporates features extracted from VGG-16 [17] to preserve the original image structure. By minimizing the differences between the enhanced and real images in the feature space, it can retain the semantic and content information.

**Final Overall Loss.** The final overall loss function for optimizing $G$ of the proposed method is defined as

$$L_G = \lambda_1 L_1 + \lambda_2 L_s + \lambda_3 L_{trans} + \lambda_4 L_{perc} + \lambda_5 L_{dc} + \lambda_6 L_d + \lambda_7 L_p. \qquad (9)$$

Due to challenges related to patient cooperation and the expertise required for capturing wide-field retinal images, it is difficult to obtain a large amount of data for training. To overcome this issue, in practice, we pre-train the model using traditional fundus images with similar features to wide-field retinal images. Subsequently, the model is fine-tuned using a wide-field fundus images dataset to achieve better performance.

## 3   Evaluation

### 3.1   Experimental Setup

**Data Sets: EyePACS-K**: A fundus image dataset, built from public datasets EyePACS [5] and Kaggle [11], contains 12,849 high-quality fundus images (12,549 from EyePACS and 300 from Kaggle) and 148 real-world fog-alike images (all from Kaggle). It is divided into a training set and a test set. The training set consists of 6,822 high-quality fundus images and 118 fog-alike images (6,940 in total). The test set consists of 6,027 high-quality fundus images and 30 real-world fog-alike images (6,057 in total). **Retivue**: This private wide-field fundus image dataset contains 26 fundus images. It is divided into a training set and a test set. The training set consisted of 9 high-quality images and 8 real-world wild-field fundus images with stray-light defects (17 in total). The test set consists of 3 high-quality images and 6 real-world fundus images (9 in total). For more details, please refer to Table 1.

**Table 1.** Data set of our experiment. The high-quality fundus images are considered as the *source domain*, and the real-world fog-alike images are as the *target domain*.

| Dataset | Total | Training Set | | | Test Set | | |
|---|---|---|---|---|---|---|---|
| | | Source | Target | Total | Source | Target | Total |
| EyePACS-K | 12,849 | 6,822 | 118 | 6,940 | 6,027 | 30 | 6,057 |
| Retivue | 26 | 9 | 8 | 17 | 3 | 6 | 9 |

**Parameter Settings:** The entire framework is implemented by PyTorch. Experiments are conducted under Ubuntu 16.04 with an NVIDIA GeForce GTX 1080Ti. The networks are trained for 200 epochs using Adam optimizer with a learning rate initialized to 0.0002 for the first 100 epochs and gradually decays to zero for the next 100 epochs. The original images are resized to $256 \times 256$, and the batch size is set to 4. Dropout is applied in U-Net to prevent over-fitting.

In the loss function, we set $\lambda_1 = \lambda_2 = 100$ for $L_1$ and $L_s$. The transmission and perception loss use weights $\lambda_3 = \lambda_4 = 50$. If $\lambda_4$ is too large, the model may pay more attention to image details and ignore the overall structure. The dark channel loss is less significant, we set $\lambda_5 = 10$. For too large $\lambda_5$, the model might pay more attention to the stray light region, leading to stronger boundary effect. Other two least significant parameters are set as $\lambda_6 = \lambda_7 = 1$.

### 3.2 Evaluation Metrics

- **Case 1**: Degraded stray-light images with ground truth (high-quality images): Peak Signal to Noise Ratio (PSNR) [8], Structural SIMilarity (SSIM) [18] and linear index of Fuzziness [1] (Fuzzy) are used to evaluate synthetic stray-light images. The PSNR and SSIM increase with the increasing similarity between the enhanced results and the ground truth. Fuzzy measures the degree of blur of an image, the lower the value, the clearer the image.
- **Case 2**: Real stray-light images without ground truth: Natural Image Quality Evaluator (NIQE) [12] is applied. The NIQE measures image quality by evaluating the degree of distortion of the image, and a low NIQE score indicates that the image quality has been effectively improved.

### 3.3   Results

**(1) Ablation Study.** To validate the effectiveness of different modules such as haze density estimation module (HDEM), detail restoration module (DRM) and loss functions, the ablation study is conducted, as shown in Table 2.

The haze density estimation module can guide the network to deal with the severity of stray-light in different locations. At the same time, the detail repair module can effectively extract the potential detail information of the image. In addition, the adopted dark channel losses ($L_{dc}$) and perceived losses ($L_{perc}$) contribute to the efficient optimization process of the model.

**Table 2.** The results of the ablation study of our proposed method RSL-GAN.

| HFE | HDEM | DRM | $L_{perc}$ | $L_{dc}$ | EyePACS-K [5,11] | | | Retivue (private) | | |
|---|---|---|---|---|---|---|---|---|---|---|
| | | | | | PSNR↑ | SSIM↑ | Fuzzy↓ | PSNR↑ | SSIM↑ | Fuzzy↓ |
| | | | | | 33.5122 | 0.8088 | 0.3952 | 32.2250 | 0.7968 | 0.2734 |
| ✓ | | | | | 34.0265 | 0.9319 | 0.3944 | 33.1855 | 0.9162 | 0.2546 |
| ✓ | ✓ | | | | 35.0259 | 0.9391 | 0.3979 | 33.8105 | 0.9162 | 0.2599 |
| ✓ | ✓ | ✓ | | | 34.9922 | 0.9509 | 0.3943 | 34.4254 | 0.9242 | 0.2474 |
| ✓ | ✓ | ✓ | ✓ | | 35.4342 | 0.9512 | 0.3863 | 34.2347 | 0.9341 | 0.2266 |
| ✓ | ✓ | ✓ | | ✓ | 35.5314 | 0.9522 | 0.3937 | 34.7747 | 0.9203 | 0.2303 |
| ✓ | ✓ | ✓ | ✓ | ✓ | **35.9077** | **0.9536** | **0.3836** | **34.8821** | **0.9368** | **0.2223** |

**(2) Comparison to the Existing Methods.** We compare our approach with two traditional image enhancement algorithms and three deep learning-based methods on degraded and real image datasets respectively.

**Results on Degraded Fundus Images.** Stray-light images are generated from real high-quality subset degradation to evaluate the effect of the model on the composite images. Table 3 shows PSNR ans SSIM for different methods on two datasets. Our proposed methods achieve the best performance.

Further, two groups of visual examples of fundus image are listed in Fig. 5. Neither CLAHE [15] nor DCP [7] can effectively handle non-uniform stray-light. While deep learning methods can deal with unevenly distributed stray-light better. The results of CycleGAN [24] and pix2pix [9] are prone to artifacts, reducing the clarity of blood vessels. D4 [22] is sharper and have better color retention, but it is less effective for removing severe regions of stray light. ArcNet [11] has a better influence on color of blood vessels while eliminating stray-light. With better contrast than ArcNet, it can be seen that our method can achieve the best performance for eliminating stray-light.

**Results on Real Images.** The real stray-light images subset of the dataset is used to evaluate the effect of the model on real stray-light images. Table 4 compares NIQE of different image enhancement algorithms on real stray-light images. Our proposed methods consistently achieve the best performance.

**Table 3.** Comparisons of different methods on degraded images, and original high-quality images are considered as ground truth.

| Methods | EyePACS-K [5,11] | | | Retivue (private) | | |
|---|---|---|---|---|---|---|
| | PSNR↑ | SSIM↑ | Fuzzy↓ | PSNR↑ | SSIM↑ | Fuzzy↓ |
| degraded image | 15.4373 | 0.7158 | 0.5638 | 19.3473 | 0.7963 | 0.3907 |
| CLAHE [15] | 14.7816 | 0.7021 | 0.5587 | 15.7068 | 0.6003 | 0.3996 |
| DCP [7] | 18.2043 | 0.8639 | 0.3927 | 20.8768 | 0.8868 | 0.2750 |
| CycleGAN [25] | 27.6675 | 0.7544 | 0.4321 | 29.2242 | 0.7088 | 0.2476 |
| pix2pix [9] | 32.6998 | 0.8461 | 0.4052 | 29.4857 | 0.8017 | 0.2892 |
| D4 [22] | 30.1012 | 0.9289 | 0.3898 | 32.6998 | 0.8461 | 0.3051 |
| ArcNet [11] | 34.0265 | 0.9319 | 0.3944 | 33.1855 | 0.9162 | 0.2546 |
| RSL-GAN(Ours) | **35.9077** | **0.9536** | **0.3836** | **34.8821** | **0.9368** | **0.2223** |

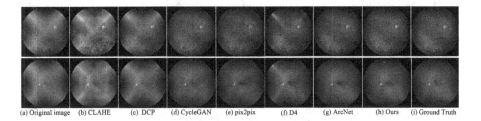

(a) Original image   (b) CLAHE   (c) DCP   (d) CycleGAN   (e) pix2pix   (f) D4   (g) ArcNet   (h) Ours   (i) Ground Truth

**Fig. 5.** Result of synthetic degraded fundus images.

Figure 6 visually compares enhanced results of different methods. The CLAHE [15] have obvious distortion, while DCP [7] can only deal with the atomization in uniform distribution effectively, and it is limited for uneven distribution of stray-light. CycleGAN [24] and pix2pix [9] perform well in mild cases of stray-light, but not satisfactory in severe cases. D4 [22] is effective at preserving image detail, but in areas with severe stray light, its removal effect is not complete. The ArcNet [11] can effectively eliminate the uneven stray-light, but it may have a significant impact on blood vessel color. However, our method can better deal with both issues. It can not only maintain a higher resolution, but also retain and restore the information of fundus blood vessels while removing stray-light.

Compared to the previous methods, our proposed method can not only eliminate stray light from the wild-filed fundus images, but also pay more attention to locations-sensitive severity of stray light. As a result, it is prone to recover more details from stray-light images, and make the image enhancement results more in line with the perceptual characteristics of human visual system. These improvements make our approach even better at improving image quality.

**Table 4.** Comparisons of different methods on real-world fog-alike images subset (similar to stray-light, but no high-quality images as the ground truth.) in terms of NIQE.

| Methods | EyePACS-K [5,11] | Retivue (private) |
|---|---|---|
| Degrade Images | 6.5904 | 5.2456 |
| CLAHE [15] | 5.5808 | 6.3588 |
| DCP [7] | 5.9279 | 5.5190 |
| CycleGAN [25] | 5.8887 | 5.0660 |
| pix2pix [9] | 5.3629 | 4.6307 |
| D4 [22] | 5.0678 | 4.6494 |
| ArcNet [11] | 5.5002 | 4.7621 |
| RSL-GAN(Ours) | **4.9667** | **4.5011** |

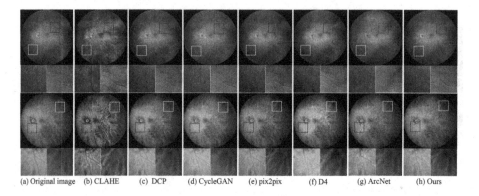

(a) Original image  (b) CLAHE  (c) DCP  (d) CycleGAN  (e) pix2pix  (f) D4  (g) ArcNet  (h) Ours

**Fig. 6.** Results comparison of different methods on real stray-light images subset.

## 4    Conclusions

We propose a GAN-based wide-field fundus image enhancement framework to remove straying-light, while preserving the lesion area and retinal structure. In addition, a degeneration model is introduced to generate low-quality fundus images to obtain sufficient low-high quality training image pairs. Experiments show that our proposed method can effectively eliminate the influence of stray-light and maintain the clarity of structural information such as blood vessels.

**Acknowledgement.** This work is supported by Xinjiang Production and Construction Corps Science and Technology Project, Science and Technology Development Program in Major Fields (2022AB021), and National High Level Hospital Clinical Research Funding (BJ-2022-120, BJ-2023-104).

# References

1. Bai, X., Zhou, F., Xue, B.: Image enhancement using multi scale image features extracted by top-hat transform. Opt. Laser Technol. **44**(2), 328–336 (2012)
2. Cai, B., Xu, X., Jia, K., Qing, C., Tao, D.: DehazeNet: an end-to-end system for single image haze removal. IEEE Trans. Image Process. **25**(11), 5187–5198 (2016)
3. Cao, L., Li, H., Zhang, Y.: Retinal image enhancement using low-pass filtering and $\alpha$-rooting. Signal Process. **170**, 107445 (2020)
4. Cheng, P., Lin, L., Huang, Y., Lyu, J., Tang, X.: I-SECRET: importance-guided fundus image enhancement via semi-supervised contrastive constraining. In: de Bruijne, M., et al. (eds.) MICCAI 2021. LNCS, vol. 12908, pp. 87–96. Springer, Cham (2021). https://doi.org/10.1007/978-3-030-87237-3_9
5. Fu, H., et al.: Evaluation of retinal image quality assessment networks in different color-spaces. In: Shen, D., et al. (eds.) MICCAI 2019. LNCS, vol. 11764, pp. 48–56. Springer, Cham (2019). https://doi.org/10.1007/978-3-030-32239-7_6
6. Goodfellow, I., et al.: Generative adversarial nets. In: Advances in Neural Information Processing Systems 27 (2014)
7. He, K., Sun, J., Tang, X.: Single image haze removal using dark channel prior. IEEE Trans. Pattern Anal. Mach. Intell. **33**(12), 2341–2353 (2010)
8. Hore, A., Ziou, D.: Image quality metrics: PSNR vs. SSIM. In: 20th International Conference on Pattern Recognition, pp. 2366–2369. IEEE (2010)
9. Isola, P., Zhu, J.Y., Zhou, T., Efros, A.A.: Image-to-image translation with conditional adversarial networks. In: Proceedings of the IEEE Conference on Computer Vision and Pattern Recognition, pp. 1125–1134 (2017)
10. Lee, K.G., Song, S.J., Lee, S., Yu, H.G., Kim, D.I., Lee, K.M.: A deep learning-based framework for retinal fundus image enhancement. PLoS ONE **18**(3), e0282416 (2023)
11. Li, H., et al.: An annotation-free restoration network for cataractous fundus images. IEEE Trans. Med. Imaging **41**(7), 1699–1710 (2022)
12. Mittal, A., Soundararajan, R., Bovik, A.C.: Making a ąřcompletely blindąś image quality analyzer. IEEE Signal Process. Lett. **20**(3), 209–212 (2012)
13. Peli, E., Peli, T.: Restoration of retinal images obtained through cataracts. IEEE Trans. Med. Imaging **8**(4), 401–406 (1989)
14. Qian, R., Tan, R.T., Yang, W., Su, J., Liu, J.: Attentive generative adversarial network for raindrop removal from a single image. In: Proceedings of the IEEE Conference on Computer Vision and Pattern Recognition, pp. 2482–2491 (2018)
15. Reza, A.M.: Realization of the contrast limited adaptive histogram equalization (CLAHE) for real-time image enhancement. J. VLSI Signal Process. Syst. Signal Image Video Technol. **38**, 35–44 (2004)
16. Shen, Z., Fu, H., Shen, J., Shao, L.: Modeling and enhancing low-quality retinal fundus images. IEEE Trans. Med. Imaging **40**(3), 996–1006 (2020)
17. Simonyan, K., Zisserman, A.: Very deep convolutional networks for large-scale image recognition. arXiv preprint arXiv:1409.1556 (2014)
18. Wang, Z., Bovik, A.C., Sheikh, H.R., Simoncelli, E.P.: Image quality assessment: from error visibility to structural similarity. IEEE Trans. Image Process. **13**(4), 600–612 (2004)
19. Wu, J., et al.: Template mask based image fusion built-in algorithm for wide field fundus cameras. In: Antony, B., Fu, H., Lee, C.S., MacGillivray, T., Xu, Y., Zheng, Y. (eds.) OMIA 2022. LNCS, vol. 13576, pp. 173–182. Springer, Cham (2022). https://doi.org/10.1007/978-3-031-16525-2_18

20. Xiong, L., Li, H., Xu, L.: An enhancement method for color retinal images based on image formation model. Comput. Methods Programs Biomed. **143**, 137–150 (2017)
21. Yang, B., Zhao, H., Cao, L., Liu, H., Wang, N., Li, H.: Retinal image enhancement with artifact reduction and structure retention. Pattern Recogn. **133**, 108968 (2023)
22. Yang, Y., Wang, C., Liu, R., Zhang, L., Guo, X., Tao, D.: Self-augmented unpaired image dehazing via density and depth decomposition. In: Proceedings of the IEEE/CVF Conference on Computer Vision and Pattern Recognition, pp. 2037–2046 (2022)
23. Yao, L., Lin, Y., Muhammad, S.: An improved multi-scale image enhancement method based on Retinex theory. J. Med. Imaging Health Inform. **8**(1), 122–126 (2018)
24. You, Q., Wan, C., Sun, J., Shen, J., Ye, H., Yu, Q.: Fundus image enhancement method based on CycleGAN. In: 41st Annual International Conference of the IEEE Engineering in Medicine and Biology Society (EMBC), pp. 4500–4503. IEEE (2019)
25. Zhu, J.Y., Park, T., Isola, P., Efros, A.A.: Unpaired image-to-image translation using cycle-consistent adversarial networks. In: Proceedings of the IEEE International Conference on Computer Vision, pp. 2223–2232 (2017)

# Training-Free Region Prediction with Stable Diffusion

Yuma Honbu and Keiji Yanai[✉]

The University of Electro-Communications, Tokyo, Japan
{honbu-y,yanai}@mm.inf.uec.ac.jp

**Abstract.** Semantic segmentation models require a large number of images with pixel-level annotations for training, which is a costly problem. In this study, we propose a method called StableSeg that infers region masks of any classes without needs of additional training by using an image synthesis foundation model, Stable Diffusion, pre-trained with five billion image-text pair data. We also propose StableSeg++, which uses the pseudo-masks generated by StableSeg to estimate the optimal weights of the attention maps, and can infer better region masks. We show the effectiveness of the proposed methods by the experiments on five datasets.

**Keywords:** foundation model · zero-shot segmentation · Stable Diffusion

## 1 Introduction

In recent years, developments in deep learning have significantly improved performance in semantic segmentation. However, standard semantic segmentation models require pixel-wise annotation data which is costly to obtain. To solve it, some methods to reduce annotation costs are proposed such as weakly supervised segmentation, in which segmentation models are trained using only image labels without pixel-level annotation, and zero-shot segmentation, in which segmentation models adapt unseen classes without additional training data on out-of-distribution data. However, these methods are limited to perform segmentation for the classes within the trained domain.

For these two years, vision-language foundation models which are pre-trained with a huge amount of image-text pair data have received much attention, since they enables us to handle arbitrary classes for various vision tasks. In fact, the method to apply a visual foundation model, CLIP [16], into semantic segmentation tasks has been proposed recently, which enabled training-free segmentation for any classes [25]. However, other visual foundation models have not been applied to semantic segmentation tasks so far. Particularly, the methods to apply text-to-image synthesis foundation models such as DALLE-2 [17], Imagen [19] and Stable Diffusion [18] into segmentation tasks have not been proposed so far.

In this study, we propose a method to apply a text-to-image synthesis foundation model, Stable Diffusion, into region prediction tasks, which needs no additional training. Since Stable Diffusion is pre-trained with five billion image-text pairs, our method, StableSeg, can estimate the region masks of arbitrary classes without training. Our

S. Rudinac et al. (Eds.): MMM 2024, LNCS 14557, pp. 17–31, 2024.
https://doi.org/10.1007/978-3-031-53302-0_2

method can make both training and annotation costs zero. Note that we call not "region segmentation" but "region prediction", since we assume that the category names of all the regions contained in a target image are given.

The main contributions of this paper are as follows:

- We have confirmed that cross-attention maps of a conditional input of the trained U-Net of Stable Diffusion model can be used for training-free region prediction.
- We propose the methods on combining self-attention with cross-attention and on prompt engineering to improve region prediction with Stable Diffusion.
- We propose StableSeg++ in which we estimate the optimal weights of cross-attention maps and train a supervised semantic segmentation model with pseudo-masks generated by StableSeg with the estimated map weights.
- We made extensive experiments to show the effectiveness of the proposed method and analyze cross-attention maps of Stable Diffusion's U-Net.

## 2   Related Work

### 2.1   Zero-Shot Segmentation

In general, zero-shot tasks do not use any training data for unseen target classes, although they need training data for seen source classes. In case of zero-shot segmentation, it is assumed that pixel-wise annotation for seen source classes which belong to the same domain as the unseen target classes is available, and segmentation masks of unseen classes are estimated by taking account of similarity of word-embeddings of class names between seen and unseen classes.

In recent years, large-scale vision-language foundation models have attracted much attention. Some of them are trained on the similarity of large data sets of image-text pairs, while others are trained as text-to-image synthesis models. The trend of using pre-trained vision-language foundation models to solve various tasks in the zero-shot manner has become popular in these two years.

Among them, CLIP [16] is the first large-scale visual language model the pre-trained model of which is open to the public. Although originally CLIP was proposed for zero-shot image classification, it can be also used for zero-shot segmentation. MaskCLIP [25] achieved per-pixel classification by removing a global average pooling layer from the CLIP image encoder without any additional training. In addition, MaskCLIP+ achieved high performance zero-shot segmentation by training DeepLabV2 [3] with the mask generated by MaskCLIP as pseudo-training masks.

The other CLIP-based method, called CLIPSeg [13], has been also proposed. The encoder of CLIPSeg uses the originally trained CLIP Visual Transformer. For the decoder, a Transformer-type architecture is proposed that incorporates the skip connection of the U-net structure, which finally outputs a segmentation mask of the binary segmentation of the text region. More than 340,000 phrases and their corresponding hand-crafted annotation masks, called the PhraseCut dataset [21], are provided to train this decoder. An extension of this dataset is used to train the decoder to build a model that can segment target regions from any texts. While these zero-shot methods train their segmentation modules using externally annotated data, our method uses self-supervised learning to achieve segmentation without the use of external data.

## 2.2    Open-Vocabulary Segmentation

Open vocabulary segmentation aims to segment object regions corresponding to arbitrary texts. ZS3Net [1] uses word embedding to train a segmentation module to segment object regions even for unseen classes. This method is based on a segmentation module that uses word embedding to segment object regions even in unseen classes. GroupVit [23] is a method to learn to group segmentation masks using textual features by adding group tokens to image tokens to achieve grouping of regions in the transformer structure.

In recent years, many methods [6, 11, 12] have been used to align text encoders with segmented regions using pre-trained CLIP. However, these methods require modules for extracting mask regions which are trained on large amounts of data corresponding to text and pixel level annotations. In contrast to this problem, our proposed method uses pseudo masks with pre-trained knowledge to train the segmentation modules, thus eliminating the use of external data.

## 2.3    Diffusion Model

In recent years, diffusion models (DMs) have achieved great success in image generation tasks: in the denoising diffusion probabilistic model (DDPM) proposed by Ho *et al.* [8], Gaussian noise is continuously added to the input image. The method then reconstructs the original image by estimating the noise at each step from the image using a neural network, which is repeated many times to gradually remove the noise. This method brought a rapid evolution of the diffusion model.

In particular, Stable Diffusion, which is pre-trained with five billion image-text pair dataset called LAION-5B [20], has attracted attention recently as an open foundation model for text-to-image synthesis, since it can produce high-quality images in line with a textual input. Stable Diffusion is based on the Latent Diffusion Model (LDM) [18] in which an input image is compressed into a latent vector by a Variational Autoencoder (VAE) [9] and then Gaussian noise is added. The model then removes the noise using the U-net architecture, to which various conditions can be added, and restores it to the image using a VAE decoder. In Stable Diffusion, LDM is conditioned on text embeddings encoded by the CLIP text encoder [16].

In the method called "prompt-to-prompt" proposed by Hertz *et al.* [7], the cross-attention maps of the attention layers used in the denoising U-Net of the diffusion model are used to control the spatial layout and the shape of the synthesized images as shown in the top row of Fig. 1. This allows various image manipulations by changing only text prompts while keeping the shape. If replacing words in a prompt is reflected in the generated image with the shape kept, this is achieved by injecting the attention map of the original image and overriding the attention map of the target. Adding words was achieved by injecting only the attention map corresponding to the unchanged part of the prompt. Inspired by this work, we found that the attention maps represented rough object shapes of the corresponding words as shown in the bottom row of Fig. 1, and expected that aggregated attention maps from several layers of the denoising U-Net could be transferred to semantic segmentation tasks.

Peekaboo by Burgert et al. [2] is a zero-shot segmentation method using Stable Diffusion. First, a learnable alpha mask is considered as the segmentation image, and this

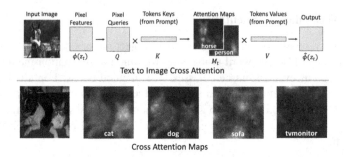

**Fig. 1.** Text to image cross attention in Transformer layer in Stable Diffusion.

mask is represented as an implicit function modeled by a neural network. Then, the alpha-synthesized image and the text prompts associated with the image regions to be segmented are iteratively optimized using a matching loss called dream loss. As a result of the optimization, the alpha mask converges to an optimal segmentation mask. In this model, there is no retraining of the diffusion model, only the neural network of the implicit function representation is trained. However, this method suffers from the problem that the optimization process is performed for each image, which takes about two minutes per image. Our proposed method, StabeSeg, is much faster and less expensive because it performs only one-step inference of the diffusion model and does not optimize another network.

## 3 Method

### 3.1 StableSeg

In our method, we focus on the Transformer blocks used in the U-net of Stable Diffusion, and we perform training-free region prediction using the cross-attention mechanism given a condition vector. There are several Transformer blocks in the U-net($\phi$) of Stable Diffusion. Each Transformer block has the self-attention between input features and the cross-attention with the conditional vector $v_t \in \mathbb{R}^{B \times C \times L}$, where $C$ is the number of classes to be detected, $L$ is the dimensions of the embedding encoded by the CLIP text encoder, and $B$ is the number of batches. Note that we assume the prompt text corresponding to a class name to be detected is provided to the CLIP text encoder one by one in the same as CLIP-based zero-shot image classification.

In the cross-attention block, the noise $z_T \in \mathbb{R}^{B \times C \times H/8 \times W/8}$ at time $T$ is recovered with multiple blocks $\phi$ in U-net. In the attention layer in each block, a linear transformation layer $l_Q$ is applied to $\phi(z_T)$ to obtain Query, $Q$, and Key, $K$, and Value, $V$, are calculated from a text embedding, $v_t$, via two linear transformation layers, $l_V$, and $l_K$. Then, the inner product of Key and Query is scaled by $\sqrt{d}$, and Softmax function is applied. We treat it as $\text{AttentionMap} \in \mathbb{R}^{B \times C \times H/2^i \times W/2^i}$. The product of this attention map and Value is used as the feature for the next layer. The $i$ varies depending on the extracted attention layer. This calculation is shown in the following equations, which is shown in the upper row of the Fig. 1 as well.

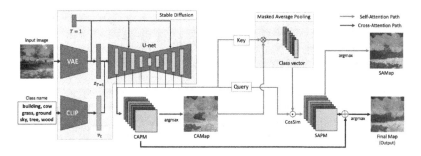

**Fig. 2.** The architecture of **StableSeg**.

$$Q = l_Q(\phi(z_T)), K = l_K(v_t), V = l_V(v_t) \qquad (1)$$

$$\text{AttentionMap} = \text{Softmax}(\frac{QK^T}{\sqrt{d}}) \qquad (2)$$

The U-net of the Stable Diffusion consists mainly of three blocks: Down block, Middle block, and Up block. Totally, it consists of 16 Cross/Self layers over these three blocks. Specifically, the Down block has two 1/8-scale ($64 \times 64$), two 1/16-scale ($32 \times 32$), and two 1/32-scale ($16 \times 16$) versions of the input image ($512 \times 512$), while the Up block has three 1/8-scale, three 1/16-scale, three 1/32-scale, and one 1/64-scale implemented in the Middle block. In the proposed method, we extract attention maps from all the Down/Middle/Up blocks and upsample them to the unified size as Cross-Attention Probability Maps (CAPM $\in \mathbb{R}^{B \times C \times H/8 \times W/8}$). Then, we obtain a Cross-Attention class Map (CAMap) by taking argmax over the class dimensions, $C$, on each pixel. CAMap can be regarded as a set of the estimated regions masks over all the classes. CAPM and CAMap are represented by the following equations and shown in the red lines of Fig. 2.

$$\text{CAPM} = (1/16) \sum_{i=1}^{16} \text{AttentionMap}_i \qquad (3)$$

$$\text{CAMap} = \underset{C}{\text{argmax}}(\text{CAPM}) \qquad (4)$$

Furthermore, we propose Self-Attention Refinement (SAR), which obtains Self-Attention Probability Map (SAPM) using Key and Query in the self-attention layer with CAMap, and refines the segmentation result by adding SAPM to CAPM as shown in the green lines in Fig. 2. At first, the Key features used in the self-attention layer are transformed into feature maps, and a average vector for each class is computed using Masked Average Pooling with Key features and each class mask in CAMap. This vector is the representative feature vector of each class, $v_c$. Next, the cosine similarity between the class vector and the self-attention Query features is calculated to create an Attention Map for each class. We call an aggregation of these attention maps as a Self-Attention Probability Map (SAPM $\in \mathbb{R}^{B \times C \times H \times W}$). This calculation is expressed by the following equations:

$$v_c = \text{MAP}(\text{Reshape}(Key_i), (\text{CAMap} == c)) \qquad (5)$$

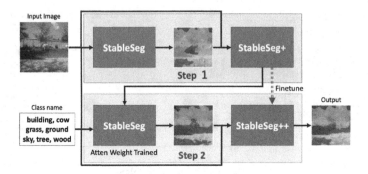

**Fig. 3.** The processing flow of **StableSeg++**.

$$\mathrm{SAPM}_c = (1/16) \sum\nolimits_{i=1}^{16} \mathrm{CosSim}(\mathrm{Reshape}(Query_i), v_c) \tag{6}$$

where $\mathrm{MAP}$ represents the Masked Average Pooling.

Finally, the final segmentation map (Final Map) is obtained by summing the CAPM and SAPM and calculating $\mathrm{argmax}$ as show in Fig. 2. This model is called **StableSeg**. By using the attention maps with only one noise removal step ($t = 1$), this processing takes only about 2 s per image.

The other problem with Stable Diffusion is that prompts are divided into tokens by the tokenizer, and separate attention maps are generated for each token. For example, "bedclothes" is divided into "bed" and "clothes". For this reason, StableSeg uses prompt engineering (p-eng). First, each token is converted into a feature using the text encoder, and the token feature is generated as a single feature by summing all the features one class name was divided into. In addition, we generate better text features by using "a photo of [prompt]", since the segmentation target is a real-world image.

## 3.2  StableSeg++

**StableSeg++** improves the accuracy of the StableSeg segmentation by using the pseudo-masks generated by StableSeg to find the optimal weights for the multiple attention maps in the StableSeg inference. The processing flow is shown in Fig. 3.

As shown in Step 1 of Fig. 3, first, a pseudo-mask is generated using StableSeg, and then this pseudo-mask is refined using the Fully Connected CRF (DenseCRF) [10]. Then DeepLabV3+ [4] is used to train the refined pseudo-masks as training data. We call this process "**StableSeg+**", the idea of which is the same as MaskCLIP+ [25].

In addition, next, as shown in Step 2 of Fig. 3, the mask inferred with StableSeg+ is used as training data to search for the optimal weights of attention maps of StableSeg. In the vanilla StableSeg, we simply summed them without weights as represented in Eq. (3), since we had no training data to estimate the optimal weights. Assuming that the output masks by StableSeg+ are pseudo-training data, we can estimate the optimal weights to maximize the match between the masks estimated by StableSeg+ and the pseudo-training masks estimates by StableSeg with weight optimization. Finally, DeepLabV3+ is trained again using the mask generated by StableSeg with the estimated weights of attention maps. We call this process "**StableSeg++**". In this way, the accuracy of the segmentation model can be improved without using any external annotation data by the step training.

## 4 Experiments

### 4.1 Experimental Setup

**StableSeg** uses the open Stable Diffusion model, which has been trained on five billion text-image pairs, and uses only $t = 1$ for the time embedding to reproduce the last

**Table 1.** Quantitative evaluation on six datasets (mIoU).

|                  | PAS-20 | PC-59 | A-150 | FoodPix | FoodSeg |
|------------------|--------|-------|-------|---------|---------|
| StableSeg (ours) | **50.3** | **36.2** | 23.6 | **63.3** | **49.1** |
| MaskCLIP [25]    | 44.7   | 25.5  | **26.0** | 33.2 | 37.0 |

**Table 2.** Ablation studies on SAR (mIoU).

|           | PAS-20 | PC-59 | A-150 | FoodPix | FoodSeg |
|-----------|--------|-------|-------|---------|---------|
| w/ SAR    | **50.3** | **36.2** | **23.6** | 63.3 | 49.1 |
| w/o SAR   | 47.2   | 31.0  | 19.4  | 53.8 | 39.3 |
| only Self | 47.1   | 33.6  | 22.5  | **65.4** | **50.0** |

denoising process in Stable Diffusion. The processing time for a single image is less than 2 s because we use the attention extracted during the denoising process only for $t = 1$. No noise is mixed into an input image, and the VAE encoder is used to compress it into latent space. Note that the size of an input image is $512 \times 512$, and the size of attention maps are $8 \times 8, 16 \times 16, 32 \times 32$ and $64 \times 64$. Because we require an accurate CAMap to estimate SAPM, we use the cross-attention maps with the size of 8 and 16 to estimate CAPM in the experiments, unless otherwise specified. For Self-Attention, we used Query, Key features of the self-attention maps with all the size to estimate SAPM in the experiments. As a baseline method, we used MaskCLIP [25], which was based on a vision-language foundation model, CLIP, and needed no training. As a publicly available pre-trained CLIP model for MaskCLIP, we used the CLIP-ViT-B/16 model. Note that we assume all the class names (including "background") of the regions contained in each input image are known in the experiments. Therefore, the experiments here is region predictions rather than semantic segmentation.

We used five segmentation datasets; the Pascal VOC (PAS-20) [5], consisting of 20 classes of general objects, Pascal Context (PC-59) [14], which consists of 59 classes of generic objects, ADE20K (A-150) [24], which is annotated with a total of 150 classes in three categories of objects, object parts, and stuff in pixel units, UECFoodPix Complete (FoodPix) [15], which consists of 100 classes of food classes, and FoodSeg103 (FoodSeg) [22], which consists of 103 food classes. Although all the datasets contain pixel-wise annotations for all the images, we used them for only evaluation. For evaluation, we used a mean Intersection over Union (mIoU) metric.

## 4.2 Comparison with the Baseline on Five Datasets

Table 1 shows the results compared to MaskCLIP [25] on each dataset. StableSeg outperformed MaskCLIP regarding PAS-20, PC-59, FoodPix and FoodSeg datasets, while StableSeg were inferior to MaskCLIP regarding A-150. From these results, StableSeg achieved much higher accuracy on food image datasets which was fine-grained datasets. One of the reason we expect is that this might come from the difference of the size of the trained datasets. In fact, Stable Diffusion was trained with 4 billion text-image pairs, while CLIP was trained with 400 million text-image pairs.

Figure 4 shows some results on the Pascal VOC dataset (PAS-20). In case of MaskCLIP, although the position of the object was captured, it can be seen that there was a lot of noise and many false areas. On the other hand, StableSeg was able to segment the target regions with less noise than MaskCLIP. However, MaskCLIP was able to segment the detailed parts better than StableSeg, suggesting that each method has its advantages and disadvantages for general object datasets.

For the fine-grained food dataset, FoodSeg, as shown in Fig. 5, StableSeg produced higher quality region masks than MaskCLIP. In MaskCLIP, unlike the general object datasets such as PAS-20, there were some food classes that cannot be segmented. StableSeg, on the other hand, was able to segment the target regions appropriately for most of the classes. This is because StableSeg has a higher diversity of training data than CLIP [16] in Stable Diffuison, suggesting that StableSeg also shows strong generalization performance in the food domain.

**Table 3.** Ablation studies on prompt engineering (p-eng).

| p-eng | "a photo of a" | PAS-20 |
|-------|----------------|--------|
| ✓ | ✓ | **50.3** |
| ✓ |  | 43.4 |
|  | ✓ | 46.8 |
|  |  | 46.0 |

**Table 4.** Ablation studies on size of Self/Cross-Attention Maps.

| | PAS-20 | | | PC-59 | | |
|-------|------|------|------|------|------|------|
| cross | self | | | | | |
| | 16 | 32 | 64 | 16 | 32 | 64 |
| 16 | 49.1 | 50.1 | 50.3 | 35.1 | 35.9 | **36.2** |
| 32 | 50.0 | 51.1 | 51.3 | 34.4 | 35.2 | 35.5 |
| 64 | 50.3 | 51.3 | **51.4** | 33.2 | 33.9 | 33.9 |
| | A-150 | | | | | |
| 16 | 23.0 | 23.5 | **23.6** | | | |
| 32 | 22.5 | 22.9 | 22.9 | | | |
| 64 | 21.8 | 22.1 | 22.0 | | | |
| | FoodPix | | | FoodSeg | | |
| 16 | 62.3 | 63.2 | 63.3 | 46.5 | 48.3 | **49.1** |
| 32 | 63.0 | 63.9 | 64.0 | 46.0 | 47.9 | 48.8 |
| 64 | 63.7 | 64.5 | **64.5** | 45.3 | 47.1 | 47.7 |

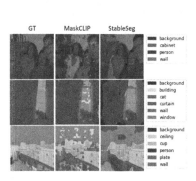

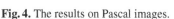

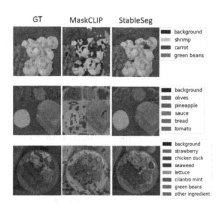

**Fig. 4.** The results on Pascal images.          **Fig. 5.** The results on food images.

### 4.3    Qualitative Results in Diverse Classes

Figure 6 shows the result of segmenting user-specified classes with StableSeg. Stable Diffusion learned the relationships between five billion image-text pairs to generate targeted attention maps for each word. Appropriate regions are segmented for proper nouns such as "Oculus" and "Batman". We found that regions corresponding to the pairs of nouns and adjectives such as "red car" and "blue car" were successfully segmented . We think that StableSeg is helpful to predict the regions corresponding to the special classes such as proper nouns and combined words of adjectives and nouns.

### 4.4    Self-attention Refinement and Prompt-Engineering

Table 2 shows ablation studies on SAR, which indicates that refining region masks using Self-Attention maps improved the region masks generated by only Cross-Attention maps. The reason is that Self-Attention is trained to specialize on similarities between

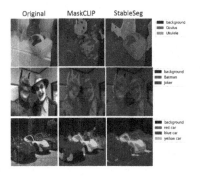

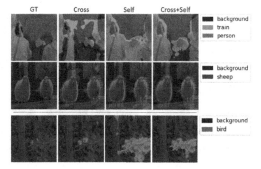

**Fig. 6.** The results for various classes.     **Fig. 7.** Estimated regions from cross/self/cross+self attention maps.

pixels, so using a vector of the target region would have resulted in a better probability map. The ablation studied for prompt-engineering (p-eng) is shown in Table 3. We found that better results were obtained by splitting the text and adding "a photo of" before the class name.

Figure 7 shows CAMaps (Cross), a SAMaps (Self), and Final Maps (Cross+Self) for three sample images. Note that the top two rows show successful results, while the bottom row shows a failure result. The results in the top two rows show that Cross extracted the incomplete person or sheep areas. In Self, on the other hand, person and sheep shapes are segmented along the contour well. In Cross+Self, the extraction of the high part of the target area, which is a feature of Cross, and the extraction of the part that extracts the contour of the target area, which is a feature of Self, are added together, and as results region prediction was successfully achieved.

On the other hand, in the bottom row, although in Cross the complete bird regions were extracted, in Self and Cross+Self too large regions were detected as bird regions.

### 4.5   Evaluation on Different Sizes of Attention Maps

Table 4 shows the experimental results on the five datasets for the different size of the Cross/Self-Attention Maps. Among 8, 16, 32, and 64 map sizes, in case of the size of 8 when selecting one size, the best results were achieved for all the datasets. Therefore, the size 8 should be always used. In the table, (cross, self) = (64, 32) means all the maps with a size of 64 or less are used for the Cross-Attention map, and all the maps with a size of 32 or less are used for the Self-Attention map. The results show that the optimal size of cross/self differed for each dataset. The best mIoU achieved with (cross, self) = (64, 64) for FoodPix, with (64, 64) for PAS-20, and with (16, 64) for PC-59, A-150 and FoodSeg.

Since the proposed method creates Self-Attention maps based on Cross-Attention maps, it is presumed that Self is influenced by the size of Cross. The best Cross size is 16 for A-150, PC-59, and FoodSeg. The smaller the cross size is, the more semantic parts are extracted, while detailed parts such as contours are not extracted. As shown in Fig. 5, the A-150, PC-59, and FoodSeg datasets often contain more than one class

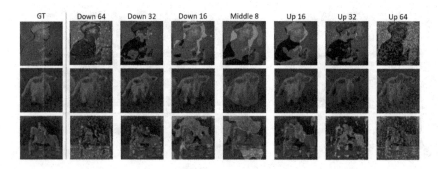

**Fig. 8.** Example of region segmentation using Cross-Attention Map with different sizes (**Down, Middle,** and **Up** indicate the position of the U-net, and **64, 32, 16,** and **8** indicate the size of one side of the Attention Map.)

per image, which may be due to the importance of determining the semantic regions of each image. On the other hand, PAS-20 and FoodPix have a small number of classes per image, so that even if weak semantic areas are extracted by cross, the accuracy is not reduced, and the accuracy can be improved by extracting detailed areas. From this consideration, it is considered optimal to set the size of cross to a small value for images with a large number of classes per image, and to set the size of cross to a small value for images with a small number of classes per image.

The results show that the optimal size of cross/self is different for each dataset, with the maximum value of (cross, self) = (64, 32) for FoodPix, (64, 64) for PAS-20, (16,64) for PC-59 and A-150 and FoodSeg. Since the proposed method creates Self-Attention maps based on Cross-Attention maps, it is presumed that Self is influenced by the size value of cross. The cross size value is 16 for A-150, PC-59, and FoodSeg. The smaller the cross size, the more semantic parts are extracted, while detailed parts such as contours are not extracted. As shown in Table 5, the A-150, PC-59, and FoodSeg datasets often contain more than four classes per image, which may be due to the importance of determining the semantic regions of each image. On the other hand, PAS-20 and FoodPix have a small number of classes per image, so even if weak semantic areas are extracted by Cross, the accuracy is not reduced, and the accuracy can be improved by extracting detailed areas. From this consideration, it is concluded that optimal to set the size of Cross to a small size for the images with a large number of classes per image, and to set the size of Cross to a larger size for images with a small number of classes per image.

The mIoU score tends to decrease as the number of classes per image increases. Comparing FoodPix and PAS-20, which have a similar number of classes per image, we found that the mIoU of FoodPix, which is a food dataset, was significantly higher. In addition, the mIoU score is competitive with FoodSeg, which has about twice as many classes as PAS-20, suggesting that StableSeg has a strong robustness in the fine-grained food domain.

Figure 8 shows the results using Cross-Attention Map with different size. Attention maps extracted at all sizes (64, 32, 16, and 8) were averaged and visualized for Down, Middle, and Up blocks. It can be seen that the larger size such as Down64, Down32, Up64, and Up32 extracted more details and produced clean region masks that follow the contours of the target regions, although there were many incorrect parts. On the other hand, when the size was small, as in Middle8, the semantic part of the target region was considered to be extracted due to the large number of feature channels and the expressiveness of Key and Query, which created the attention map. Although this reduced the number of false parts extracted, it is found that regions were segmented that ignored the shape of the target regions.

**Table 5.** Number of classes per image in each dataset.

| A-150 | PC-59 | FoodSeg | FoodPix | PAS-20 |
|-------|-------|---------|---------|--------|
| 9.5   | 5.8   | 4.6     | 2.7     | 2.5    |

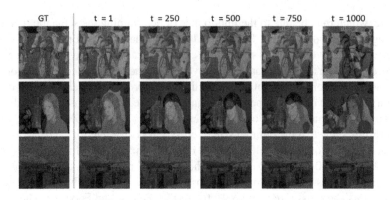

GT      t = 1      t = 250      t = 500      t = 750      t = 1000

**Fig. 9.** Estimated regions when time embedding is changed.

**Table 6.** Quantitative evaluation with PAS-20 in different time embeddings, $t$

| $t$ | 1 | 250 | 500 | 750 | 1000 |
|---|---|---|---|---|---|
| mIoU | **49.9** | 46.4 | 42.9 | 38.8 | 31.3 |

**Table 7.** Quantitative evaluation of Stable-Seg++ (mIoU).

| | | PAS-20 | PC-59 |
|---|---|---|---|
| MaskCLIP | init | 44.7 | 37.9 |
| | + | 53.4 | 40.5 |
| StableSeg | init | 51.4 | 33.9 |
| | + | 55.6 | 34.6 |
| | ++ | 59.1 | 36.6 |

### 4.6   Evaluation with Different Time Embeddings

When the time embedding ($t$) of StableSeg was varied, the results were also varied as shown in Fig. 9 and Table 6. In actual Stable Diffusion, the larger $t$ is, the closer to Gaussian noise is used to denoise the image, so it is presumed that an attention map is obtained that roughly determines the position of the object to be generated. On the other hand, as $t$ becomes smaller, an image closer to the real image is denoised, and thus an attention map is estimated that is more consistent with the shape of the object present in the real image. This shows that, quantitatively, mIoU is higher when $t = 1$. Therefore, StableSeg can obtain an attention map more suitable for real images by compressing the original image into latent space with VAE and then passing it through $t = 1$ embedding and U-net without mixing noise. Furthermore, it enabled fast and high quality segmentation map estimation in only one step.

### 4.7   Evaluation of StableSeg++

We compared StableSeg++ with MaskCLIP+ [25]. In this case, MaskCLIP+ was set to Annotation-Free setting and MaskCLIP created pseudo-masks using the class names in an input image.

The results are shown in Table 7. In this experiments, StableSeg "init" used all cross/self attention maps which is different from the condition of Sect. 4.2. StableSeg+

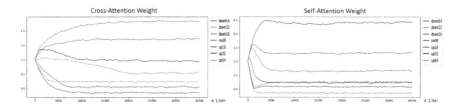

**Fig. 10.** Transition of the estimated weights of CA/SA maps using **the pseudo masks**.

means that the DeepLabV3+ trained with the masks created by StableSeg and CRF. StableSeg++ means that the DeepLabV3+ finetuned with the pseudo-mask of attention map weighted StableSeg. The dataset used for evaluation is PAS-20 and PC-59. The results show that StableSeg++ outperformed MaskClip+ for PAS-20, and vice verse for PC-59. The effectiveness of StableSeg++ was also verified, suggesting that the pseudo-mask generation using DeepLabV3+ twice was effective.

In this experiment, we explain the results of a weight search for the attention maps for StableSeg++. To estimate the optimal weights, groundtruth masks are needed. We used the masks estimated by StableSeg as pseudo masks, and the groundtruth masks of the PASCAL VOC for comparison. The weights were all initially set to 1, the optimizer updated the weights using Adam, and the attention maps for each class were trained using Binary Cross Entropy Loss until the loss is converged.

Figure 10 show the transition of the weights over the training iterations in case of using pseudo masks. The results show that the down/up layer with a size of 16 is the most important when focusing on the weights of the Cross-Attention maps, while the down/up layer with a size of 64 is almost ignored. Since Self-Attention uses the similarity between Key features and Query features to estimate regions, it is assumed that the size with the most features expressing more semantic regions was given the most importance.

## 5  Conclusions

We proposed a novel training-free region prediction method, **StableSeg**, that effectively used the attention maps and pre-trained knowledge in the Stable Diffusion model. In the experiments, we evaluated the method on a variety of datasets and found that it had the potential to predict the regions corresponding to any texts. The method was found to be generalizable to specific domains such as food images, and is considered to be highly robust to various domains including proper nouns and combination words of adjectives and nouns. We also proposed **StableSeg++**, a model that iteratively performed self-supervised learning of DeepLabV3+ and the weights of the attention maps using pseudo-masks generated by StableSeg. StableSeg++ improved segmentation quality and achieved better region prediction compared to the original StableSeg.

**Limitation:** The current StableSeg assumes that all the class names (including "background") of the regions contained in each input image are known. This is why we call StableSeg as "a region prediction model" rather than a semantic segmentation model.

The one idea to remove this limitation is combining CLIP-based zero-shot image classification [16] to estimate class candidates. Despite this limitation, we believe that the idea to use the attention maps of Stable Diffusion for region prediction is enough insightful to stimulates many new research directions.

**Acknowledgments.** This work was supported by JSPS KAKENHI Grant Numbers, 21H05812, 22H00540, 22H00548, and 22K19808.

# References

1. Bucher, M., Vu, T., Cord, M., Pérez, P.: Zero-shot semantic segmentation. In: Proceedings of CVF/IEEE Computer Vision and Pattern Recognition (2019)
2. Burgert, R., Ranasinghe, K., Li, X., Ryoo, M.S.: Peekaboo: text to image diffusion models are zero-shot segmentors. In: Proceedings of arXiv:2211.13224 (2022)
3. Chen, L., Papandreou, G., Kokkinos, I., Murphy, K., Yuille, A.L.: Deeplab: smantic image segmentation with deep convolutional nets, atrous convolution, and fully connected CRFs. IEEE Trans. Pattern Anal. Mach. Intell. **40**(4), 834–848 (2018)
4. Chen, L.-C., Zhu, Y., Papandreou, G., Schroff, F., Adam, H.: Encoder-decoder with atrous separable convolution for semantic image segmentation. In: Ferrari, V., Hebert, M., Sminchisescu, C., Weiss, Y. (eds.) ECCV 2018. LNCS, vol. 11211, pp. 833–851. Springer, Cham (2018). https://doi.org/10.1007/978-3-030-01234-2_49
5. Everingham, M., Eslami, S.M.A., Van Gool, L., Williams, C.K.I., Winn, J., Zisserman, A.: The pascal visual object classes challenge: a retrospective. Int. J. Comput. Vision **111**(1), 98–136 (2015)
6. Ghiasi, G., Gu, X., Cui, Y., Lin, T.: Scaling open-vocabulary image segmentation with image-level labels. In: Avidan, S., Brostow, G., Cissé, M., Farinella, G.M., Hassner, T. (eds.) ECCV 2022. LNCS, vol. 13696, pp. 540–557. Springer, Cham (2022). https://doi.org/10.1007/978-3-031-20059-5_31
7. Hertz, A., Mokady, R., Tenenbaum, J., Aberman, K., Pritch, Y., Cohen-Or, D.: Prompt-to-prompt image editing with cross attention control. arXiv preprint arXiv:2208.01626 (2022)
8. Ho, J., Jain, A., Abbeel, P.: Denoising diffusion probabilistic models. In: Advances in Neural Information Processing Systems, vol. 33 (2020)
9. Kingma, P., Welling, M.: Auto-encoding variational Bayes. In: Proceedings of International Conference on Machine Learning (2014)
10. Krähenbühl, P., Koltun, V.: Efficient inference in fully connected CRFs with gaussian edge potentials. In: Advances in Neural Information Processing Systems (2011)
11. Li, B., Weinberger, K.Q., Belongie, S., Koltun, V., Ranftl, R.: Language-driven semantic segmentation. In: Proceedings of International Conference on Learning Representation (2022)
12. Liang, F., et al.: Open-vocabulary semantic segmentation with mask-adapted clip. arXiv preprint arXiv:2210.04150 (2022)
13. Lüddecke, T., Ecker, A.S.: Image segmentation using text and image prompts. In: Proceedings of CVF/IEEE Computer Vision and Pattern Recognition, pp. 7086–7096 (2022)
14. Mottaghi, R., et al.: The role of context for object detection and semantic segmentation in the wild. In: Proceedings of CVF/IEEE Computer Vision and Pattern Recognition (2014)
15. Okamoto, K., Yanai, K.: UEC-FoodPIX complete: a large-scale food image segmentation dataset. In: Proceedings of ICPR Workshop on Multimedia Assisted Dietary Management (2021)
16. Radford, A., et al.: Learning transferable visual models from natural language supervision. arXiv preprint arXiv:2103.00020 (2021)

17. Ramesh, A., Dhariwal, P., Nichol, A., Chu, C., Chen, M.: Hierarchical text-conditional image generation with clip Latents. arXiv preprint arXiv:2204.06125 (2022)
18. Rombach, R., Blattmann, A., Lorenz, D., Esser, P., Ommer, B.: High-resolution image synthesis with latent diffusion models. In: Proceedings of CVF/IEEE Computer Vision and Pattern Recognition, pp. 10684–10695 (2022)
19. Saharia, C., et al.: Photorealistic text-to-image diffusion models with deep language understanding. arXiv preprint arXiv:2205.11487 (2022)
20. Schuhmann, C., et al.: Laion-5b: an open large-scale dataset for training next generation image-text models. arXiv preprint arXiv:2210.08402 (2022)
21. Wu, C., Lin, Z., Cohen, S., Bui, T., Maji, S.: Phrasecut: language-based image segmentation in the wild. In: Proceedings of CVF/IEEE Computer Vision and Pattern Recognition, pp. 7086–7096 (2020)
22. Xiongwei, W., Xin, F., Ying, L., Ee-Peng, L., Steven, H., Qianru, S.: A large-scale benchmark for food image segmentation. arXiv preprint arXiv:2105.05409 (2021)
23. Xu, J., et al.: GroupViT: semantic segmentation emerges from text supervision. In: Proceedings of CVF/IEEE Computer Vision and Pattern Recognition, pp. 18134–18144 (2022)
24. Zhou, B., Zhao, H., Puig, X., Fidler, S., Barriuso, A.: Scene parsing through ade20k dataset. In: Proceedings of CVF/IEEE Computer Vision and Pattern Recognition (2017)
25. Zhou, C., Loy, C.C., Dai, B.: Extract free dense labels from clip. In: Avidan, S., Brostow, G., Cissé, M., Farinella, G.M., Hassner, T. (eds.) ECCV 2022. LNCS, vol. 13688, pp. 696–712. Springer, Cham (2022)

# Mitigating Fine-Grained Hallucination by Fine-Tuning Large Vision-Language Models with Caption Rewrites

Lei Wang[1], Jiabang He[2], Shenshen Li[2], Ning Liu[3], and Ee-Peng Lim[1(✉)]

[1] Singapore Management University, Singapore, Singapore
{lei.wang.2019,eplim}@smu.edu.sg
[2] University of Electronic Science and Technology of China, Chengdu, China
[3] Beijing Forestry University, Beijing, China

**Abstract.** Large language models (LLMs) have shown remarkable performance in natural language processing (NLP) tasks. To comprehend and execute diverse human instructions over image data, instruction-tuned large vision-language models (LVLMs) have been introduced. However, LVLMs may suffer from different types of object hallucinations. Nevertheless, LVLMs are evaluated for coarse-grained object hallucinations only (i.e., generated objects non-existent in the input image). The fine-grained object attributes and behaviors non-existent in the image may still be generated but not measured by the current evaluation methods. In this paper, we thus focus on reducing fine-grained hallucinations of LVLMs. We propose *ReCaption*, a framework that consists of two components: rewriting captions using ChatGPT and fine-tuning the instruction-tuned LVLMs on the rewritten captions. We also propose a fine-grained probing-based evaluation method named *Fine-Grained Object Hallucination Evaluation* (*FGHE*). Our experiment results demonstrate that ReCaption effectively reduces fine-grained object hallucination for different LVLM options and improves their text generation quality. The code can be found at https://github.com/Anonymousanoy/FOHE.

**Keywords:** Hallucination Mitigation · Large Vision-Language Models

## 1 Introduction

Large language models (LLMs), such as GPT-3 [4] and ChatGPT [17], have demonstrated impressive performance in a wide range of natural language processing (NLP) tasks [19]. To extend LLMs to comprehend and execute both text-only and multi-modal (i.e., vision + text) instructions, new multi-modal large language models have been introduced, exemplified by GPT-4 [18]. Despite the impressive capabilities of GPT-4 in understanding and processing multi-modal information, the underlying mechanisms responsible for these exceptional abilities remain unclear due to its black-box nature.

To shed light on this mystery, recent research endeavors have focused on extending text-only LLMs to comprehend visual inputs by incorporating vision-language models (LVMs) into text-only LLMs. The resultant model is called the large vision-language model (LVLM). One LVLM research direction involves using vision modality models to provide textual description for visual information, followed by employing

© The Author(s), under exclusive license to Springer Nature Switzerland AG 2024
S. Rudinac et al. (Eds.): MMM 2024, LNCS 14557, pp. 32–45, 2024.
https://doi.org/10.1007/978-3-031-53302-0_3

closed-source LLMs, such as ChatGPT, to address multi-modal tasks such as visual QA and image caption generation. The LVLM examples using this approach include Visual ChatGPT [28], MM-REACT [33], and HuggingGPT [22]. Nevertheless, this approach requires good alignment across modalities to understand specific multi-modal instructions.

Another alternative approach focuses on instruction-tuned large vision-language models (LVLMs), e.g., LLaVA [14], MiniGPT-4 [36], mPLUG-Owl [34], and Instruct-BLIP [6], which extend language-only open-source LLMs (e.g., FlanT5 [5] and LLaMA [24]) to encompass visual reasoning abilities and instruction execution abilities by training LVLMs on text-image pairs and multi-modal instructions. Instruction-tuned LVLMs demonstrate outstanding capabilities in solving diverse multi-modal tasks [1,15,21]. Despite the success of instruction-tuned LVLMs, these models are prone to hallucination, which compromises both model performance and user experience in real-world use cases [9]. Similar to text-only LLMs [2,9], LVLMs may generate text descriptions that include non-existent or inaccurate objects in the target image also known as object hallucination [3,9,20].

To evaluate object hallucination in instruction-tuned LVLMs, several evaluation methods (e.g., CHAIR [20], POPE [12] have been proposed. However, these methods only focus on coarse-grained object hallucination but not the hallucinated object attributes and behaviors. We call the latter *fine-grained object hallucination*. Fine-grained object hallucination refers to the phenomenon wherein LVLMs generate captions that include not only non-existent or erroneous objects but also inaccurate object attributes and behaviors.

Consider the example in Fig. 1. This example illustrates that MiniGPT-4, an instruction-tuned LVLM, generates fine-grained hallucinations. The input image depicts two airplanes: a smaller one in flight and a larger one parked on the runway, accompanied by clouds. Multi-object hallucination occurs when the generated text mistakenly introduces erroneous or irrelevant relations between objects. Object attribute hallucination refers to generating incorrect attributes for a particular object. Vanilla MiniGPT-4 generates a description such as "small, white, and blue turboprop airplane", even though the color of the turboprop is unknown. Object behavior hallucination pertains to describing incorrect actions for objects. In this example scenario, the smaller airplane is flying, but MiniGPT-4 incorrectly states it is on the runway.

In this paper, we make key contributions to address fine-grained object hallucination. First, we introduce *ReCaption*, a framework that enables instruction-tuned LVLMs to reduce fine-grained object hallucination by fine-tuning them on additional rewritten captions derived from curated high-quality image captions.

ReCaption consists of two components: 1) rewriting captions using ChatGPT and 2) additional training of instruction-tuned LVLM on the rewritten captions. To develop the first component, we employ a two-stage prompting strategy to guide ChatGPT to generate high-quality image-text pairs. In the first stage, we utilize a prompt to tell ChatGPT to extract verbs, nouns, and adjectives from the original input caption. In the second stage, the extracted verbs, nouns and adjectives are merged into a list which is used in another ChatGPT prompt to generate a rewritten image caption that covers the list of words. This caption rewriting process is repeated multiple times, creating a diverse collection of captions that still retain the core content of the original caption. At

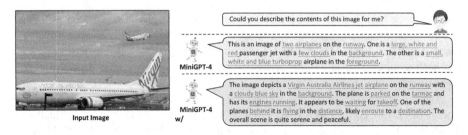

**Fig. 1.** An illustrative example is presented to compare the output of MiniGPT-4 and MinitGPT-4 with ReCaption. The generated caption by MiniGPT-4 contains words in blue that are inconsistent with the given image. In contrast, MinitGPT-4 with ReCaption demonstrates a superior ability to generate words in red that align more closely with the image at a fine-grained level. The words marked with an underline represent objects. The words marked with an dotted underline denote attributes and behaviors. (Color figure online)

the end of the first stage, we obtain a set of high-quality image-caption pairs. The second component of ReCaption performs fine-tuning of the instruction-tuned LVLM using the above set of image-caption pairs to strengthen the model's fine-grained alignment between visual and text modalities.

To better evaluate the proposed method, we introduce a new evaluation method called *Fine-Grained Object Hallucination Evaluation (FGHE)*. This method assesses how well any LVLM performs in minimizing fine-grained object hallucination by incorporating another evaluation method POPE with the measurement of hallucinated object attributes and behaviors. Similar to POPE, FGHE converts object hallucination evaluation into a set of binary classification tasks, prompting instruction-tuned LVLMs with simple Yes-or-No questions about the probed objects (e.g., *"Is there a car in the image?"*) and about the attributes/behaviors of objects (e.g., *"Is the man's clothing blue in the picture?"*). We evaluate the fine-grained hallucination reduction of ReCaption using POHE and FGHE. Our evaluation results demonstrate that any LVLM adopting the ReCaption framework can effectively reduce fine-grained object hallucination.

## 2   Related Work

### 2.1   Large Vision-Language Models

As text-only LLMs [4,17,24] show very good results across NLP tasks [19], there are many works extending LLMs to comprehend visual inputs, and to the development of large vision-language models (LVLMs). Two primary paradigms have been pursued in this line of research. The first paradigm involves representing visual information through textual descriptions. Closed-source LLMs, such as ChatGPT [17], are used to establish connections between vision modality models, enabling subsequent handling of multimodal tasks. Several notable approaches following this paradigm include Visual ChatGPT [28], MM-REACT [33], and HuggingGPT [22].

The second paradigm centers around training LVLMs using vision-language instructions. MultiInstruct [32] engages in vision-language instruction tuning, containing various multi-modal tasks involving visual comprehension and reasoning.

LLaVA [14] employs self-instruct [27] to generate instructions and optimize the alignment network's and LLM's model parameters. MiniGPT-4 [36] integrates a visual encoder derived from BLIP-2 [11] and trains the model using image captions generated by ChatGPT, ensuring that these captions are longer than the training data of BLIP-2. mPLUG-Owl [34] equips LLMs with multimodal abilities through modularized learning, enabling LLMs to support multiple modalities. Lastly, InstructBLIP [6] enhances the cross-modal alignment network, empowering the LLM to generate meaningful semantic descriptions for a given image. By exploring these paradigms, researchers enhance the ability of LLMs to process visual information, thus expanding their applicability and effectiveness in various vision-language tasks.

### 2.2 Hallucnation in Large Vison-Language Mdoels

A recent survey [9] has thoroughly analyzed studies examining hallucinations in various tasks, including text summarization [8, 16], dialogue generation [23, 29], and vision-language generation [10, 13, 20, 25, 26, 30, 35]. Specifically, within the domain of vision-language generation, object hallucination can be further classified as intrinsic and external hallucinations. Intrinsic hallucinations in vision-language generation refer to generated captions that contain incorrect or non-existent objects in the given image. On the other hand, external hallucinations in vision-language generation refer to generated captions that contain irrelevant objects. To comprehensively investigate the phenomenon of object hallucination in LVLMs, POPE [12] endeavors to conduct an extensive empirical examination of object hallucinations across various LVLMs, and provides an evaluation benchmark for evaluating VLVMs. While previous works mainly focus on hallucinations about the presence or absence of objects, LVLMs may also generate more fine-grained erroneous or incomplete descriptions for target images, specifically regarding incorrect attributes associated with objects. Therefore, this paper aims to mitigate and evaluate finer-grained hallucinations within LVLMs.

## 3   ReCaption Framework

In this section, we describe the design of our ReCaption framework, highlighting its key components and strategies. ReCaption is LVLM-agnostic and can be used to reduce fine-grained object hallucination of any LVLMs. There are two components in ReCaption, caption rewriting and training of instruction-tuned LVLM. The latter is conducted using the rewritten captions.

### 3.1   Caption Rewriting

The caption rewriting component aims to create good quality captions of training images. To generate rewritten image-text pairs, we first randomly select image-text pairs from the cc_sbu_align dataset, which is curated by MiniGPT-4 [36]. As shown in Fig. 2, we use a two-stage prompting strategy on the caption of each selected image to generate multiple different captions that convey the essential elements of the original caption but with varied details. We use ChatGPT [17] for both stages because it performs well in

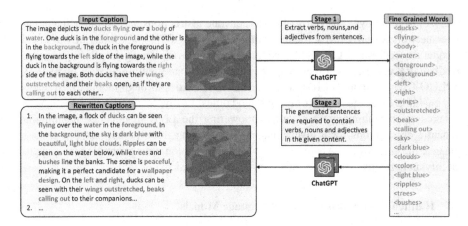

**Fig. 2.** Illustration of rewriting image captions using ChatGPT. Stage 1: Keyword Extraction Prompt (i.e., "Derive verbs, nouns, and adjectives from sentences"), directs ChatGPT to generate verbs, nouns, and adjectives (highlighted in brown) from the original caption. Stage 2: Caption Generation Prompt (i.e., "The resulting sentences must encompass verbs, nouns, and adjectives aligned with the provided content") guides ChatGPT to generate a rewritten caption. By repeating this prompt, multiple rewritten captions will be generated. (Color figure online)

both stages. We believe other similar LLMs can also be used. This two-stage prompting aims to preserve the original semantics associated with the corresponding image, which is important for enriching the fine-grained alignment between image and text. Note that we provide one demonstration example for each stage to assist LLMs in improving their keyword extraction and caption generation abilities. In the following, we elaborate on the proposed two-stage prompting strategy.

*Stage 1: Keyword Extraction.* This stage aims to preserve essential information from the caption corresponding to a specific image. In generating descriptions for an input image, LVLMs may produce inaccurate or irrelevant objects, object attributes, and object behaviors. Since verbs, nouns, and adjectives in captions often denote objects, object attributes, and object behaviors, we devise a prompt to facilitate the extraction of these pertinent words from the original caption. Through this prompting, the extracted keywords preserve the caption's essential information. Below is the prompt template for the first-stage prompting:

```
Extract verbs, nouns, and adjectives from [X],
```

where $X$ denotes the original caption of the input image. Figure 2 depicts the first-stage prompting directly extracts nouns (e.g., "ducks"), adjectives (e.g., "right"), and verbs (e.g., "flying") from the caption. The extracted keywords, denoted as $\{x_1^k, x_2^k, ..., x_n^k\}$, capture essential information in the caption, where $k$ means keywords.

*Stage 2: Caption Generation.* The second stage is to rewrite the original caption, conditioned on the extracted words obtained in stage 1. The prompt template for second-stage prompting can be denoted as follows:

```
The generated sentences are required to contain verbs, nouns and
adjectives in the given content: [x₁ᵏ, x₂ᵏ, ..., xₙᵏ].
```

Through prompting ChatGPT, we can obtain a new version of the caption, denoted as $X'$. To keep the characteristic of randomness in the LLM, we use a temperature ratio of 1.0. Further, we prompt ChatGPT $R$ times to generate diverse captions while preserving essential information from the original caption. As shown in Fig. 2, each rewritten caption has different details while contextually relevant to the original caption. This caption generation prompt basically elicits the imagination and rewriting abilities of ChatGPT.

## 3.2 Additional Tuning

By producing $R$ diverse rewritten captions for each original image caption, we can now enhance the implicit fine-grained alignment between the input images and captions. In our experiments, we set $R = 5$. These rewritten image-caption pairs are model-agnostic to language model architecture, thereby enabling their seamless adaptation to various LVLM models, including MiniGPT-4 [36], LLaVA [14], mPLUG-Owl [34], and MultiModal-GPT [7], with minimal changes. Our later experimental analysis reveals that a small number of rewritten image-text pairs can significantly reduce fine-grained object hallucination and improve caption generation quality. Model tuning using such a small set of pairs entails negligible additional computational cost compared to training of the original LVLM. The training loss over the images with their rewritten captions can be formulated as follows:

$$\mathcal{L} = \sum_{i=1}^{M} \sum_{j=1}^{R} \mathcal{L}_{\mathrm{CE}}(X'_{i,j}, \hat{X}_i), \tag{1}$$

where $X'_i$ represents a rewritten caption for image-caption pair $i$, $\hat{X}_i$ is generated caption of image-caption pair $i$, $R$ is the total number of rewritten captions for the same image, $M$ denotes the total number of training image-caption examples, and $\mathcal{L}_{\mathrm{CE}}$ is the cross-entropy loss.

## 4 Hallucination Evaluation of LVLMs

To evaluate the effectiveness of ReCaption framework, we use an existing evaluation dataset and method introduced by the POPE work [12]. As POPE only evaluates coarse-grained hallucination, we introduce another dataset with specific evaluation method. In this section, we will fist introduce POPE dataset and method. We then introduce our proposed dataset and its evaluation method also known as *Fine-Grained Object Hallucination Evaluation (FGHE)*. We leave out the Caption Hallucination Assessment with Image Relevance (CHAIR) metric [20] which suffers from prompt sensitivity and inaccuracies as reported in [12].

**Table 1.** Statistics of FGHE Evaluation Dataset.

| Type | Questions | Multi-Object | Attribute | Behavior |
|------|-----------|--------------|-----------|----------|
| Yes | 100 | 47 | 45 | 8 |
| No | 100 | 51 | 42 | 7 |
| Total | 200 | 98 | 87 | 15 |

*POPE Dataset and Evaluation Method.* In POPE [12], the Polling-based Object Probing Evaluation (POPE) evaluation method was proposed to improve the evaluation of object hallucination in LVLMs. The basic idea behind POPE is to evaluate hallucination by asking LVLMs simple Yes-or-No questions concerning the probed objects. For instance, a sample question could be: "Is there a car in the image?". By including existent or non-existent object into a question of an input image, the evaluation method obtains a question with "yes" or "no" answer. In our experiments, we use popular non-existent objects in the no-questions (i.e., questions with "no" answer). Finally, the POPE dataset involves 3000 questions for the captions of 500 images. By treating a LVLM's answers to the questions as a binary classification task, we obtain the Accuracy, Precision, Recall and F1 scores of the LVLM on this dataset. The higher the scores, the less hallucination. POPE has been adopted by LVLM-eHub, an evaluation benchmark for LVLMs [31].

*FGHE Dataset and Evaluation Method.* **FGHE** follows the binary classification approach of POPE to evaluate LVLMs. However, unlike POPE, FGHE requires a different set of binary questions to measure fine-grained hallucination. The FGHE dataset consists of 50 images and 200 binary questions divided into three categories: (a) *multiple-object* question which verifies the relationships between multiple objects in the image; (b) *attribute* question which verifies an attribute of an object in the image; and (c) *behavior* question which verifies a behavior or an object in the image. Figure 3 presents an illustrative comparison between probing questions in FGHE and POPE. All the above questions are manually defined by human annotators on a subset of 50 images from the validation set of MSCOCO dataset. As shown in, the FGHE dataset consists of 100 yes-questions and 100 no-questions. Among the yes-questions, 47, 45 and 8 are multi-object, attribute and behavior questions. Among the no-questions, 51, 42 and 7 are multi-object, attribute and behavior questions. Table 3 only displays few behavior questions as some of them are counted towards multiple objects. Similar to POPE, we finally employ Accuracy, Precision, Recall, and F1 score of all questions as the evaluation metrics (Table 1).

## 5   Experiment

### 5.1   Experimental Setup

*Model Settings.* Since the proposed ReCaption is model-agnostic to instruction-tuned LVLMs, we can enrich any LVLMs with ReCaption. In this paper, we choose four open-sourced representative instruction-tuned LVLMs for evaluation: mPLUG-Owl [34], LLaVA [14], Multimodal-GPT [7], and MiniGPT-4 [36].

Question:Is there a plane in the image?
Please answer yes or no. Yes.

Question:Is there a person near the wheel
of the plane? Please answer yes or no. No.

Question:Is there a cup in the image?
Please answer yes or no. No.

Question:Is there a letter 'entourage' in the
picture? Please answer yes or no. Yes.

Question:Is there a tennis racket? Please
answer yes or no. Yes.

**(a) POPE**

Question:Is the athlete holding a tennis
racket? Please answer yes or no. Yes.

**(b) FOHE**

**Fig. 3.** Examples of evaluation data of POPE (a) and FGHE (b).

*Implementation Details.* A total of 500 image-caption pairs are randomly selected from a high-quality well-aligned image-text pair dataset named cc_sbu_align, which is curated by MiniGPT-4 [36], for our study. To increase the diversity of the data, we generated 5 rewritten versions for each original caption. We employed the AdamW optimizer with a beta value of (0.9, 0.98) for optimization purposes. The learning rate and weight decay were set to 0.0001 and 0.1, respectively. During the training process, we initiated a warm-up phase consisting of 2, 000 steps, after which we applied the cosine schedule to decay the learning rate. As for the input image, it was randomly resized to dimensions $224 \times 224$. We additionally fine-tune LVLMs 20 epochs with the batch size 256, and the learning rate is set to 0.00002.

## 5.2  Main Results

**LVLMs with ReCaption have Reduced Hallucination.** Overall, a remarkable improvement in performance can be observed across various LVLMs and evaluation metrics when the LVLM is used with ReCaption, as compared to the original LVLM. For instance, Mini-GPT4 with ReCaption demonstrates a significant improvement of F1 over Mini-GPT4 without ReCaption (7% improvement on POPE and 10% on FGHE). Among the LVLMs, mPLUG-Owl with ReCaption enjoys the least improvement in F1 score (with 1.32% improvement on POPE but more substantial 3.71% improvement on FGHE).

**Table 2.** Results of LVLMs w/o ReCaption and LVLMs w/ ReCaption on POPE and FGHE datasets. The best results are denoted in bold.

| Model | POPE | | | | FGHE | | | |
|---|---|---|---|---|---|---|---|---|
| | Accuracy | Precision | Recall | F1 Score | Accuracy | Precision | Recall | F1 Score |
| mPLUG-Owl | 59.37 | 55.51 | 94.34 | 69.89 | 55.56 | 57.52 | 84.42 | 68.42 |
| w/ ReCaption | **61.79** | **57.11** | **94.61** | **71.22** | **62.22** | **62.26** | **85.71** | **72.13** |
| | (+2.42) | (+1.60) | (+0.27) | (+1.33) | (+6.66) | (+4.74) | (+1.29) | (+3.71) |
| LLaVA | 50.00 | 50.00 | **100.00** | 66.67 | 56.83 | 56.52 | **100.00** | 72.22 |
| w/ ReCaption | **61.87** | **57.33** | 92.80 | **70.88** | **67.63** | **65.42** | 89.74 | **75.68** |
| | (+11.87) | (+7.33) | (−7.80) | (+4.21) | (+10.80) | (+8.90) | (−10.26) | (+3.46) |
| MultiModal-GPT | 50.00 | 50.00 | **100.00** | 66.67 | 57.56 | 56.93 | **100.00** | 72.56 |
| w/ ReCaption | **62.40** | **57.60** | 93.93 | **71.41** | **75.83** | **72.31** | 83.73 | **77.60** |
| | (+12.40) | (+7.60) | (−6.07) | (+4.74) | (+18.27) | (+15.38) | (−16.27) | (+5.04) |
| MiniGPT-4 | 54.23 | 58.24 | 31.97 | 41.29 | 53.01 | 59.62 | 63.27 | 61.39 |
| w/ ReCaption | **57.69** | **62.54** | **39.67** | **48.55** | **60.24** | **62.12** | **83.67** | **71.30** |
| | (+3.46) | (+4.30) | (+7.70) | (+7.26) | (+7.23) | (+2.50) | (+20.40) | (+9.91) |

The results above demonstrate that our proposed ReCaption framework enhances the generation quality of LVLMs. It effectively reduce both coarse-grained and fine-grained hallucinations. The improvement on fine-grained hallucinations can be attributed to a strong alignment between images and text descriptions at the fine-grained level. Additionally, the proposed ReCaption approach is model-agnostic, allowing for effortless integration as a plug-and-play component during the training of LVLMs.

**LVLMs with ReCaption Reduce Over-Confidence.** As shown in Table 2, based on the Recall of POPE and FGHE, it is evident that mPLUG-Owl, LLaVA, and MultiModel-GPT show a strong inclination to respond with the affirmative answer "Yes". For instance, both LLaVA and MultiModel-GPT provide "Yes" responses to most questions. The mPLUG-Owl also achieve a high recall rate of 94.34% on POPE and 84.42% on FGHE. It suggests that certain LVLMs exhibit a high degree of over-confidence and struggle to accurately identify objects, object attributes, and object behaviors in the given images. With ReCaption, LLaVA and MultiModal-GPT reduces their over-confidence.

**FGHE Serves as a Different Hallucination Evaluation Compared to POPE.** Table 2 reveals different performances of LVLMs evaluated on POPE and FGHE, suggesting that LVLMs possess varying degrees of coarse-grained and fine-grained object hallucinations. For example, the mPLUG-Owl model attains an F1 score of 69.89% on POPE but only 68.42% on FGHE, indicating that mPLUG-Owl may excel in identifying objects rather than attributes or behaviors within a given image. Among the LVLMs, MultiModal-GPT performs the best for both coarse-grained and fine-grained object hallucinations.

**Table 3.** Evaluation over different hallucination categories in terms of F1 score of FGHE.

| Method | Multi-Object | Attribute | Behavior |
|---|---|---|---|
| mPLUG-Owl | 72.36 | 68.14 | 61.59 |
| w/ ReCaption | **74.23** | **71.55** | **67.90** |
| | (+1.87) | (+3.41) | (+6.31) |
| LLaVA | 74.26 | 71.49 | 66.45 |
| w/ ReCaption | **76.70** | **75.18** | **71.82** |
| | (+2.44) | (+3.69) | (+5.37) |
| MultiModal-GPT | 74.82 | 70.56 | 68.13 |
| w/ ReCaption | **76.84** | **78.92** | **74.22** |
| | (+2.02) | (+8.36) | (+6.09) |
| MiniGPT-4 | 63.30 | 60.16 | 56.72 |
| w/ ReCaption | **71.02** | **72.16** | **67.93** |
| | (+7.72) | (+12.00) | (+11.21) |

### 5.3  Further Analysis

*Break-Down Study of Fine-Grained Object Hallucination.* FGHE includes three types of binary questions, namely multiple objects, object attributes, and object behaviors. Table 2 depicts the F1 Scores of FGHE, combining the three types of questions. We further examine the hallucination organized by the three types of questions using the F1 scores.

Table 3 presents the results, demonstrating that adding ReCaption to any LVLM reduces fine-grained hallucinations measured by the F1 scores computed over three question categories. This finding suggests that ReCaption is effective for reducing fine-grained hallucination generation by enriching alignment between images and texts through additional training with rewritten captions. Furthermore, ReCaption is also more effective in reducing hallucination in object attributes and behaviors than in multiple objects. One possible reason is that the rewritten captions may contain more verb and adjective keywords, thereby reducing hallucination, especially in object attributes and behaviors.

*Varying Rewriting Strategies.* Figure 4 depicts a comparison between our proposed two-stage caption rewriting strategy and a simple rewriting method (called Rewrite) employing ChatGPT. We select MiniGPT-4 as the model of choice since the data used for rewriting is based on image-text pairs collected by MiniGPT-4. The latter uses the prompt "Rewrite the following image description" to generate a new caption without any constraint. The performance of this simple rewriting technique exhibits some improvements compared to the original LVLMs without rewrites. However, this improvement is relatively minor. In contrast, our proposed two-stage rewriting prompting approach yields substantial improvements, highlighting multiple diverse rewritten captions with the same keywords to guide the LVLM in establishing a more robust alignment between images and captions.

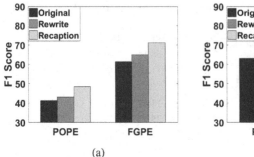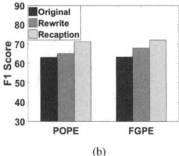

(a)                                    (b)

**Fig. 4.** Performance comparison of MiniGPT-4 training with different caption rewriting strategies. Original means MiniGPT-4 without rewriting. Rewrite denotes MiniGPT-4 with a simple rewriting prompting (e.g., "Rewrite the following image description"). ReCaption means MiniGPT-4 with our proposed strategy.

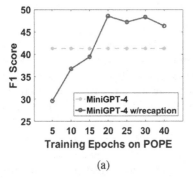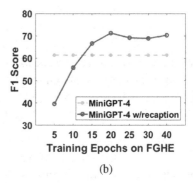

(a)                                    (b)

**Fig. 5.** F1 Score Curves for MiniGPT-4 with and without ReCaption over training epochs. The evaluation datasets used are POPE (a) and FGHE (b).

*Number of Training Epochs.* We examine the impact of the number of additional training epochs on hallucination reduction of LVLMs with ReCaption. Figure 5 presents the F1 Score curves in relation to the training epochs for MiniGPT-4 with ReCaption on POPE and FGHE. The figure illustrates that the F1 Score of MiniGPT-4 with ReCaption increases as the number of training epochs increases. This suggests that conducting further training using this limited number of rewritten data pairs can enhance the fine-grained alignment between images and text, consequently reducing the generation of hallucinations by LVLMs.

*Number of Images Used for Rewriting.* Figure 6a illustrates the impact of the number of images utilized for rewriting in additional training of LVLMs, including MiniGPT-4, mPLUG-Owl, and LLaVA. All are evaluated in terms of F1 Score of FGHE. In the figure, a value of 50 indicates LVLMs trained with ReCaption incorporating 50 distinct images and their respective rewritten captions. The remaining hyperparameters, such as the number of rewritten captions per image, remain unchanged and fixed. The results

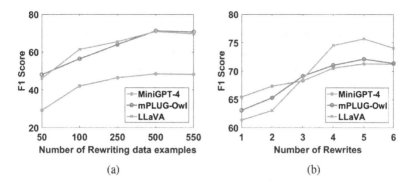

**Fig. 6.** F1 Score Curves of three LVLMs with our ReCaption as the number of images used for rewriting and the number of rewrites per image increase. We evaluate hallucination degree of models using FGHE.

strongly indicate that our proposed ReCaption consistently enhances the performance of all LVLMs as more images are employed for rewriting.

*Number of Rewrites per Image.* In Fig. 6b, we observe the effect of the number of rewritten captions per image on LVLMs, such as MiniGPT-4, mPLUG-Owl, and LLaVA. We use the F1 Score of FGHE to evaluate them. Increasing the number of rewritten captions per image is expected to enhance the alignment between input images and output captions. This is because more output captions containing the same keywords are utilized for training the mapping from image to text. The findings strongly support that our proposed ReCaption consistently improves the hallucination reduction ability of all LVLMs as more rewritten captions per image are used during training.

*Case Study.* In the appendix, we show more examples of image caption generation using four vanilla LVLMs and these LVLMs with our ReCaption. Overall, adding ReCaption to LVLMs yields superior quality captions with fewer hallucinated objects, object attributes, and object behaviors. Conversely, based on the provided examples, Mini-GPT4 performs poorly in generating accurate descriptions of the given image, while MultiModal-GPT and LLaVA are prone to generating irrelevant content. Despite vanilla mPlug-Owl displaying a relatively higher-quality caption generation compared to the other three LVLMs, ReCaption still allows it to generate more accurate image captions.

## 6  Conclusion

In this paper, we aim to address the issue of fine-grained object hallucinations in instruction-tuned large vision-language models (LVLMs). We introduced a framework called ReCaption, which comprises two components: caption rewriting using ChatGPT and fine-tuning of LVLMs based on the rewritten captions. To evaluate the effectiveness of our approach, we proposed a fine-grained probing-based evaluation dataset and method called Fine-Grained Object Hallucination Evaluation (FGHE). Our experimental results demonstrate that ReCaption can effectively mitigate fine-grained object hallucinations across various LVLMs, thereby enhancing text generation quality. Future work could focus on refining and expanding the evaluation metrics to support more comprehensive evaluation of LVLM performance in hallucination reduction. One should also explore more effective rewriting techniques to enrich alignment between unnatural images, such as invoices and cartoon pictures, and longer output text.

# References

1. Agrawal, H., et al.: nocaps: novel object captioning at scale. In: 2019 IEEE/CVF International Conference on Computer Vision, ICCV, pp. 8947–8956. IEEE (2019)
2. Bang, Y., et al.: A multitask, multilingual, multimodal evaluation of chatgpt on reasoning, hallucination, and interactivity. CoRR abs/2302.04023 (2023). https://doi.org/10.48550/arXiv.2302.04023. https://doi.org/10.48550/arXiv.2302.04023
3. Biten, A.F., Gómez, L., Karatzas, D.: Let there be a clock on the beach: educing object hallucination in image captioning. In: IEEE/CVF Winter Conference on Applications of Computer Vision, WACV (2022)
4. Brown, T.B., et al.: Language models are few-shot learners. In: NeurIPS (2020)
5. Chung, H.W., et al.: Scaling instruction-finetuned language models. arXiv preprint arXiv:2210.11416 (2022)
6. Dai, W., et al.: Instructblip: Towards general-purpose vision-language models with instruction tuning. arXiv preprint arXiv:2305.06500 (2023)
7. Gong, T., et al.: Multimodal-gpt: A vision and language model for dialogue with humans. arXiv preprint arXiv:2305.04790 (2023)
8. Huang, Y., Feng, X., Feng, X., Qin, B.: The factual inconsistency problem in abstractive text summarization: a survey. arXiv preprint arXiv:2104.14839 (2021)
9. Ji, Z., et al.: Survey of hallucination in natural language generation. ACM Comput. Surv. **55**(12), 1–38 (2023)
10. Lee, S., Park, S.H., Jo, Y., Seo, M.: Volcano: mitigating multimodal hallucination through self-feedback guided revision. arXiv preprint arXiv:2311.07362 (2023)
11. Li, J., Li, D., Xiong, C., Hoi, S.: Blip: Bootstrapping language-image pre-training for unified vision-language understanding and generation. In: International Conference on Machine Learning, pp. 12888–12900. PMLR (2022)
12. Li, Y., Du, Y., Zhou, K., Wang, J., Zhao, W.X., Wen, J.R.: Evaluating object hallucination in large vision-language models. arXiv preprint arXiv:2305.10355 (2023)
13. Liu, F., Lin, K., Li, L., Wang, J., Yacoob, Y., Wang, L.: Aligning large multi-modal model with robust instruction tuning. arXiv preprint arXiv:2306.14565 (2023)
14. Liu, H., Li, C., Wu, Q., Lee, Y.J.: Visual instruction tuning (2023)
15. Lu, P., et al.: Learn to explain: multimodal reasoning via thought chains for science question answering. In: The 36th Conference on Neural Information Processing Systems (NeurIPS) (2022)
16. Maynez, J., Narayan, S., Bohnet, B., McDonald, R.: On faithfulness and factuality in abstractive summarization. In: Proceedings of the 58th Annual Meeting of the Association for Computational Linguistics, pp. 1906–1919 (2020)
17. OpenAI: Introducing chatgpt (2022). https://openai.com/blog/chatgpt
18. OpenAI: GPT-4 technical report. CoRR abs/2303.08774 (2023)
19. Qin, C., Zhang, A., Zhang, Z., Chen, J., Yasunaga, M., Yang, D.: Is chatgpt a general-purpose natural language processing task solver? arXiv preprint arXiv:2302.06476 (2023)

20. Rohrbach, A., Hendricks, L.A., Burns, K., Darrell, T., Saenko, K.: Object hallucination in image captioning. In: EMNLP (2018)
21. Schwenk, D., Khandelwal, A., Clark, C., Marino, K., Mottaghi, R.: A-OKVQA: a benchmark for visual question answering using world knowledge. In: ECCV 2022, vol. 13668, pp. 146–162. Springer (2022)
22. Shen, Y., Song, K., Tan, X., Li, D., Lu, W., Zhuang, Y.: Hugginggpt: solving AI tasks with chatgpt and its friends in huggingface. CoRR abs/2303.17580 (2023)
23. Shuster, K., Poff, S., Chen, M., Kiela, D., Weston, J.: Retrieval augmentation reduces hallucination in conversation. EMNLP (2021)
24. Touvron, H., et al.: Llama: open and efficient foundation language models. arXiv preprint arXiv:2302.13971 (2023)
25. Wang, J., et al.: An llm-free multi-dimensional benchmark for mllms hallucination evaluation. arXiv preprint arXiv:2311.07397 (2023)
26. Wang, J., et al.: Evaluation and analysis of hallucination in large vision-language models. arXiv preprint arXiv:2308.15126 (2023)
27. Wang, Y., et al.: Self-instruct: aligning language model with self generated instructions. arXiv preprint arXiv:2212.10560 (2022)
28. Wu, C., Yin, S., Qi, W., Wang, X., Tang, Z., Duan, N.: Visual chatgpt: talking, drawing and editing with visual foundation models. CoRR abs/2303.04671 (2023)
29. Wu, Z., et al.: A controllable model of grounded response generation. In: Proceedings of the AAAI Conference on Artificial Intelligence, vol. 35 (2021)
30. Xiao, Y., Wang, W.Y.: On hallucination and predictive uncertainty in conditional language generation. In: Proceedings of the 16th Conference of the European Chapter of the Association for Computational Linguistics: Main Volume, pp. 2734–2744 (2021)
31. Xu, P., et al.: Lvlm-ehub: a comprehensive evaluation benchmark for large vision-language models (2023)
32. Xu, Z., Shen, Y., Huang, L.: Multiinstruct: improving multi-modal zero-shot learning via instruction tuning. arXiv preprint arXiv:2212.10773 (2022)
33. Yang, Z., et al.: MM-REACT: prompting chatgpt for multimodal reasoning and action. CoRR abs/2303.11381 (2023)
34. Ye, Q., et al.: mplug-owl: modularization empowers large language models with multimodality (2023)
35. Yin, S., et al.: Woodpecker: hallucination correction for multimodal large language models. arXiv preprint arXiv:2310.16045 (2023)
36. Zhu, D., Chen, J., Shen, X., Li, X., Elhoseiny, M.: Minigpt-4: enhancing vision-language understanding with advanced large language models (2023)

# GDTNet: A Synergistic Dilated Transformer and CNN by Gate Attention for Abdominal Multi-organ Segmentation

Can Zhang[1], Zhiqiang Wang[2], Yuan Zhang[1(✉)], Xuanya Li[3], and Kai Hu[1(✉)]

[1] Key Laboratory of Intelligent Computing and Information Processing
of Ministry of Education, Xiangtan University, Xiangtan 411105, China
{yuanz,kaihu}@xtu.edu.cn

[2] Key Laboratory of Medical Imaging and Artificial Intelligence of Hunan Province,
Xiangnan University, Chenzhou 423000, China

[3] Baidu Inc., Beijing 100085, China

**Abstract.** As one of the key problems in computer-aided medical image analysis, learning how to model global relationships and extract local details is crucial to improve the performance of abdominal multi-organ segmentation. While current techniques for Convolutional Neural Networks (CNNs) are quite mature, their limited receptive field makes it difficult to balance the ability to capture global relationships with local details, especially when stacked onto deeper networks. Thus, several recent works have proposed Vision Transformer based on a self-attentive mechanism and used it for abdominal multi-organ segmentation. However, Vision Transformer is computationally expensive by modeling long-range relationships on pairs of patches. To address these issues, we propose a novel multi-organ segmentation framework, named GDTNet, based on the synergy of CNN and Transformer for mining global relationships and local details. To achieve this goal, we innovatively design a Dilated Attention Module (DAM) that can efficiently capture global contextual features and construct global semantic information. Specifically, we employ a three-parallel branching structure to model the global semantic information of multiscale encoded features by Dilated Transformer, combined with global average pooling under the supervision of Gate Attention. In addition, we fuse each DAM with DAMs from all previous layers to further encode features between scales. Extensive experiments on the Synapse dataset show that our method outperforms ten other state-of-the-art segmentation methods, achieving accurate segmentation of multiple organs in the abdomen.

**Keywords:** CT images · Abdominal multi-organ segmentation · CNN and Transformer

---

C. Zhang and Z. Wang—These authors contributed equally to this paper.

S. Rudinac et al. (Eds.): MMM 2024, LNCS 14557, pp. 46–57, 2024.
https://doi.org/10.1007/978-3-031-53302-0_4

# 1    Introduction

Abdominal multi-organ segmentation is one of the main challenges in medical image segmentation [5]. It aims to accurately capture the shape and volume of target organs and tissues through pixel classification, assisting physicians in analyzing organs and inter-organ relationships more accurately for disease assessment and diagnostic treatment. With the increasing use of abdominal multi-organ segmentation, the need for highly accurate and robust segmentation has become increasingly crucial.

Improving the performance of multi-organ segmentation relies on learning how to model global relationships and extract local details. Over the past decades, Convolutional Neural Networks (CNNs) have greatly contributed to the development of dense prediction tasks. They have been widely employed in various image segmentation tasks due to their excellent ability to extract image features. The advent of encoder-decoder-based networks, such as Fully Convolutional Networks (FCNs) [12], U-shaped structured networks such as U-Net [13] and its variants [9,23], has further propelled the success of CNN models in computer vision tasks. However, CNNs have obvious drawbacks in obtaining global information due to their limited receptive field and inherent inductive bias [4]. If a larger receiving field is obtained by constant stacking and downsampling, it will cause the training of the deep network to suffer from the feature reuse problem [14]. In addition to this fact, the gradual decrease in spatial resolution often leads to the loss of crucial local information, which is essential for dense prediction tasks.

In recent years, considering the excellent achievements of Transformer in Natural Language Processing (NLP) [16], research members have also begun to try to introduce it into computer vision tasks for mitigating the shortcomings of CNNs. Building upon this, Dosovitskiy et al. introduced the first Vision Transformer (ViT) for image recognition, which relies on a purely self-attention mechanism [4]. The Transformer primarily leverages the Multi-head Self-Attention (MSA) mechanism to efficiently establish long-term dependencies among token sequences and capture global context. While this approach yielded competitive results on ImageNet [3], it relies on pre-training on large external datasets. And the squared complexity of the number of patches will consume a lot of computational resources.

In order to overcome this weakness, DeiT [15] was proposed, providing various training strategies, including knowledge distillation, which contributed to improved Transformer performance. In addition, adaptations of the Transformer, such as Swin Transformer [11] and Pyramid Vision Transformer [19], attempted to reduce the computational complexity of Vision Transformer by exploiting the window mechanism. Because of this, it does not model long distance dependencies at the beginning, but instead divides the feature map into many windows. These are similar to CNNs that require a certain number of layers to be superimposed in order to model long-distance relationships, which makes Swin Transformer somewhat weaker than Vision Transformer would have been. Another notable adaptation, MISSFormer [8], is a location-free hierarchical

U-shape Transformer that solves the feature recognition constraint and scale information fusion problems using Transformer.

On the other hand, TransWnet [20] employs a row and column attention mechanism to capture the global semantic information across multiple scales. The channels are processed through a cross-like global receptive field. Other similar approaches are LeVit-Unet [21] and AFTer-UNet [22]. However, most of these methods simply add the combined CNN-Transformer models to conventional U-shaped encoder-decoder skip connections, limiting their ability to fully exploit the multivariate information generated by hierarchical encoders. Since the semantic divide between different scale layers is not taken into account, they cannot well combine the advantages of CNN with Transformer. While Transformer excels in modeling the global context, its self-attention mechanism leads to a lack of low-level features, making it inadequate in capturing fine-grained details. To address this limitation, researchers have explored the combination of CNN and Transformer to encode both local and global features [7,18]. For instance, Chen et al. proposed TransUnet [2], which leverages CNNs to extract low-level features and utilizes Transformers to model global information. TransUnet achieved a new record in CT abdominal multi-organ segmentation.

In this paper, we propose a novel and efficient combination of CNN and Transformer called GDTNet, aiming to further explore their potential for abdominal multi-organ segmentation in a powerful manner. Our approach leverages the strength of CNN in capturing local details hierarchically, while also exploiting the self-attention mechanism of Transformer to mine global semantic information. Inspired by the Focus Module in YOLOv5 [24], we design the Dilated Attention Module (DAM) for long-distance information collection, while being able to address the weakening of the Transformer by the window mechanism. Specifically, we first model the semantic information of the whole feature map by Dilated Transformer, which effectively expands the receptive field of the Transformer without increasing the computational complexity. In images, the relationship between visual tokens and their surrounding context typically depends on the content itself. To capture this, we incorporate a Gate Attention module into DAM to obtain spatial and level-aware weights via a linear layer. Supervised by Gate Attention, the output of Dilated Transformer is combined with the global semantic information collected through Global Average Pooling (GAP) to serve as the output of the DAM layer. This output is then fused with the information from all previous DAM layers and concatenated with the local information mined by the encoder to form the input to the decoder.

To validate the effectiveness of DAM, extensive ablation experiments are conducted on the publicly available abdominal multi-organ dataset Synapse, where we focus on comparing the performance improvement of each sub-module on the network. Experimental results show that DAM can effectively extract feature map information. For the overall network performance, our GDTNet outperforms ten state-of-the-art methods in terms of average Dice Similarity Coefficients, achieving high segmentation accuracy. In summary, our contributions are listed as follows:

- We propose a novel CNN-Transformer synergistic approach for mining global relationships and local details, named GDTNet, to effectively perform abdominal multi-organ segmentation.
- We innovatively design a Dilated Attention Module that captures intra- and inter-scale feature information through a three-parallel branch structure, combining GAP with Dilated Transformer to expand the effective receptive field and fully extract global semantic information under the supervision of Gate Attention.
- We conduct extensive ablation experiments on the abdominal multi-organ dataset Synapse to validate the effectiveness of the key components of our method. Moreover, the results show that our method outperforms ten state-of-the-art methods and achieves accurate abdominal multi-organ segmentation.

## 2    Methods

In this section, we will first introduce the general flow and specific structure of GDTNet, then introduce our proposed Dilated Attention Module (DAM) in detail, and finally analyze the features and advantages of this module. It is worth noting that the DAM is able to model hierarchical multi-scale information relevance and quickly collect global semantic information.

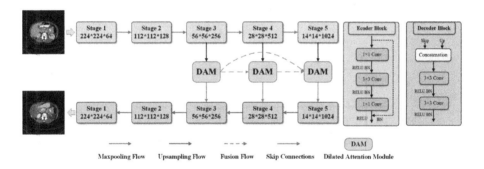

**Fig. 1.** The framework for GDTNet, where the Dilated Attention Module is used to simultaneously capture intra- and inter-scale feature information.

### 2.1    Overall Network Architecture

Figure 1 shows an overview of the proposed GDTNet. First, local information is extracted by encoder-decoder architecture in layers, and in the skip connection part, we add our proposed Dilated Attention Module (DAM). Given an input image $X^{H \times W \times C}$ with spatial dimensions of $H \times W$ and $C$ channels, we use a convolutional layer to transform its channel number to 64 as the input to the encoder. Each encoder block consists of two $1 \times 1$ convolutions and one $3 \times 3$

convolution, and here we use residual connections to prevent overfitting of the deep network. The corresponding decoder uses two $3 \times 3$ convolutional blocks and no longer uses residual connections. The number of channels in the encoder increases by a factor of two at a time as the depth of the network increases, and the resolution of the feature map decreases accordingly, while the decoder does the opposite. As the CNN gradually deepens, different granular features are extracted at different stages.

However, the above operation has a limitation: the receptive field of CNN limits its ability to extract global semantic information, which can only be captured after stacking multiple layers of CNNs. But this information is very important in medical image segmentation. In general, ordinary U-shaped architectures fail to fuse the information between multi-scale feature maps well. Therefore, considering the Transformer's strong advantage in capturing long-distance relationships, we propose the Dilated Attention Module (DAM) to learn global information within modeling scales as well as dependencies between different scales. Then we link the DAM at each level to further interact across multiple scales for richer feature representations. Finally, the output of the DAM is concatenated with the corresponding CNN block of the encoder as a skip connection and sent to the decoder for final prediction. In the following, we detail the concrete implementation details of the Dilated Attention Module.

## 2.2 Dilated Attention Module

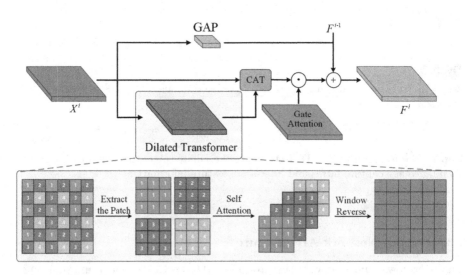

**Fig. 2.** Operational flowchart of Dilated Attention Module (DAM), which performs feature extraction through a three-parallel branching structure. "CAT" denotes the connection operation.

Figure 2 shows the flowchart of the operation of the Dilated Attention Module (DAM). Let $i$ refers to the corresponding CNN layer number, $N$ refers to the total number of modules, and $X_C^i$ denotes the output of each CNN branch in each layer, the feature map $F_{DAM}^i$ can be expressed as:

$$\begin{cases} [X_C^i, DT(X_C^i)] \times G_i + GAP(X_C^i), & i=1 \\ DWConv(F_{DAM}^{i-1}) + [X_C^i, DT(X_C^i)] \times G_i + GAP(X_C^i), & i=2,\ldots,N \end{cases} \quad (1)$$

where $[\cdot]$ denotes the concatenation function. We use Global Average Pooling (GAP) to directly assign actual category meanings to each channel to improve interpretability. GAP does not have to consume high computational, and can strengthen the connection between the categories and the feature map. DWConv$(\cdot)$ represents the deepwise convolution. Instead of pooling, we implement the downsampling by using DWConv, which divides each channel of the input features into a group, and each time the convolved features are connected to output features by the channel. Compared to pooling, deep convolution is learnable and structure-aware, while it has lower computational complexity when compared to regular convolution.

**Dilated Transformer.** To overcome the limitation of using Global Average Pooling (GAP) without learnable parameters, we design the Dilated Transformer (DT), which is a global-scaled multi-head self-attention mechanism. This allows for information interactions across different units in a dilated manner. To achieve this, we divide the whole feature map into windows with a window size denoted as $w$. We select a patch at every distance of $w$ and form a new window cell by combining all the selected patches. Subsequently, we perform self-attention within this window. Specifically, each patch within a window is combined with patches in the same position from other windows to create a new cell. This process continues until all the patches within the window have been utilized. In practice, we set the step size $w$ to 7. As a result, we can still access the semantic information of the whole feature map, while overcoming the drawbacks of high computational cost and limited fine-grained feature extraction typically associated with conventional transformers. In Eq. (1), we denote the Dilated Transformer module as $DT(\cdot)$.

**Gate Attention.** In images, selecting all the memorized information would result in redundancy or incorrect information. In general, the relationship between visual tokens and their surrounding context depends on the content itself. Leveraging this understanding, the Gate Attention module obtains spatial and level-aware weights through a linear layer. These weights are then multiplied pixel-by-pixel with the feature map, enabling control over the aggregation of queries from different levels. Specifically, for each layer of input $X_C^i$, assuming it has $c$ channels, we apply a split operation to change its channel number to $c+1$. The feature map on the $c+1$th channel serves as the value for the Gate Attention. This allows the Gate Attention units to selectively retain what they deem important based on the content itself. In Eq. (1), the Gate Attention module is denoted as $G_i$.

**Overall.** We concatenate the output of the Dilated Transformer module with another copy of the CNN branch and multiply it with Gate Attention. Finally, the result obtains the output of the current DAM layer by adding the result of GAP and the output of the previous DAMs. That is:

$$F_{out}^i = \sum_{k=1}^{i} Dwon(F_{DA}^k) \tag{2}$$

where $Dwon(\cdot)$ denotes the downsampling operation. In this case, each DAM layer benefits from the feature information learned by all the previous DAMs, enabling improved fusion of semantic information across different scales.

### 2.3 Discussion and Analysis

The Vision Transformer performs self-attention over the whole feature map, collecting global semantic information at the beginning to capture the pairwise relationships between patches, which results in a squared complexity of the number of patches. Since the resolution of images processed in the medical image segmentation domain is generally high, such primitive self-attention will no longer be suitable for tasks related to medical image segmentation. Later proposed Swin Transformer consists of two parts: window attention and shift window attention mechanism, which significantly reduces the time complexity of using fine-grained patches and provides the Vision Transformer with previously unavailable local information. However, it is also because of this shifted window-based self-attention mechanism that it does not model the global semantic information at the beginning, but divides the feature map into many windows, similar to the CNN that requires a certain number of layers of superposition to model the remote information, which inevitably makes Swin Transformer weaker to a certain extent than the original advantage of Transformer. In contrast, our designed Dilated Attention Module can effectively enlarge the receptive field in the self-attention mechanism, but the number of involved patches does not increase.

This means that the Dilated Attention Module can retain the advantage of the Transformer in rapidly modeling long-range information interactions and model dependencies between long-range patches in a relatively more dynamic and efficient manner without additional computational cost.

## 3   Experimental Results and Discussion

### 3.1   Dataset

We train and test the model on the publicly available Synapse dataset, which performs 30 abdominal CT scans on 8 abdominal organs (Aorta, Gallbladdr,

Left Kidney, Right Kidney, Liver, Pancreas, Spleen, Stomach. Synapse). Each consists of 148 to 241 axial slices (depending on body size). A total of 3779 axial CT images are obtained with a resolution size of $512 \times 512$. They are randomly divided into 18 for training and 12 for testing [1,2]. We use mean Dice Similarity Coefficient (DSC) and mean Hausdorff Distance (HD) for performance evaluation.

## 3.2  Implementation Details

We train our GDTNet with Pytorch on an NVIDIA RTX A5000 GPU with 24 GB of video memory. The input image size is $224 \times 224$, the initial learning rate is 0.05, the maximum training epoch is 200, the batch size is 8, and the SGD optimizer with a momentum of 0.9 and a weight decay of 1e-4 is used.

## 3.3  Results and Discussion

We conduct experiments in two aspects: (1) For verifying the effectiveness of the proposed DAM, we conduct ablation experiments of Dilated Transformer, Gate Attention, and Global Average Pooling within the Dilated Attention Module. (2) For verifying the high accuracy of the model, we compare GDTNet to other existing advanced segmentation methods.

**Ablation Experiments.** This section verifies the effectiveness of the current design by conducting ablation experiments on the design of DAM, tuning the Dilated Transformer, Gate Attention, and Global Average Pooling modules within the module. The experimental results are shown in Table 1, where the best results are in bold.

**Table 1.** Ablation experiments on the Synapse dataset. GA, DT, and GAP denote the Gate Attention, Dilated Transformer, and Global Average Pooling, respectively. B denotes Baseline.

| Ver | Method | DSC ↑ | HD ↓ | Aorta | Gallbladdr | Kidney (L) | Kidney (R) | Liver | Pancreas | Spleen | Stomach |
|-----|--------|-------|------|-------|-----------|-----------|-----------|-------|----------|--------|---------|
| No.1 | B | 82.08 | 19.24 | 88.22 | 68.18 | 88.09 | 85.87 | 95.03 | 61.14 | 90.01 | 79.82 |
| No.2 | B+GA | 82.11 | 25.48 | 89.60 | 66.90 | 81.12 | 80.40 | 95.16 | 67.42 | 92.43 | 83.85 |
| No.3 | B+DT | 83.47 | 19.23 | 89.28 | 72.03 | 87.24 | 83.96 | 95.26 | 67.50 | 91.38 | 81.08 |
| No.4 | B+GAP | 82.75 | 16.39 | 89.43 | 62.74 | 85.51 | 84.48 | 94.58 | 70.28 | 92.52 | 82.45 |
| No.5 | B+DT+GA | 84.47 | 14.65 | 89.62 | 72.79 | 87.22 | 85.69 | 95.34 | 70.90 | 93.08 | 81.11 |
| No.6 | B+DT+GAP | 84.73 | 16.23 | **89.68** | **74.19** | 87.35 | 85.67 | 95.58 | **72.04** | 92.74 | 80.62 |
| No.7 | B+GA+GAP | 84.40 | 24.99 | 89.55 | 73.72 | 88.11 | 84.69 | 95.08 | 68.11 | 92.02 | **83.93** |
| No.8 | B+GA+DT+GAP *in sequential* | 83.46 | 21.31 | 89.01 | 74.01 | 84.34 | 81.87 | 95.29 | 68.00 | 92.54 | 82.61 |
| No.9 | B+GA+DT+GAP *in parallel* | **85.06** | **13.16** | 88.48 | 73.24 | **90.30** | **86.48** | **96.12** | 70.87 | **93.64** | 81.36 |

As shown in Table 1, the inclusion of DT, GA, and GAP modules individually brings improvements in the segmentation results of the network. The network enhancement is most obvious with the addition of the DT module, a result that validates the effectiveness of the DT module in modeling long-range dependencies. If these modules are combined two by two, better results can be achieved, in particular the combination of Dilated Transformer and Global Average Pooling achieves the best results in the segmentation of Aorta, Gallbladdr, and Pancreas. This result verifies that combining the DT module with GAP enables fine segmentation of small targets and efficient learning of relationships between different targets. The combination of GA and GAP yields the best results for Stomach, indicating that using GAP for global semantic information extraction is effective. However, since GAP lacks trainable parameters, the information it captures is simple and straightforward. To overcome this limitation, Gate Attention is employed to allow the network to focus on target structures of different shapes and sizes. In addition, we try to concatenate the three modules in a sequential manner. The output of the Dilated Transformer is directly multiplied with Gate Attention to add the output to its own global average pooling, and this result is pooled with the output features of the other DAM layers as the output of this layer. While simple serialization provides some enhancement with respect to the baseline network, it blurs the relationship of the three modules to each other's constraints and therefore has an impact on the final result.

Experimental results show that the network model achieves the best segmentation results when all three modules are used simultaneously and when all three belong to parallel branches. This means that our proposed Dilated Attention Module can effectively handle various abdominal multi-organ segmentation problems.

**Table 2.** Comparison with state-of-the-art methods on the Synapse dataset. The best results are highlighted in bold fonts.

| Method | Year | DSC ↑ | HD ↓ | Aorta | Gallbladdr | Kidney (L) | Kidney (R) | Liver | Pancreas | Spleen | Stomach |
|---|---|---|---|---|---|---|---|---|---|---|---|
| U-Net [13] | 2015 | 76.85 | 39.70 | 89.07 | 69.72 | 77.77 | 68.60 | 93.43 | 53.98 | 86.67 | 75.58 |
| Att-UNet [10] | 2018 | 77.77 | 36.02 | **89.55** | 68.88 | 77.98 | 71.11 | 93.57 | 58.04 | 87.30 | 75.75 |
| R50 ViT [19] | 2020 | 71.29 | 32.87 | 73.73 | 55.13 | 75.80 | 72.20 | 91.51 | 45.99 | 81.99 | 73.95 |
| TransUnet [2] | 2021 | 77.48 | 31.69 | 87.23 | 63.13 | 81.87 | 77.02 | 94.08 | 55.86 | 85.08 | 75.62 |
| UCTransNet [17] | 2021 | 78.23 | 26.74 | 88.86 | 66.97 | 80.18 | 73.17 | 93.16 | 56.22 | 87.84 | 79.43 |
| SwinUNet [1] | 2021 | 79.12 | 21.55 | 85.47 | 66.53 | 83.28 | 79.61 | 94.29 | 56.58 | 90.66 | 76.60 |
| MISSFormer [8] | 2021 | 81.96 | 18.20 | 86.99 | 68.65 | 85.21 | 82.00 | 94.41 | 65.67 | 91.92 | 80.81 |
| HiFormer [6] | 2022 | 80.69 | 19.14 | 87.03 | 68.61 | 84.23 | 78.37 | 94.07 | 60.77 | 90.44 | 82.03 |
| ScaleFormer [7] | 2022 | 82.86 | 16.81 | 88.73 | **74.97** | 86.36 | 83.31 | 95.12 | 64.85 | 89.40 | 80.14 |
| TransWnet [20] | 2023 | 84.56 | 14.12 | 89.48 | 74.52 | 88.76 | 84.92 | 95.47 | 67.49 | **93.71** | **82.37** |
| GDTNet | Ours | **85.06** | **13.16** | 88.48 | 73.24 | **90.30** | **86.48** | **96.12** | **70.87** | 93.64 | 81.36 |

**Comparison with Other Methods.** To validate the overall segmentation performance of the proposed GDTNet, we compare GDTNet with three types of advanced segmentation methods, including pure CNN-based segmentation methods, such as U-Net [13], Att-UNet [10], pure Transformer-based segmentation methods, such as Vision Transformer [19], SwinUNet [1], MISSFormer [8], and

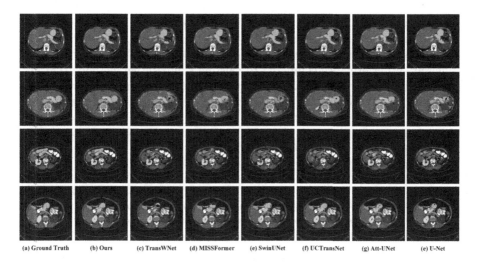

**Fig. 3.** The visual comparison of the proposed method with the state-of-the-art approaches on the Synapse dataset.

the methods based on the combination of CNN and Transformer, such as TransUNet [2], UCTransNet [17], HiFormer [6], ScaleFormer [7], and TransWnet [20]. For a fair comparison, their initially released code and released settings are used in the experiments. The experimental results are shown in Table 2, where the best results are in bold.

It can be found that in general, our GDTNet outperforms all other existing methods in terms of average Dice Similarity Coefficient and average Hausdorff Distance, while being able to achieve the best segmentation results on most abdominal organs. In addition, Fig. 3 shows the visual segmentation results of different models for multiple organs of the abdomen. It can be seen that our GDTNet is able to provide more accurate segmentation results on almost all organs compared to other state-of-the-art methods.

## 4  Conclusion

In this paper, we propose a novel complementary CNN and Transformer coupled segmentation method, called GDTNet, for abdominal multi-organ segmentation from CT images. It introduces a new module Dilated Attention Module (DAM) in the skip-connection part of the traditional encoder-decoder, which utilizes a three-parallel branch structure supervised by Gate Attention to quickly converge global semantic information and perform multi-scale feature fusion. To address the issue of weak deep semantic information present in the U-shaped structure, we fuse each output layer of DAM. In other words, each DAM layer uses information from all past DAM layers. We show the advantages of the GDTNet, which makes full use of multi-scale features in the encoding stage and narrows the semantic gap between the layers. Compared to Vision Transformer, our proposed

Dilated Transformer uses fewer parameters and captures the global information of the entire image at the beginning. Finally, the experimental evaluation on the publicly available Synapse dataset shows that our GDTNet achieves higher segmentation performance compared to existing comparative methods that are CNN-based, Transformer-based, or a combination of both. In future, we plan to reproduce our GDTNet on the open source deep learning platform paddle.

**Acknowledgments.** This work was supported in part by the National Natural Science Foundation of China under Grants 62272404 and 62372170, in part by the Open Project of Key Laboratory of Medical Imaging and Artificial Intelligence of Hunan Province under Grant YXZN2022004, in part by the Natural Science Foundation of Hunan Province of China under Grants 2022JJ30571 and 2023JJ40638, in part by the Research Foundation of Education Department of Hunan Province under Grant 21B0172, in part by the Innovation and Entrepreneurship Training Program for China University Students under Grant 202210530002, and in part by the Baidu Program.

# References

1. Cao, H., et al.: Swin-unet: Unet-like pure transformer for medical image segmentation. In: Karlinsky, L., Michaeli, T., Nishino, K. (eds.) ECCV 2022. LNCS, vol. 13803, pp. 205–218. Springer, Cham (2022). https://doi.org/10.1007/978-3-031-25066-8_9
2. Chen, J., et al.: Transunet: transformers make strong encoders for medical image segmentation (2021)
3. DENG, J.: A large-scale hierarchical image database. Proc. IEEE Comput. Vision Pattern Recogn. **2009** (2009)
4. Dosovitskiy, A., et al.: An image is worth 16x16 words: transformers for image recognition at scale (2021)
5. Gibson, E., et al.: Automatic multi-organ segmentation on abdominal CT with dense v-networks. IEEE Trans. Med. Imaging **37**(8), 1822–1834 (2018)
6. Heidari, M., et al.: Hiformer: hierarchical multi-scale representations using transformers for medical image segmentation. In: Proceedings of the IEEE/CVF Winter Conference on Applications of Computer Vision (WACV), pp. 6202–6212 (January 2023)
7. Huang, H., et al.: Scaleformer: revisiting the transformer-based backbones from a scale-wise perspective for medical image segmentation (2022)
8. Huang, X., Deng, Z., Li, D., Yuan, X.: Missformer: an effective medical image segmentation transformer (2021)
9. Jin, Q., Meng, Z., Pham, T.D., Chen, Q., Wei, L., Su, R.: Dunet: a deformable network for retinal vessel segmentation. Knowl.-Based Syst. **178**, 149–162 (2019)
10. Lian, S., Luo, Z., Zhong, Z., Lin, X., Su, S., Li, S.: Attention guided u-net for accurate iris segmentation. J. Vis. Commun. Image Represent. **56**, 296–304 (2018)
11. Liu, Z., et al.: Swin transformer: hierarchical vision transformer using shifted windows. In: Proceedings of the IEEE/CVF International Conference on Computer Vision (ICCV), pp. 10012–10022, October 2021
12. Long, J., Shelhamer, E., Darrell, T.: Fully convolutional networks for semantic segmentation. In: Proceedings of the IEEE Conference on Computer Vision and Pattern Recognition (CVPR), June 2015

13. Ronneberger, O., Fischer, P., Brox, T.: U-net: convolutional networks for biomedical image segmentation. In: Navab, N., Hornegger, J., Wells, W.M., Frangi, A.F. (eds.) MICCAI 2015. LNCS, vol. 9351, pp. 234–241. Springer, Cham (2015). https://doi.org/10.1007/978-3-319-24574-4_28

14. Srivastava, R.K., Greff, K., Schmidhuber, J.: Highway networks (2015)

15. Touvron, H., Cord, M., Douze, M., Massa, F., Sablayrolles, A., Jegou, H.: Training data-efficient image transformers & distillation through attention. In: Meila, M., Zhang, T. (eds.) Proceedings of the 38th International Conference on Machine Learning. Proceedings of Machine Learning Research, vol. 139, pp. 10347–10357. PMLR, 18–24 July 2021

16. Vaswani, A., et al.: Attention is all you need. In: Guyon, I., et al. (eds.) Advances in Neural Information Processing Systems, vol. 30. Curran Associates, Inc. (2017)

17. Wang, H., Cao, P., Wang, J., Zaiane, O.R.: Uctransnet: rethinking the skip connections in u-net from a channel-wise perspective with transformer. In: Proceedings of the AAAI Conference on Artificial Intelligence, vol. 36, mo. 3, pp. 2441–2449 (2022)

18. Wang, H., et al.: Mixed transformer u-net for medical image segmentation. In: ICASSP 2022–2022 IEEE International Conference on Acoustics, Speech and Signal Processing (ICASSP), pp. 2390–2394 (2022)

19. Wang, W., et al.: Pyramid vision transformer: a versatile backbone for dense prediction without convolutions. In: Proceedings of the IEEE/CVF International Conference on Computer Vision (ICCV), pp. 568–578, October 2021

20. Xie, Y., Huang, Y., Zhang, Y., Li, X., Ye, X., Hu, K.: Transwnet: integrating transformers into CNNs via row and column attention for abdominal multi-organ segmentation. In: ICASSP 2023–2023 IEEE International Conference on Acoustics, Speech and Signal Processing (ICASSP), pp. 1–5 (2023)

21. Xu, G., Wu, X., Zhang, X., He, X.: Levit-unet: make faster encoders with transformer for medical image segmentation (2021)

22. Yan, X., Tang, H., Sun, S., Ma, H., Kong, D., Xie, X.: After-unet: axial fusion transformer unet for medical image segmentation. In: Proceedings of the IEEE/CVF Winter Conference on Applications of Computer Vision (WACV), pp. 3971–3981 (January 2022)

23. Zhou, Z., Siddiquee, M.M.R., Tajbakhsh, N., Liang, J.: Unet++: redesigning skip connections to exploit multiscale features in image segmentation. IEEE Trans. Med. Imaging **39**(6), 1856–1867 (2020)

24. Zhu, X., Lyu, S., Wang, X., Zhao, Q.: Tph-yolov5: improved yolov5 based on transformer prediction head for object detection on drone-captured scenarios (2021)

# Fine-Grained Multi-modal Fundus Image Generation Based on Diffusion Models for Glaucoma Classification

Xinyue Liu[2], Gang Yang[1,2(✉)], Yang Zhou[3], Yajie Yang[3], Weichen Huang[2], Dayong Ding[3], and Jun Wu[4]

[1] MOE Key Lab of DEKE, Renmin University of China, Beijing, China
yanggang@ruc.edu.cn
[2] School of Information, Renmin University of China, Beijing, China
[3] Vistel Inc., Beijing, China
[4] School of Electronics and Information,
Northwestern Polytechnical University, Xi'an, China

**Abstract.** With the emergence of Foundation Model, the generation quality and generalisation ability of image generation method have been further improved. However, medical image generation is still a challenging and promising task. Recently, diffusion-based models are more prominent in multi-modal image generation for its flexibility. Therefore, in order to solve the problem of lack of high-quality medical images and high annotation costs, we propose a fine-grained multi-modal fundus image generation method based on foundation models to research an efficient way of data augmentation. First, we adopt optic fundus images, fundus vessel images and class textual information to form a weakly supervised fine-tuning dataset. Then, based on the Stable-Diffusion and Control-Net model, we fine-tune our method by LoRA model to generate high-resolution fundus images of special diseases in a targeted manner. Furthermore, we use these synthetic fundus images in conjunction with existing datasets for data augmentation or model fine-tuning to improve performance in the glaucoma classification task. Extensive experiments have shown that our method produces high quality medical fundus images and can be well applied to real-world medical imaging tasks. Moreover, experimental results show that we are able to generate fundus images that act as an augmentation, meaning that the generation of foundation models is effective in certain domains.

**Keywords:** Foundation Model · Multi-Modal AIGC · Glaucoma Classification

## 1 Introduction

In the realm of medical imaging, the low volume of available data and challenges in acquiring it pose a persistent issue, hindering the adoption of artificial intelligence in the field and limiting progress in associated tasks. As a type of medical image, fundus image data faces the same issue. Therefore, expanding the amount of fundus image data is necessary. We attempt to generate fundus images by using foundation models to explore a feasible solution to this problem.

Generative modeling is a particular focus for researchers, with particular attention given to the concept of content generated through Artificial Intelligence, known as AIGC [5]. In the field of image generation, prominent generative models include VAE [12], flow-based Generative Model [7], GAN [1,27], and Diffusion Models [6,19]. Among these generative models, Diffusion Models, especially the Stable Diffusion Model, have demonstrated superior performance. As a foundation model, the Stable Diffusion Model comprises a significant quantity of image data, allowing for a robust basis for producing area-specific images. Besides, the quantity and quality of the generated images are critical factors for effectively deploying image generation techniques to medical imaging applications. Therefore, we employed the Stable Diffusion Model along with its associated model, ControlNet [26], to generate a significant number of high-quality glaucomatous and non-glaucoma fundus images.

Extensive experiments indicate that our method generates medical images that perform well in both qualitative and quantitative aspects. In the qualitative aspect, fundus images generated by our method feature distinct structures and precisely defined lesions. In the quantitative aspect, the results of glaucoma classification are improved after data augmentation using the images generated by our method. Through the experiment, the conclusion can be drawn that our method is able to generate high-quality fundus images that can be applied to practical medical tasks. Our contributions in this paper are as follows:

- We propose an effective method based on foundation models for fundus image generation. This method can generate vast high-quality fundus images with distinct structures and precisely defined lesions flexibly.
- Our method has been proven to be an effective approach for glaucoma classification. Extensive experiments on multiple datasets show that data augmentation using the filtered synthetic images can improve the performance of glaucoma classification.
- Our work demonstrates that foundation models can support the generation of fundus images. We initially explore ways to improve the performance of medical image generation by studying the generation of glaucoma fundus images.

## 2 Related Work

### 2.1 Foudation Models

Recently, as computing power continues to advance, foundation models have been becoming increasingly popular. In the field of language modeling, OpenAI has proposed a large-scale language pre-training model, GPT [4,15,16]. This model is equipped with strong linguistic correlation capabilities and is able to learn more information from context. In the multi-modal domain, CLIP [14] is proposed as a pre-trained neural network model for matching images and text. DALL·E [17,18] combines CLIP and Diffusion Models to advance the excellence of image generation for text-contextualized problems. Besides, the emergence of the Latent Diffusion Model and the Stable Diffusion Model have greatly advanced the field of image generation.

Stable Diffusion is a text-to-image diffusion model working in the same principle as Latent Diffusion Models (LDMs) [19] that can generate realistic images with given prompts. Dhariwal et al. [6] achieved the first time in the field of image generation using Diffusion Models to generate images with better results than GAN models. However, since the diffusion model has to be calculated at the pixel level, the computational effort is enormous. The Latent Diffusion Model is proposed aiming at addressing the limitation of high computational resource consumption in the diffusion model. By transferring the training and inference processes of pixel-based Diffusion Models (DMs) to a compressed lower-dimension latent space, LDMs are able to compute more efficiently. Due to the advantages of LDMs over pixel-based DMs, the model has been widely employed in multiple tasks, including unconditional image generation [24], text-to-image synthesis [11,20,21], video synthesis [2] , and super-resolution [13].

## 2.2  AIGC

Artificial Intelligence Generated Content (AIGC) is an AI technology that generates relevant content with appropriate generalization capabilities based on AI methods by learning from existing data. After the Generative Adversarial Network (GAN) [8] model was proposed, GAN-based image generation methods have become increasingly popular. Zhu et al. [27] propose CycleGAN which achieved domain migration task in the absence of paired examples. Brock et al. [3] propose BigGAN, a model that uses orthogonal regularisation, truncation tricks, etc., to achieve the goal of synthesizing image samples with higher resolution and more diversity, and achieves excellent generation results on ImageNet. In addition to GAN-based image generation methods, VAE-based image generation has also been attempted. Oord et al. [12] propose the VQ-VAE model based on VAE for high-resolution image synthesis. In this work, the authors also propose the use of the codebook mechanism to encode images into discrete vectors, which inspired countless subsequent works, including the Stable Diffusion Model.

With the proposal of the Latent Diffusion Model and Stable Diffusion Model, Diffusion Models have developed into the most popular and best generative model for producing images. While Stable Diffusion is capable of producing high-quality images, it lacks the ability to generate images conditionally. Zhang et al. [26] propose the ControlNet to improve the control and versatility of DMs in various applications. Thus, DMs, like the Stable Diffusion Model are able to be enhanced by enabling conditional inputs such as edge maps and segmentation maps with or without prompts. ControlNet clones the weights of a large diffusion model into a 'trainable copy', which retains the network's ability to learn from billions of images, and a 'locked copy', which is trained on a task-specific dataset to learn conditional control. In this way, foundation models can retain their ability to generalize when trained on a specific domain.

Previous work on fundus image generation has been mainly based on GAN models. Guo et al. [9] propose a method for generating fundus images based on a "Combined GAN" (Com-GAN), which consists of two sub-networks: im-WGAN and im-CGAN. Fundus images are generated using the trained model, and then the generated images are evaluated qualitatively and quantitatively, and the images are added to the original image set for data augmentation. Shenkut et al. [22] propose a method for synthesizing fundus images (Fundus GAN) using Generative Adversarial Networks (GANs). This method achieves image-to-image conversion between vascular trees and fundus images via unsupervised generative attention networks.

## 3 Methodology

### 3.1 Overview

To address the challenges arising from limited data availability and the complexities associated with acquiring medical images, we present a three-phase glaucoma classification method grounded in diffusion models. An overview of our approach is illustrated in Fig. 1. The method is systematically partitioned into three distinct phases: the data construction phase, the generation phase, and the classification phase. In the data construction stage, we construct the necessary data by leveraging existing fundus image datasets. Employing two refinements of the Stable Diffusion Model, high-resolution fundus images are generated. To further enhance the fidelity of vascular information in the generated fundus images, the ControlNet model is refined based on the fine-tuned Stable Diffusion Model and the vascular images are seamlessly integrated into the fundus images. In oder to obtain images of higher quality, these generated images are filtered subsequently. The resulting high-quality synthetic images are then utilized for data augmentation and fine-tuning, thereby amplifying the effectiveness of glaucoma classification.

### 3.2 Fine-Grained Multi-model Fundus Image Generation Method

**Data Construction Phase.** In our methodology, three types of data are employed: a text prompt, fundus images, and vascular mask maps. Initially, the term 'fundus' serves as the text prompt. Despite training the generation phase separately for glaucoma and non-glaucoma images, our experiments reveal that maintaining consistency with the 'fundus' prompt yields superior training performance compared to using 'glaucoma' and 'non-glaucoma' prompts individually. Consequently, for optimal generation of high-quality fundus images during this phase, we exclusively employ the coarse-grained term 'fundus' as the text prompt. In the second data type, the fundus images used for training are derived

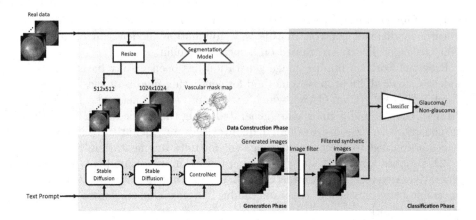

**Fig. 1.** The method of fine-grained multi-modal fundus image generation based on diffusion models for glaucoma classification. Solid lines indicate data stream, while dotted lines indicate parameter stream.

from resizing an existing fundus image dataset. These resized images, formatted to dimensions of $512 \times 512$ and $1024 \times 1024$, are instrumental in training our model effectively. The third data type involves the use of vascular mask maps, obtained through a segmentation model, MMF-Net [25]. To enhance the vascular information within the generated fundus image, the vascular mask maps are segmented from the fundus images using the MMF-Net, with these maps then incorporated into the generation stage.

**Generation Phase.** In the generation phase, the Stable Diffusion Model and ControlNet are leveraged to produce high-quality fundus images for both glaucoma and non-glaucoma cases. Notably, to enhance accuracy in generating fundus images for these distinct conditions, separate training processes are employed for glaucomatous and non-glaucoma fundus images. Given the absence of highly resolution pre-trained generative models and the nuanced nature of fundus image generation, the image generation phase is divided into two essential processes: the generation of high-resolution images and the incorporation of vascular information.

Consistent use of the term 'Fundus' as the text prompt in Fig. 1 is maintained to ensure coherence throughout the generation phase. To fine-tune the Stable Diffusion Model, the LoRA [10] method is used. During the fine-tuning process, the parameters of the original model are frozen and only the newly added model weights are trained to fine-tune. The first Stable Diffusion Model (SD1) is fine-tuned using low-resolution fundus images ($512 \times 512$), initialized with Stable Diffusion v1.5[1], yielding a model capable of generating low-resolution fundus images. Subsequently, to obtain a model (SD2) capable of generating

---

[1] https://huggingface.co/runwayml/stable-diffusion-v1-5.

high-resolution fundus images, the parameters of SD1 initialize the second Stable Diffusion Model, which is then fine-tuned using high-resolution fundus images (1024 × 1024).

Nevertheless, the fundus images generated by SD2 exhibit limitations in terms of vascular information. To address this shortfall, we leverage the parameters of SD2 to initialize ControlNet. The corresponding vascular mask maps are then supplied to ControlNet for fine-tuning, specifically focusing on enhancing vascular details. It is notable that during the training stage, a strict one-to-one correspondence is maintained between the fundus map and the vascular mask map. In contrast, during the inference stage, there are two approaches to generate fundus images, text-prompt-based generation and image-based generation. For our text-prompt-based generation, only the vessel mask map is necessary, and noise is randomly sampled. On the other hand, for our image-based generation, both the fundus images and the vessel mask maps are required, although they do not necessarily need to maintain a one-to-one correspondence. ControlNet, integral to the generation phase, plays an important role in producing fine-grained fundus images by addressing the limitations in vascular information.

**Classification Phase.** Due to the substantial volume of generated fundus images in the generation phase, a filtering process is necessary. Two principal filtering techniques are employed, namely, quality modeling and manual filtering. In the quality modeling approach, the generated images undergo classification on a scale from 1 to 5 based on their quality, with 1 indicating the highest and 5 indicating the lowest quality. Images classified as grade 5, signifying low quality, are subsequently filtered out. In the manual filtering approach, generated images demonstrating poor quality visually undergo manual filtering for exclusion. Using both techniques, fundus images that are produced with notably low quality are excluded.

To enhance the effectiveness of glaucoma classification utilizing the filtered synthetic fundus images, we employ a data augmentation method in two distinct ways. Firstly, we combine images from the original dataset with the filtered synthetic images in a proportional mix, utilizing this amalgamated dataset to train the classifier. Secondly, we exclusively utilize the filtered synthetic fundus images for initial classifier training, followed by fine-tuning with the original existing fundus images.

Our approach capitalizes on the robust image generation capabilities of Stable Diffusion, serving as the foundational model enriched with potential medical domain insights. By incorporating ControlNet's precision in controlling image generation, we segment the fundus image generation process into distinct phases. Commencing with the generation of low-resolution fundus images, followed by high-resolution counterparts, and culminating with the incorporation of vascular information, our method meticulously controls detail generation to produce realistic fundus images. Moreover, the utilization of filtered synthetic images enhances our methodology, whether through data augmentation or fine-tuning for glaucoma classification. Subsequent experiments will show that our methodology is capable of producing high-quality, fine-grained fundus images.

# 4    Experiments

## 4.1    Experimental Settings

**Datasets.** Glaucomatous and non-glaucoma images used to train the image generation model are acquired from the fundus image dataset REFUGE[2]. To generate the vascular mask map, we segment the vascular maps using the vascular segmentation model. REFUGE, RIMONE_DL, and Drishti-GS [23] datasets are used to train and test the medical image classification models. Besides, since the XH dataset is a privately owned dataset rather than a public dataset, the data labeling of this dataset is coarse and not up to the standard of test data. Consequently, it is only intended for use in training the classifier, but not for testing the classifier. All these datasets are described in detail in Table 1.

**Table 1.** Fundus Images Related Datasets

| Dataset | Glaucoma | | | Non-glaucoma | | |
|---|---|---|---|---|---|---|
| | train | validate | test | train | validate | test |
| REFUGE | 40 | 40 | 40 | 360 | 360 | 360 |
| RIMONE_DL | 120 | – | 52 | 219 | – | 94 |
| Drishti-GS | 32 | – | 38 | 18 | – | 13 |
| XH | 728 | – | – | 4424 | – | – |

The REFUGE dataset is composed of 1200 images (training, validate and test set). The training, test and validation set are all composed of 40 glaucomatous and 360 non-glaucoma images.

The RIMONE_DL dataset includes 485 images (training and test set). The training set contains 219 non-glaucoma images and 120 glaucomatous images, and the test set contains 94 non-glaucoma images and 52 glaucomatous images.

The Drishti-GS dataset is composed of 101 images (training and test set). The training set of Drishti-GS contains 32 glaucomatous images and 18 nonglaucoma images. The test set includes 38 glaucomatous images and 13 nonglaucoma images.

The XH dataset is a private dataset which contains 5125 images (training set). The training set contains 4424 non-glaucoma images and 728 glaucomatous images. Due to the previously mentioned labeling issues with the XH dataset, it is only used for training the classifiers and not for testing.

Additionally, the filtered synthetic fundus images are composed into a dataset and named as SD-Generated dataset (SDG), which contains 1.6k glaucoma fundus images and 7k normal fundus images.

---

[2] https://refuge.grand-challenge.org/.

**Metrics.** Our proposed method undergoes evaluation through a combination of visual inspection and quantitative metrics. The classification effectiveness of the classifier is assessed using sensitivity (Sen), specificity (Spe), F1-measure, and the Area Under the Receiver Operating Characteristic Curve (AUC). Sen and Spe are key indicators describing the classifier's accuracy in reporting the presence or absence of a disease. Sen represents the true positive rate, while Spe denotes the true negative rate. F1-measure and AUC offer a comprehensive evaluation, considering both Sen and Spe. A higher F1-measure indicates an improved balance and overall performance of Sen and Spe, while a higher AUC signifies the ease with which the classifier identifies the point where Sen and Spe are balanced. In glaucoma classification, the interplay between Spe and Sen is crucial. Consequently, the F1-measure and AUC metrics hold greater significance than Spe and Sen alone, providing a more holistic assessment of the classification model's overall performance.

**Implementation Details.** Given that the images in the REFUGE dataset are originally 2056 × 2124 in size, we initiate our preprocessing by resizing these images to dimensions of 512 × 512 and 1024 × 1024. Additionally, the original blood vessel mask undergoes expansion using a 10 × 10 kernel, magnifying it tenfold. This expansion aims to preserve the recognizability of blood vessel contours after fine blood vessels are inevitably downsized fourfold during the downsampling process. Moreover, after experimenting with various prompts, we ultimately establish the input model prompt for 'fundus' to achieve optimal fundus image generation. Glaucoma and non-glaucoma fundus images are produced seperately through two models trained individually.

## 4.2 Experimental Results

Our method is compared with GAN-based methods for medical image generation. A multi-disease dataset is used to train CycleGAN to generate fundus images. Comparing the fundus image generated by CycleGAN with the fundus image generated by our method, it is clear that our method generates a better quality fundus image.

The fundus image generated by CycleGAN is shown in the first row of Fig 2. Though the training data for CycleGAN is more extensive, the fundus images generated by CycleGAN still present some issues. The optic cups and discs of the generated fundus images are rather unclear, and the blood vessels lack realism. In contrast, the fundus images produced by our method contain abundant vascular information and the lesions are distinctly visible, as demonstrated in the second and third row of Fig. 2.

The REFUGE and SD-Generated dataset (SDG) datasets are utilized to assess the validity of our filtered synthetic images in genuine medical tasks, pertaining to medical image classification. Our method is evaluated using two medical image classification models, InceptionV3 and Swin-Transformer. Initially, we determined the effectiveness of the method for classifying glaucoma using

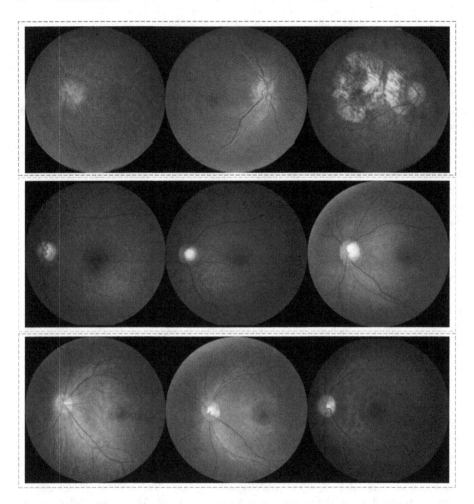

**Fig. 2.** The Fundus images produced by CycleGAN and our method. The images in the red box are generated by CycleGAN, while the images inside the yellow and blue boxes display the glaucomatous and non-glaucoma fundus images generated by our method respectively. (Color figure online)

three different existing datasets and the SDG dataset separately. Subsequently, the model trained with the SDG dataset is fine-tuned using the existing original dataset and tested the method for glaucoma classification. Finally, the SDG dataset and the existing original dataset are combined in a 75% : 25% ratio and the classifier is trained to obtain the method's classification performance. We evaluated the efficacy of the data augmentation technique on InceptionV3 and Swin-Transformer classification models utilizing the REFUGE, Drishti-GS, and Rimone_DL datasets. The classification results tested on the REFUGE, Drishti-GS, and Rimone_DL dataset are presented in Table 2 and Table 3.

**Table 2.** Performance of Our Method on InceptionV3 (Indices in %)

| Method | Rimone_DL | | | | Drishti-GS | | | | REFUGE | | | |
|---|---|---|---|---|---|---|---|---|---|---|---|---|
| | Sen | Spe | F1 | AUC | Sen | Spe | F1 | AUC | Sen | Spe | F1 | AUC |
| InceptionV3 | **92.3** | 94.7 | **93.5** | 97.2 | **100** | 46.2 | 63.2 | 93.3 | **82.5** | 91.4 | **86.7** | 95.6 |
| InceptionV3(+SDG) | 57.7 | 95.7 | 72.0 | 93.2 | 62.5 | **99.4** | **76.8** | **97.6** | 62.5 | 99.4 | 76.8 | 97.6 |
| InceptionV3(+SDG+FT) | 90.4 | 96.8 | **93.5** | 97.3 | 97.4 | 53.9 | 69.3 | 97.0 | 70.0 | **99.7** | 82.3 | 97.4 |
| InceptionV3(+SDG+Aug) | 82.7 | **97.9** | 89.6 | **97.4** | 97.4 | 46.2 | 62.7 | 95.3 | 67.5 | **99.7** | 80.5 | **97.8** |

**Table 3.** Performance of Our Method on Swin-Transformer(Indices in %)

| Method | Rimone_DL | | | | Drishti-GS | | | | REFUGE | | | |
|---|---|---|---|---|---|---|---|---|---|---|---|---|
| | Sen | Spe | F1 | AUC | Sen | Spe | F1 | AUC | Sen | Spe | F1 | AUC |
| Swin-Transformer | 90.4 | 88.3 | 89.3 | 95.7 | 57.9 | 84.6 | 68.8 | 81.8 | 75.0 | 94.4 | 83.6 | 94.8 |
| Swin-Transformer(+SDG) | 46.2 | **98.9** | 62.9 | 92.6 | 57.5 | **100** | 73.0 | **95.9** | 57.5 | **100** | 73.0 | 95.9 |
| Swin-Transformer(+SDG+FT) | 80.8 | 96.8 | 88.1 | 97.6 | 60.5 | 92.3 | 73.1 | 89.9 | 70.0 | **100** | 82.4 | 97.0 |
| Swin-Transformer(+SDG+Aug) | **98.1** | 88.3 | **92.9** | **98.2** | **73.7** | 84.6 | **78.8** | 88.5 | **80.0** | 99.7 | **88.8** | **97.1** |

Table 2 presents the experimental findings utilizing the InceptionV3 classification model, whereas Table 3 displays the experimental findings using the Swin-Transformer classification model. In Table 2 and Table 3, the initial row displays the classification results obtained from training with the current dataset. The second row reflects the classification outcomes from training using the filtered synthetic image dataset, SDG. The third row exhibits the classification performance after fine-tuning the existing dataset based on the second-row model, and the fourth row reflects the classification performance after training using the data-augmented dataset.

In Table 2, our method achieves optimal F1-measure and AUC scores for both the Rimone_DL and Drishti-GS datasets. Furthermore, our method attains optimal AUC and nearly optimal F1-measure on the REFUGE dataset. As depicted in Table 2, our method's Sen values do not exhibit the same performance as the original dataset's training results for all three datasets. However, the Spe values attain optimal values. This contrasting outcome can be explicable by the generation of certain domain bias during the fundus map generation process. In Table 3, our method achieves the optimal values of not only F1-measure and AUC but also Sen and Spe on the three datasets. Based on the metric analysis, it can be inferred that our method generates images that exhibit strong performance in the glaucoma classification task with favorable usability.

It is worth mentioning that the creation of a novel method is generally dependent on the integration of its various modules. However, our method is a holistic approach based on data augmentation and is not applicable to traditional ablation experiments. Therefore, rather than conducting module-specific ablation experiments, we compare the effects of glaucoma classification before and after data augmentation using our method. Through such comparative experiments, we have comprehensively demonstrated the validity of our method.

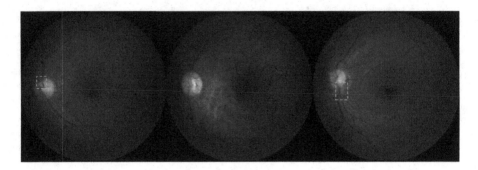

**Fig. 3.** The less well reproduced fundus images generated by our method. The dotted box highlights problematic areas.

While our method generates fundus images with some low-quality instances, notably in the optic cup/disc region with broken blood vessels, as shown in Fig. 3, these issues stem from the complexity of vessel structures and inaccuracies in the utilized vessel mask maps. Despite challenges in this specific region, the overall impact of the generated images remains satisfactory. Our ongoing work will focus on resolving these challenges in the optic cup/disc region.

In brief, the experimental results indicate an enhancement in the efficacy of models that are trained, fine-tuned, or data-augmented with our filtered synthetic fundus images.

## 5   Conclusion

We introduce a method to generate fundus images utilizing the diffusion model. The experimental findings unequivocally establish the superior quality of the medical images generated by our method. Additionally, we conduct a thorough assessment of the applicability of our method's fundus images in genuine clinical scenarios, leveraging techniques such as data augmentation and fine-tuning. Extensive experiments consistently demonstrate a notable enhancement in the performance of classification models when coupled with data augmentation or fine-tuning using fundus images generated by our method. In brief, this paper represents an initial exploration into improving the performance of medical image generation, focusing specifically on the generation of glaucoma fundus images.

**Acknowledgement.** This work is supported by fund for building world-class universities (disciplines) of Renmin University of China. The computer resources were provided by Public Computing Cloud Platform of Renmin University of China.

## References

1. Arjovsky, M., Chintala, S., Bottou, L.: Wasserstein generative adversarial networks. In: International Conference on Machine Learning, pp. 214–223. PMLR (2017)

2. Blattmann, A., et al.: Align your latents: high-resolution video synthesis with latent diffusion models. In: Proceedings of the IEEE/CVF Conference on Computer Vision and Pattern Recognition (CVPR), pp. 22563–22575, June 2023

3. Brock, A., Donahue, J., Simonyan, K.: Large scale GAN training for high fidelity natural image synthesis. In: International Conference on Learning Representations (2018)

4. Brown, T., et al.: Language models are few-shot learners. In: Advances in Neural Information Processing Systems, vol. 33, pp. 1877–1901 (2020)

5. Cao, Y., et al.: A comprehensive survey of AI-generated content (AIGC): a history of generative AI from GAN to chatGPT. arXiv abs/2303.04226 (2023)

6. Dhariwal, P., Nichol, A.: Diffusion models beat GANs on image synthesis. In: Advances in Neural Information Processing Systems, vol. 34, pp. 8780–8794 (2021)

7. Dinh, L., Sohl-Dickstein, J., Bengio, S.: Density estimation using real NVP. In: International Conference on Learning Representations (2016)

8. Goodfellow, I., et al.: Generative adversarial networks. Commun. ACM **63**(11), 139–144 (2020)

9. Guo, J., Pang, Z., Yang, F., Shen, J., Zhang, J.: Study on the method of fundus image generation based on improved GAN. Math. Probl. Eng. **2020**, 1–13 (2020)

10. Hu, E.J., et al.: Lora: low-rank adaptation of large language models. In: International Conference on Learning Representations (2021)

11. Kawar, B., et al.: Imagic: text-based real image editing with diffusion models. 2023 IEEE/CVF Conference on Computer Vision and Pattern Recognition (CVPR), pp. 6007–6017 (2022)

12. van den Oord, A., Vinyals, O., Kavukcuoglu, K.: Neural discrete representation learning. In: Advances in Neural Information Processing Systems, vol. 30 (2017)

13. Pandey, K., Mukherjee, A., Rai, P., Kumar, A.: Diffusevae: efficient, controllable and high-fidelity generation from low-dimensional latents. Trans. Mach. Learn. Res. **2022** (2022)

14. Radford, A., et al.: Learning transferable visual models from natural language supervision. In: International Conference on Machine Learning (2021)

15. Radford, A., Narasimhan, K., Salimans, T., Sutskever, I., et al.: Improving language understanding by generative pre-training (2018)

16. Radford, A., Wu, J., Child, R., Luan, D., Amodei, D., Sutskever, I., et al.: Language models are unsupervised multitask learners. OpenAI blog **1**(8), 9 (2019)

17. Ramesh, A., Dhariwal, P., Nichol, A., Chu, C., Chen, M.: Hierarchical text-conditional image generation with clip latents. arXiv abs/2204.06125 (2022)

18. Ramesh, A., et al.: Zero-shot text-to-image generation. In: International Conference on Machine Learning, pp. 8821–8831. PMLR (2021)

19. Rombach, R., Blattmann, A., Lorenz, D., Esser, P., Ommer, B.: High-resolution image synthesis with latent diffusion models. In: 2022 IEEE/CVF Conference on Computer Vision and Pattern Recognition (CVPR), pp. 10674–10685 (2021)

20. Rombach, R., Blattmann, A., Ommer, B.: Text-guided synthesis of artistic images with retrieval-augmented diffusion models. arXiv abs/2207.13038 (2022)

21. Ruiz, N., Li, Y., Jampani, V., Pritch, Y., Rubinstein, M., Aberman, K.: Dreambooth: fine tuning text-to-image diffusion models for subject-driven generation. In: 2023 IEEE/CVF Conference on Computer Vision and Pattern Recognition (CVPR), pp. 22500–22510 (2022)

22. Shenkut, D., Kumar, B.V.K.V.: Fundus GAN - GAN-based fundus image synthesis for training retinal image classifiers. In: 2022 44th Annual International Conference of the IEEE Engineering in Medicine and Biology Society (EMBC), pp. 2185–2189 (2022)

23. Sivaswamy, J., et al.: Drishti-gs: retinal image dataset for optic nerve head(ONH) segmentation. In: 2014 IEEE 11th International Symposium on Biomedical Imaging (ISBI), pp. 53–56 (2014)

24. Wang, Z., Wang, J., Liu, Z., Qiu, Q.: Binary latent diffusion. In: Proceedings of the IEEE/CVF Conference on Computer Vision and Pattern Recognition, pp. 22576–22585 (2023)

25. Yi, J., Chen, C., Yang, G.: Retinal artery/vein classification by multi-channel multi-scale fusion network. Appl. Intell. (2023)

26. Zhang, L., Rao, A., Agrawala, M.: Adding conditional control to text-to-image diffusion models. In: Proceedings of the IEEE/CVF International Conference on Computer Vision, pp. 3836–3847 (2023)

27. Zhu, J.Y., Park, T., Isola, P., Efros, A.A.: Unpaired image-to-image translation using cycle-consistent adversarial networks. In: 2017 IEEE International Conference on Computer Vision (ICCV), pp. 2242–2251 (2017)

# Adapting Pretrained Large-Scale Vision Models for Face Forgery Detection

Lantao Wang and Chao Ma[✉]

MoE Key Lab of Artificial Intelligence, AI Institute, Shanghai Jiao Tong University,
Shanghai, China
{lantao11,chaoma}@sjtu.edu.cn

**Abstract.** In the evolving digital realm, generative networks have catalyzed an upsurge in deceptive media, encompassing manipulated facial imagery to tampered text, threatening both personal security and societal stability. While specialized detection networks exist for specific forgery types, their limitations in handling diverse online forgeries and resource constraints necessitate a more holistic approach. This paper presents a pioneering effort to efficiently adapt pre-trained large vision models (LVMs) for the critical task of forgery detection, emphasizing face forgery. Recognizing the inherent challenges in bridging pre-training tasks with forgery detection, we introduce a novel parameter-efficient adaptation strategy. Our investigations highlight the imperative of focusing on detailed, local features to discern forgery indicators. Departing from conventional methods, we propose the Detail-Enhancement Adapter (DE-Adapter), inspired by 'Unsharp Masking'. By leveraging Gaussian convolution kernels and differential operations, the DE-Adapter enhances detailed representations. With our method, we achieved state-of-the-art performance with only 0.3% network adjustment. Especially when the number of training samples is limited, our method far surpasses other methods. Our work also provides a new perspective for the Uni-Vision Large Model, and we call on more fields to design suitable adapting schemes to expand the capabilities of large models instead of redesigning networks from scratch.

**Keywords:** Face Forgery Detection · Parameter-efficient Tuning · Pretrained Large-Scale Vision Models · Uni-Vision Large Model

## 1 Introduction

In the modern digital landscape, the development of generative networks [7, 13] has led to a surge in forged media online. Such deceptive content, ranging from manipulated facial imagery [2,12,20,24,43] to tampered text [41], poses significant threats to personal security and societal harmony.

To counter this issue, researchers typically devise dedicated detection networks based on their expertise in specific forgery types. Although effective for specific cases, this method struggles when confronted with the myriad of

S. Rudinac et al. (Eds.): MMM 2024, LNCS 14557, pp. 71–85, 2024.
https://doi.org/10.1007/978-3-031-53302-0_6

forgery media found online. Moreover, the resource-intensive nature of deploying specialized systems for each forgery media type renders this approach impractical. Therefore, we turn our attention to pre-trained large vision models (LVMs) [28,42], which possess vast semantic knowledge learned from large-scale images and perform exceptionally across various downstream tasks [9,33]. We aim to exploit the power of LVMs to develop a unified solution, avoiding the need to deploy multiple specialist networks.

This paper tackles the novel and critical task of *parameter efficiently adapting pre-trained large vision models for forgery detection*, with a particular emphasis on the widely influential face forgery detection task. Given the significant gap between the pre-training task and forgery detection, this challenge is formidable and non-trivial. While direct detection of facial forgery by LVMs without additional training has poor performance, fine-tuning LVMs for this purpose is computationally demanding due to the numerous parameters involved. Although partial fine-tuning offers a potential solution, it presents its own set of difficulties, such as deciding the optimal tunable parameters ratio.

More importantly, our insights into forgery detection indicate that identifying forgery markers requires a keen focus on detailed features that encapsulate local nuances. This perspective stands in stark contrast to existing parameter-efficient tuning (PET) methods for LVMs [3,5,22], which may not fully address the intricacies of forgery. Inspired by the traditional image processing technique 'Unsharp Masking' [38], we propose a lightweight Detail-Enhancement Adapter (DE-Adapter). By employing Gaussian convolution kernels and differential operations, the DE-Adapter captures detail-enhanced representations which are then integrated into the original representation, thereby 'sharpen' detail information. Additionally, our method is versatile and compatible with a wide range of LVM architectures, including both Convolutional networks and Transformers.

The main contributions of this work are summarized as follows:

- We establish a new paradigm, namely solving the forgery detection problem by maximizing the use of pre-trained knowledge from LVMs. Numerous studies have already proven that LVMs possess strong capabilities; hence, we advocate for more domains to design suitable schemes. The idea is to utilize LVMs to solve problems rather than redesigning networks entirely.
- We propose a lightweight Detail-Enhanced Adapter (DE-Adapter). By employing Gaussian convolution kernels and differential operations, the DE-Adapter captures detail-enhanced representations which are then integrated into the original representation, thereby 'sharpen' detail information, the DE-Adapter empowers LVMs to excel at detecting face forgeries.
- Extensive experiments have proven the effectiveness of our method. We can make LVMs achieve state-of-the-art results with only a very small number of parameters trained, especially when the number of training datasets is limited, our method far surpasses other methods. Moreover, our method is applicable to LVMs with different structures, including Convolutional networks and Transformers.

## 2    Related Work

### 2.1    Face Forgery Detection

Face forgery detection is a critical task in computer vision and image processing, with the objective of identifying manipulated or forged facial images or
videos. Typically, face forgery detection methods rely on human prior knowledge
when formulating model architectures. The prior knowledge typically includes
visual cues that are crucial to identifying the differences between real and fake
images, such as noise statistics [14], spatial domain [25,44], and frequency information [4,21,31,35]. Zhou *et al.* [45] attempted to add a side branch to the
image classification backbone, focusing on local noise patterns under the assumption that these patterns differ between real and fake images. Zhao *et al.* [44]
redesigned the spatial attention module to enhance the network's capability of
extracting subtle forgery traces from local regions. Qian *et al.* [35] and Miao
*et al.* [31] proposed frequency-aware models utilizing Discrete Cosine Transform (DCT) and Discrete Wavelet Transform (DWT) tools to extract frequency
details. While these methods have demonstrated admirable performance, they
have not fully harnessed the potential of LVMs. These methods also require
extensive data for training from scratch, which can be a limitation.

### 2.2    Parameter-Efficient Transfer Learning

Over the past few years, transfer learning has surged in prominence, leading
to an increased dominance of large-scale foundation models within the field of
deep learning. In this context, PET has garnered attention due to its effectiveness and efficiency. Current PET methods can be categorized into three groups.
Firstly, Adapter [5] inserts a trainable bottleneck block into the LVMs for downstream tasks adaptation. Secondly, Prompt Learning [22] adds several trainable
tokens to the input sequences of LVM blocks. Lastly, Learning weight Decomposition [17,18] breaks down the learning weight into low-rank metrics, training
only the low-rank portion. Moreover, researchers have begun applying the PET
paradigm to specific task that may not initially seem suitable for LVMs. For
example, Pan *et al.* [33] proposed a new Spatio-Temporal Adapter (ST-Adapter)
to adapt LVMs, lacking temporal knowledge, to dynamic video content reasoning. Huang *et al.* [19] introduced ensemble adapters in the Vision Transformer
(ViT) for robust cross-domain face-anti-spoofing. In these instances, PET not
only matches full fine-tuning performance but also enables LVM adaptation to
incompatible downstream tasks. In this work, we address the challenging problem
of adapting LVMs for face forgery detection, which demands fine-grained local
features rather than category-level differences inherent in the original LVMs.

## 3    Methodology

### 3.1    Preliminary: Adapter

We first define the face forgery detection task as an image classification problem. The LVM can be separated into a backbone and a classifier. To efficiently

use the LVM for downstream tasks, a straightforward approach is to freeze the backbone during training and insert a trainable lightweight module into it. This plugged-in module is designed to learn task-specific features, enhancing the original representations. Essentially, it modulates the original hidden features [15].

Formally, given an image $x \in \mathbb{R}^{H \times W \times 3}$ as input, the backbone extracts the image features as $h_l \in \mathbb{R}^{H' \times W' \times C}$, where $h_l$ is the output of the $l$-th block in the backbone, $\Delta h_l$ denotes the task-specific representation, $(H, W)$ and $(H', W')$ represents the size of the input image and the features respectively, $C$ denotes the channel dimension.

$$h_{l+1} \leftarrow h_l + \alpha \cdot \Delta h_l, \tag{1}$$

where $\alpha$ is a scale factor. After that, the classifier is applied on the image feature to output the prediction.

To introduce task-specific representations, the simplest solution is the basic Adapter [5]. When constructing $\Delta h$, the adapter module is designed as a bottleneck structure. It includes a down-projection layer with parameters $W_{down} \in \mathbb{R}^{C \times c'}$, an up-projection layer with parameters $W_{up} \in \mathbb{R}^{c' \times C}$ and an activation function. Here, $c'$ represents the middle dimension and satisfies $c' \ll C$. The task-specific feature learning process can be expressed as:

$$\Delta h_l = f(h_l \cdot W_{down}) \cdot W_{up}, \tag{2}$$

where $f(\cdot)$ denotes the activation function.

The basic Adapter [5] is lightweight and flexible, facilitating the design of different structures to achieve optimal task-specific representations. We refer to this architecture in our design, adhering to two principles: (1) parameter efficiency to minimize the cost of parameters used for face forgery detection, and (2) suitability for the task at hand. The design should enable the LVM to effectively extract detail information relevant to face forgery detection.

### 3.2    The Design of Our Detail-Enhanced Adapter(DE-Adapter)

Given our understanding of face forgery, we recognize the critical importance of the detail information. A pretrained LVM is proficient at extracting category semantic information, enabling high performance on classification tasks like ImageNet [6]. However, without incorporating detailed information, fine-tuning the LVM with existing PET methods does not yield satisfactory results, as validated by Table 2. Thus, introducing the detail information is pivotal to adapting the LVM for face forgery detection with minimal parameter fine-tuning.

By incorporating the detail information, a type of task-specific representation, into the backbone to enhance the original representation, this process resembles the 'sharpening' operation in traditional image processing. Drawing inspiration from the 'Unsharp Masking' technique [38] used in conventional image processing, we obtain detail-enhanced representations through Gaussian convolution kernels combined with differential operations. This processed detail-enhanced representation is then introduced into the original representation to achieve a 'sharpening' effect.

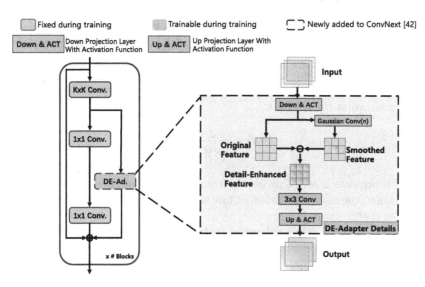

**Fig. 1.** The architecture of our proposed Detail-Enhanced Adapter. Taking the ConvNext [42] network as an example, we obtain detail-enhanced representations through Gaussian convolution kernels combined with differential operations. This processed detail-enhanced representation is then introduced into the original representation to achieve a 'sharpening' effect.

A classical 'Unsharp Masking' techniques [38] can be described by the equation:

$$\hat{I} = I + \lambda(I - G(I)) \tag{3}$$

where $\hat{I}$ denotes the enhanced image, $I$ denotes the original image, an unsharp mask is represented by $(I - G(I))$ where $G$ denotes a Gaussian filter and an amount coefficient by $\lambda$ controls the volume of enhancement achieved at the output.

Indeed, the Eq. 3 bears a striking resemblance in Eq. 1. Intuitively, we can treat $(I - G(I))$ part as a task-specific representation. However, in contrast to traditional 'Unsharp Masking' technique, we need to extract detail information of varying scales from the feature map generated by the preceding block to enrich the original representation of the deeper block. This process requires more than just simple operations on the image. Consequently, we cannot merely employ a static Gaussian convolution kernel but must dynamically adjust the kernel based on the depth of the block.

We design a circular concentric Gaussian Convolution (GC) based on the one-dimensional (1D) Gaussian function:

$$f_{1d}(d) = A \cdot exp(-\frac{(d - \mu)^2}{2\sigma^2}) \tag{4}$$

where $A = (\sqrt{2\pi}\sigma)^{-1}$ represents the coefficient terms, $d$ represents the distance between points and the center $\mu$ within the sample grids, $\sigma$ determines the distribution of the convolutional kernel, $\mu$ denotes the extreme point, where the function takes on the highest value. For simplicity, we put $\mu$ at the center of the kernel. We set $\sigma$ as a trainable parameter to accommodate multi-scale features generated by blocks at varying depths.

Furthermore, to prevent extreme values and a lack of receptive field, we apply Max normalization to the mask:

$$G(d) = \frac{f_{1d}(d)}{\max_{d \in D} f_{1d}(d)} \tag{5}$$

where $D$ represents a set of distances from the center of the kernel. With Max normalization, the modified Gaussian function always reaches its maximum value of 1 at the center point.

To better extract detail-enhanced features, we incorporate a 3*3 convolutional layer. Simultaneously, to maintain the lightweight nature of the model as much as possible, the input feature $h$ is first downsampled before performing the 'Unsharp Masking' operation and then upsampled again after completion to restore its original shape. Taking the ConvNext [42] network as an example, our proposed Detail-Enhanced Adapter (DE-Adapter) is depicted in Fig. 1. The computation process can be formulated as follows:

$$\Delta h_l = f(CONV(h_l^d - G_\sigma(h_l^d)) \cdot W_{up}) \tag{6}$$

where $CONV$ denotes the 3*3 convolutional layer, $h_l^d = f(h_l \cdot W_{down})$ represents the original feature after downsampling, $G_\sigma(\cdot)$ denotes the Gaussian Convolution with a trainable parameter $\sigma$.

### 3.3   The DE-Adapter Integration Scheme

When integrating the proposed DE-Adapter module into the backbone, the integration scheme also plays a pivotal role, influencing knowledge transfer performance. To explore an effective integration scheme, both the position adaptation in the LVM and the insertion form of the adapter warrant careful consideration [15]. The position determines which layer the hidden representation $h$ is to be adapted in the LVM, while the insertion form decides how to set the input to the DE-Adapter to compute the task-specific representation $\Delta h$.

Combining design dimension from these two perspectives, we meticulously design and assess five different integration schemes. Using the inverted residual blocks of ConvNext [29] as an example, Fig. 2 illustrates the proposed five variants of integration designs. The DE-Adapter can be flexibly inserted into every block in ConvNet, introducing only a minimal number of parameters. From our empirical studies, we found that: (1) Compared to the sequential architecture, the parallel architecture is more adept at extracting task-specific features. (2) When adapting convolutional networks on tasks with substantial domain shifts, the radical mismatch of the receptive field in $\Delta h$ and $h$ might result in inferior transfer performance. Therefore, we select the inverse residual parallel approach

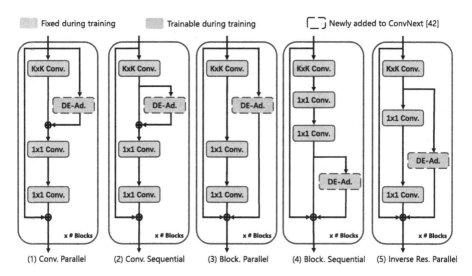

**Fig. 2.** Illustrations of five integrating designs of DE-Adapter to ConvNext. The schemes differ regarding the position of the modified representation and corresponding insertion form. DE-Ad. denotes the proposed DE-Adapter.

that simultaneously extracts task-specific features and maintains the same receptive field. Our ablation study in Table 5 further verifies this analysis.

Besides, it has been experimentally proven that other LVMs such as Vision Transformer [10] and even Swin Transformer [28] are also applicable.

## 4 Experiments

### 4.1 Experimental Settings

**Datasets and Metrics.** In this research, we utilize three extensively recognized deepfake datasets, including FaceForensics++ (FF++) [36], Celeb-DeepFake (Celeb-DF) [26] and DFDC [8]. The FF++ dataset is the most frequently employed dataset in this field, encompassing 1,000 original videos and 4,000 corresponding fake videos. The content within FF++ is compressed into two distinct versions: high quality (C23) and low quality (C40). The Celeb-DF dataset incorporates 590 real videos and employs the advanced DeepFake algorithm [26] to generate a substantial collection of 5,639 high-quality forgery videos. The DFDC dataset presents a unique challenge, housing an extensive array of 128,154 facial videos. These videos, originating from 960 diverse subjects, have been subjected to a variety of manipulations and perturbations, adding to the complexity of the dataset. For a rigorous comparative analysis, we report the Accuracy (ACC) and the Area Under the Receiver Operating Characteristic Curve (AUC), both of which are critical metrics in this field.

**Implementation Details.** Both our method and the re-implemented approaches are built on PyTorch [34]. All experiments were conducted using

**Table 1.** Comparison with State-of-the-art Forgery Detector Models. **Bold** and underline refer to the top and second result separately. # Params refers to the amount of tuning parameters. The symbol * indicates the result of reproduction.

| Method | #Params. | Input | FF++(C23) | | FF++(C40) | | Celeb-DF | | DFDC | |
|---|---|---|---|---|---|---|---|---|---|---|
| | | | ACC | AUC | ACC | AUC | ACC | AUC | ACC | AUC |
| MesoNet [1] | 2.8 | 256 | 83.10 | – | 70.47 | – | – | – | – | – |
| Multi-task [32] | 26.08 | 256 | 85.65 | 85.43 | 81.30 | 75.59 | – | – | – | – |
| SPSL [27] | – | 299 | 91.51 | 95.32 | 81.57 | 82.82 | – | – | – | – |
| Face X-ray [25] | – | 299 | – | 87.40 | – | 61.60 | – | – | – | – |
| Xception [36] | 20.81 | 299 | 95.73 | 96.30 | 86.86 | 89.31 | 97.90 | 99.73 | 78.87 | 89.39 |
| RFM [39] | – | 299 | 95.69 | 98.79 | 87.06 | 89.83 | 97.96 | 99.94 | 80.83 | 89.75 |
| Add-Net [46] | – | 299 | 96.78 | 97.74 | 87.59 | 91.01 | 96.93 | 99.55 | 78.71 | 89.85 |
| $F^3$-Net [35]* | 41.99 | 299 | 96.52 | 98.11 | 86.43 | 91.32 | 95.95 | 98.93 | 76.17 | 88.39 |
| MultiAtt [44] | 18.82 | 380 | <u>97.61</u> | <u>99.29</u> | <u>87.69</u> | <u>91.41</u> | 97.92 | 99.89 | 76.81 | 90.32 |
| $M^2TR$ [40]* | 40.12 | 320 | 93.22 | 97.84 | 86.09 | 87.97 | 98.76 | 99.02 | – | – |
| $F^2$Trans-B [31]* | 128.01 | 224 | 96.59 | 99.24 | 87.21 | 89.91 | <u>98.79</u> | <u>99.23</u> | <u>81.32</u> | <u>89.12</u> |
| DE-Adapter(Ours) | **0.38** | 224 | **98.01** | **99.31** | **88.92** | **93.10** | **98.93** | **99.91** | **81.59** | **90.37** |

4 Nvidia GeForce 3090 GPUs. During training, we utilized random horizontal flipping as a form of data augmentation. We employed the AdamW optimizer [30] with an initial learning rate of 1e-3 and a weight decay of 1e-3. Additionally, a step learning rate scheduler was used to adjust the learning rate over time.

## 4.2   Main Results and Analysis

**Comparison with State-of-the-Art Forgery Detector Models.** Table 1 reveals that our method not only delivers state-of-the-art results across all datasets, but does so with a mere 0.38M training parameters - significantly less than other methods by at least one to two orders of magnitude. These findings suggest that in the era of large models, fully exploiting the robust capabilities of LVMs may be more beneficial than redesigning network structures.

**Comparison with Previous PET Methods.** Table 2 compares our method with the widely used and effective PET approach [5,22], as well as three baselines: Full Finetune (FT), Partially Tuning (PT), and Linear Probing (LP). The results underscore the efficacy of our proposed DE-Adapter, corroborating our analysis that for efficient fine-tuning of LVMs to solve face forgery detection task, merely adjusting representation mapping is insufficient. It is crucial to incorporate detailed information.

**Evaluation on Limited Sample Training Dataset.** Limited Sample Training Datasets, a common real-world scenario often overlooked by the community, was used to evaluate our model's data utilization efficiency. We restricted the detector to use only a tiny fraction of the training set (like 1/512, 1/256, etc.). Three state-of-the-art detectors (Multiatt [44], $F^3$-Net [35], and $M^2TR$ [40]) were chosen as baselines.

**Table 2.** Comparison with PET methods, including Fine-Tuning (FT), Linear Probing (LP), Partially Tuning (PT), Bias Tuning (Bias), visual prompt tuning (VPT) [22] and Adaptformer [5]. **Bold** refers to the top result.

| Method | #Params. | FF++(C23) | | FF++(C40) | | Celeb_DF | | DFDC | |
|---|---|---|---|---|---|---|---|---|---|
| | | ACC | AUC | ACC | AUC | ACC | AUC | ACC | AUC |
| FT | 87.69 | 96.67 | 98.91 | 89.44 | 94.32 | 98.34 | 99.03 | 82.95 | 91.48 |
| LP | 0.01 | 68.21 | 54.19 | 58.76 | 50.79 | 84.64 | 55.44 | 61.07 | 51.39 |
| PT | 8.46 | 77.35 | 58.96 | 65.08 | 58.30 | 85.46 | 65.40 | 68.90 | 66.28 |
| VPT [22] | 0.02 | 71.91 | 55.31 | 66.51 | 52.51 | 83.89 | 66.26 | 65.73 | 56.43 |
| Bias [3] | 0.13 | 84.65 | 84.08 | 80.89 | 73.58 | 94.18 | 86.07 | 74.69 | 80.54 |
| AdaptFormer [5] | 0.09 | 86.72 | 83.19 | 81.11 | 76.25 | 95.18 | 89.27 | 71.15 | 79.69 |
| DE-Adapter | 0.38 | +1.34 | +0.40 | −0.52 | −1.22 | +0.59 | +0.88 | −1.36 | −1.11 |
| (Ours) | | **98.01** | **99.31** | **88.92** | **93.10** | **98.93** | **99.91** | **81.59** | **90.37** |

**Table 3.** Generalization across datasets in terms of AUC (%) by training on FF++. **Bold** and underline refer to the top and second result separately.

| Method | Celeb-DF | DFDC |
|---|---|---|
| RFM [39] | 57.75 | 65.63 |
| Add-Net [46] | 62.35 | 65.29 |
| $F^3$-Net [35] | 61.51 | 64.59 |
| MultiAtt [44] | <u>67.02</u> | <u>67.79</u> |
| DE-Adapter (Ours) | **67.12** | **68.84** |

As shown in Fig. 3, our method significantly outperforms these small expert networks built on human prior knowledge when training data is limited, with the gap reaching over 40% at most. These results validate our analysis - fully leveraging LVMs' pre-training knowledge while avoiding network structure rebuilds drastically reduces the need for large training data volumes. In the era of large models, it becomes possible to achieve high performance with minimal training data.

**Generalization Across Datasets.** We conduct experiments on evaluating the generalization performance to unknown forgeries. Specifically, we train the models on the FF++ dataset and evaluate their performance on Celeb-DF and DFDC. The results, shown in Table 3, demonstrate that despite introducing only a minimal number of learnable parameters, our proposed DE-Adapter outperforms the baselines in terms of generalization. This result further validates the effectiveness of DE-Adapter and underscores its potential for robust generalization.

**Universality of DE-Adapter.** We evaluate the universality of DE-Adapter by applying it to various backbones pre-trained with various strategies. Specifically, we use Vision Transformer [10], Swin Transformer [28], ConvNext Model [29] as

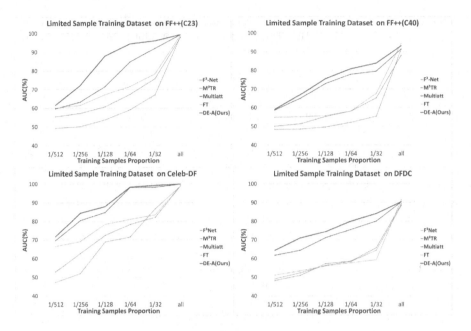

**Fig. 3.** Results of Limited Sample Training setting on various datasets. Our method significantly outperforms these small expert networks built on human prior knowledge when training data is limited, with the gap reaching over 40% at most. In the era of large models, it becomes possible to achieve high performance with minimal training data.

the backbone models. These models are trained on ImageNet-21K dataset [6] with supervised or self-supervised training. MAE [16] is adapted for self-supervised training. Table 4 presents the experiment results of different backbones and pre-training methods. We utilized the results of full fine-tuning as a benchmark for comparison. Our DE-Adapter performs comparably to full fine-tuning across all datasets, which not only attests to its ability to harness the potential of LVMs with a minimal number of parameters but also showcases the universality of our method regarding different structures and pre-training methods of LVMs.

## 4.3   Ablation Study

We provide an ablation study on each component of the DE-Adapter, including the adapter architectures and integration schemes. We use ConvNext V2 [42] as the backbone and conduct experiments on the FF++(C40) dataset.

**Effects of Detail-Enhanced Operation.** We try various Detail-Enhanced Operations in the adapter module. The results are shown in Table 5(a). Compared to the vanilla convolution baseline, a gain of +16.19% and +7.53% is obtained by using our Gaussian Mask. We have also tried other methods that extract detail information, such as LBP [23] and SRM [11]. However, the results are not satisfactory. We argue that these operations destroy the original semantic features.

**Effects of Adapting Schemes.** Table 5(b) showcases comparisons among various integration schemes. In general, the parallel approach outperforms the sequential one. We argue that the sequential method can not facilitate the extraction of task-specific features by the adapting module. Among the parallel approaches, both the convolution and block parallel methods can lead to a mismatch in the receptive fields due to differing convolution kernel sizes on either side of the parallel. Conversely, the inverse residual parallel emerges as the optimal choice due to its unique design that aligns the receptive fields.

**Table 4.** Comparison of DE-Adapter (DE-A) and full FineTuning (FT) with various backbones and different pre-training. CN denotes ConvNext backbones. Sup. denotes supervised pre-training.

| Pre-train | Backbone | Method | #Param. | FF++(C23) ACC | FF++(C23) AUC | FF++(C40) ACC | FF++(C40) AUC | Celeb-DF ACC | Celeb-DF AUC | DFDC ACC | DFDC AUC |
|---|---|---|---|---|---|---|---|---|---|---|---|
| Sup. | ViT-B | FT | 86.13 | **93.76** | **98.16** | 85.24 | 90.36 | 97.86 | 99.47 | **81.65** | **90.12** |
| | | DE-A(Ours) | 0.33 | 93.75 | 97.62 | **85.78** | **89.53** | **98.17** | **99.67** | 80.92 | 89.11 |
| | Swin-B | FT | 86.75 | **97.10** | **99.48** | **88.97** | **94.59** | **98.95** | **99.95** | **82.76** | **89.77** |
| | | DE-A(Ours) | 0.43 | 96.61 | 98.97 | 87.51 | 92.30 | 98.05 | 98.63 | 81.17 | 89.59 |
| | CN-T | FT | 27.94 | 95.62 | 98.49 | 85.93 | 88.68 | **97.81** | **98.71** | **81.16** | **89.54** |
| | | DE-A(Ours) | 0.12 | **96.34** | **99.07** | **86.59** | **92.51** | 97.79 | 98.53 | 80.10 | 89.10 |
| | CN-B | FT | 87.57 | **97.22** | **99.43** | **88.89** | **94.41** | 98.11 | 99.68 | **81.63** | 89.74 |
| | | DE-A(Ours) | 0.32 | 96.70 | 99.18 | 88.09 | 93.89 | **98.74** | **99.83** | 81.59 | **90.44** |
| MAE | ViT-B | FT | 85.80 | 96.29 | 98.51 | **87.85** | **93.12** | **98.37** | 99.39 | **81.06** | **89.01** |
| | | DE-A(Ours) | 0.32 | **96.12** | **98.76** | 87.22 | 92.59 | 98.03 | **99.58** | 80.56 | 88.15 |
| | CNV2-T | FT | 27.99 | 94.43 | 97.12 | **87.23** | 88.12 | **98.37** | **99.34** | 79.35 | **89.50** |
| | | DE-A(Ours) | 0.12 | **95.81** | **97.46** | 86.03 | **91.77** | 97.91 | 99.12 | **81.51** | 89.12 |
| | CNV2-B | FT | 87.69 | 96.67 | 98.91 | **89.44** | **94.32** | 98.34 | 99.03 | **82.95** | **91.48** |
| | | DE-A(Ours) | 0.32 | **98.01** | **99.31** | 88.92 | 93.10 | **98.93** | **99.91** | 81.59 | 90.37 |

**Table 5.** Ablation study on each component of DE-Adapter.

(a) Effects of convolution types

| Detail-Enhanced Operation Type | AUC |
|---|---|
| Linear | 76.19 |
| Vanilla 3 × 3 Conv. | 85.57 |
| LBP Conv. [23] | 84.15 |
| SRM Conv. [11] | 87.17 |
| Gaussian Mask (Ours) | **93.10** |

(b) Effects of adapting schemes

| Adapting Scheme | AUC |
|---|---|
| Conv. Parallel | 91.53 |
| Conv. Sequential | 89.72 |
| Block Parallel | 91.54 |
| Block Sequential | 90.90 |
| Inverse Residual Parallel | **93.10** |

**Visualization.** To gain deeper insights into the decision-making mechanism of our approach, we utilize Grad-CAM [37] for visualization on FF++ as displayed in Fig. 4. It is noticeable that the baseline method (an Adapter with standard 3×3 convolution) is highly vulnerable to high-frequency noise beyond the facial area in DeepFakes (DF), NeuralTextures (NT), and Real scenarios. In contrast, our method generates distinguishable heatmaps for authentic and forged faces where the highlighted regions fluctuate according to the forgery techniques, even though it solely relies on binary labels for training. For example, the heatmaps for both DeepFakes (DF) and FaceSwap (FS) concentrate on the central facial area, whereas that for Face2Face (F2F) identifies the boundary of the facial region. These results corroborate the effectiveness of the proposed DE-Adapter from a decision-making standpoint.

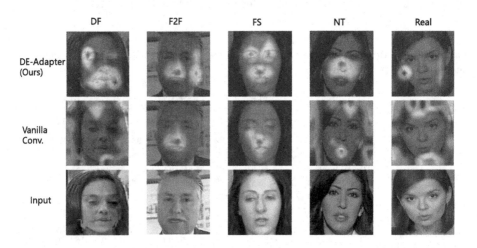

**Fig. 4.** The Grad-CAM visualization on the FF++ dataset. The first row and the second row display the proposed method result and the baseline result, respectively. It is noticeable that the baseline method is highly vulnerable to high-frequency noise beyond the facial area in DF, NT, and Real scenarios. In contrast, our method generates distinguishable heatmaps for authentic and forged faces where the highlighted regions fluctuate according to the forgery techniques, even though it solely relies on binary labels for training. For example, the heatmaps for both DF and FS concentrate on the central facial area, whereas that for F2F identifies the boundary of the facial region. These results corroborate the effectiveness of the proposed DE-Adapter from a decision-making standpoint.

## 5  Conclusion

In this paper, we propose a new paradigm to solve the forgery detection problem. Instead of redesigning small expert systems based on prior experience as before, we leverage the powerful capabilities of LVMs to address it. In order for LVMs to better adapt to such a specific downstream task as face forgery detection, inspired by the traditional image processing technique 'Unsharp Masking', we propose a lightweight Detail-Enhancement Adapter (DE-Adapter). By employing Gaussian convolution kernels and differential operations, the DE-Adapter captures detail-enhanced representations which are then integrated into the original representation, thereby 'sharpen' detail information, which enables LVMs to excel at detecting face forgery. With our method, we achieved state-of-the-art performance with only 0.3% of the training parameter volume. Especially when the number of training samples is limited, our method far surpasses other methods.

**Acknowledgement.** This work was supported in part by NSFC (62376156, 62322113), Shanghai Municipal Science and Technology Major Project (2021SHZDZX0102), and the Fundamental Research Funds for the Central Universities.

# References

1. Afchar, D., Nozick, V., Yamagishi, J., Echizen, I.: MesoNet: a compact facial video forgery detection network. In: WIFS (2018)
2. Bitouk, D., Kumar, N., Dhillon, S., Belhumeur, P.N., Nayar, S.K.: Face swapping: automatically replacing faces in photographs. ACM Trans. Graph **27**, 1–8 (2008)
3. Cai, H., Gan, C., Zhu, L., Han, S.: Tinytl: reduce memory, not parameters for efficient on-device learning. In: NeurIPS (2020)
4. Chen, S., Yao, T., Chen, Y., Ding, S., Li, J., Ji, R.: Local relation learning for face forgery detection. In: AAAI (2021)
5. Chen, S., et al.: Adaptformer: adapting vision transformers for scalable visual recognition. In: NeurIPS (2022)
6. Deng, J., Dong, W., Socher, R., Li, L., Li, K., Fei-Fei, L.: Imagenet: A large-scale hierarchical image database. In: CVPR (2009)
7. Dhariwal, P., Nichol, A.Q.: Diffusion models beat GANs on image synthesis. In: NeurIPS (2021)
8. Dolhansky, B., et al.: The deepfake detection challenge (DFDC) dataset. CoRR (2020)
9. Dong, B., Zhou, P., Yan, S., Zuo, W.: LPT: long-tailed prompt tuning for image classification. CoRR (2022)
10. Dosovitskiy, A., et al.: An image is worth 16x16 words: transformers for image recognition at scale. In: ICLR (2021)
11. Fridrich, J.J., Kodovský, J.: Rich models for steganalysis of digital images. TIFS **7**, 868–882 (2012)
12. Gao, Y., et al.: High-fidelity and arbitrary face editing. In: CVPR (2021)
13. Goodfellow, I.J., et al.: Generative adversarial networks. CoRR (2014)
14. Han, X., Morariu, V., Larry Davis, P.I., et al.: Two-stream neural networks for tampered face detection. In: CVPR Workshop (2017)
15. He, J., Zhou, C., Ma, X., Berg-Kirkpatrick, T., Neubig, G.: Towards a unified view of parameter-efficient transfer learning. In: ICLR (2022)
16. He, K., Chen, X., Xie, S., Li, Y., Dollár, P., Girshick, R.: Masked autoencoders are scalable vision learners. In: CVPR (2022)
17. He, X., Li, C., Zhang, P., Yang, J., Wang, X.E.: Parameter-efficient fine-tuning for vision transformers. CoRR (2022)

18. Hu, E.J., et al.: Lora: low-rank adaptation of large language models. In: ICLR (2022)
19. Huang, H., et al.: Adaptive transformers for robust few-shot cross-domain face anti-spoofing. In: Avidan, S., Brostow, G., Cissé, M., Farinella, G.M., Hassner, T. (eds.) ECCV 2022. LNCS, vol. 13673, pp. 37–54. Springer, Cham (2022). https://doi.org/10.1007/978-3-031-19778-9_3
20. Huang, Z., Chan, K.C.K., Jiang, Y., Liu, Z.: Collaborative diffusion for multi-modal face generation and editing. In: CVPR (2023)
21. Jia, G., et al.: Inconsistency-aware wavelet dual-branch network for face forgery detection. Trans. Biom. Behav. Ident. Sci. **3**, 308–319 (2021)
22. Jia, M., et al.: Visual prompt tuning. In: Avidan, S., Brostow, G., Cissé, M., Farinella, G.M., Hassner, T. (eds.) ECCV 2022. LNCS, vol. 13693, pp. 709–727. Springer, Cham (2022). https://doi.org/10.1007/978-3-031-19827-4_41
23. Juefei-Xu, F., Boddeti, V.N., Savvides, M.: Local binary convolutional neural networks. In: CVPR (2017)
24. Kim, K., et al.: Diffface: diffusion-based face swapping with facial guidance. CoRR (2022)
25. Li, L., et al.: Face x-ray for more general face forgery detection. In: CVPR (2020)
26. Li, Y., Yang, X., Sun, P., Qi, H., Lyu, S.: Celeb-DF: a large-scale challenging dataset for deepfake forensics. In: CVPR (2019)
27. Liu, H., et al. Spatial-phase shallow learning: rethinking face forgery detection in frequency domain. In: CVPR (2021)
28. Liu, Z., et al.: Swin transformer: hierarchical vision transformer using shifted windows. In: ICCV (2021)
29. Liu, Z., Mao, H., Wu, C.Y., Feichtenhofer, C., Darrell, T., Xie, S.: A convnet for the 2020s. In: CVPR (2022)
30. Loshchilov, I., Hutter, F.: Decoupled weight decay regularization. In: ICLR (2019)
31. Miao, C., Tan, Z., Chu, Q., Liu, H., Hu, H., Yu, N.: $F^2$trans: High-frequency fine-grained transformer for face forgery detection. TIFS (2023)
32. Nguyen, H.H., Fang, F., Yamagishi, J., Echizen, I.: Multi-task learning for detecting and segmenting manipulated facial images and videos. In: BTAS (2019)
33. Pan, J., Lin, Z., Zhu, X., Shao, J., Li, H.: St-adapter: parameter-efficient image-to-video transfer learning. In: NeurIPS (2022)
34. Paszke, A., et al.: Automatic differentiation in pytorch (2017)
35. Qian, Y., Yin, G., Sheng, L., Chen, Z., Shao, J.: Thinking in frequency: face forgery detection by mining frequency-aware clues. In: Vedaldi, A., Bischof, H., Brox, T., Frahm, J.M. (eds.) ECCV 2022. LNCS, vol. 12357, pp. 86–103. Springer, Cham (2020). https://doi.org/10.1007/978-3-030-58610-2_6
36. Rössler, A., Cet al.: Faceforensics++: learning to detect manipulated facial images. In: ICCV (2019)
37. Selvaraju, R.R., Cogswell, M., Das, A., Vedantam, R., Parikh, D., Batra, D.: Grad-CAM: visual explanations from deep networks via gradient-based localization. In: IJCV (2020)
38. Shi, Z., Chen, Y., Gavves, E., Mettes, P., Snoek, C.G.M.: Unsharp mask guided filtering. TIP **30**, 7472–7485 (2021)
39. Wang, C., Deng, W.: Representative forgery mining for fake face detection. In: CVPR (2021)
40. Wang, J., et al.: M2tr: multi-modal multi-scale transformers for deepfake detection. In: ICMR (2022)

41. Wang, Y., Xie, H., Xing, M., Wang, J., Zhu, S., Zhang, Y.: Detecting tampered scene text in the wild. In: Avidan, S., Brostow, G., Cissé, M., Farinella, G.M., Hassner, T. (eds.) ECCV 2022. LNCS, vol. 13688, pp. 215–232. Springer, Cham (2022). https://doi.org/10.1007/978-3-031-19815-1_13
42. Woo, S., et al.: Convnext v2: co-designing and scaling convnets with masked autoencoders. CoRR (2023)
43. Yao, G., et al.: One-shot face reenactment using appearance adaptive normalization. In: AAAI (2021)
44. Zhao, H., Wei, T., Zhou, W., Zhang, W., Chen, D., Yu, N.: Multi-attentional deepfake detection. In: CVPR (2021)
45. Zhou, P., Han, X., Morariu, V.I., Davis, L.S.: Two-stream neural networks for tampered face detection. In: CVPR Workshop (2017)
46. Zi, B., Chang, M., Chen, J., Ma, X., Jiang, Y.: Wilddeepfake: a challenging real-world dataset for deepfake detection. In: ACM MM (2020)

# ICDAR: Special Session on Intelligent Cross-Data Analysis and Retrieval

# Towards Cross-Modal Point Cloud Retrieval for Indoor Scenes

Fuyang Yu[1], Zhen Wang[2], Dongyuan Li[2], Peide Zhu[3], Xiaohui Liang[1(✉)],
Xiaochuan Wang[4], and Manabu Okumura[2]

[1] Beihang University, Beijing, China
liang_xiaohui@buaa.edu.cn
[2] Tokyo Institute of Technology, Tokyo, Japan
[3] Delft University of Technology, Delft, Netherlands
[4] Beijing Technology and Business University, Beijing, China

**Abstract.** Cross-modal retrieval, as an important emerging foundational information retrieval task, benefits from recent advances in multimodal technologies. However, current cross-modal retrieval methods mainly focus on the interaction between textual information and 2D images, lacking research on 3D data, especially point clouds at scene level, despite the increasing role point clouds play in daily life. Therefore, in this paper, we proposed a cross-modal point cloud retrieval benchmark that focuses on using text or images to retrieve point clouds of indoor scenes. Given the high cost of obtaining point cloud compared to text and images, we first designed a pipeline to automatically generate a large number of indoor scenes and their corresponding scene graphs. Based on this pipeline, we collected a balanced dataset called CRISP, which contains 10K point cloud scenes along with their corresponding scene images and descriptions. We then used state-of-the-art models to design baseline methods on CRISP. Our experiments demonstrated that point cloud retrieval accuracy is much lower than cross-modal retrieval of 2D images, especially for textual queries. Furthermore, we proposed ModalBlender, a tri-modal framework which can greatly improve the Text-PointCloud retrieval performance. Through extensive experiments, CRISP proved to be a valuable dataset and worth researching. (Dataset can be downloaded at https://github.com/CRISPdataset/CRISP.)

**Keywords:** Point Cloud · Cross-modal Retrieval · Indoor Scene

## 1 Introduction

With the advancement of multimodal technologies, various tasks are now benefiting from multimodality. Unlike traditional information retrieval methods, cross-modal retrieval [26] involves not only text, but also other modalities, such as images and videos, in both queries and retrieved results. Currently, the most

---

F. Yu and Z. Wang—Equal contribution.

S. Rudinac et al. (Eds.): MMM 2024, LNCS 14557, pp. 89–102, 2024.
https://doi.org/10.1007/978-3-031-53302-0_7

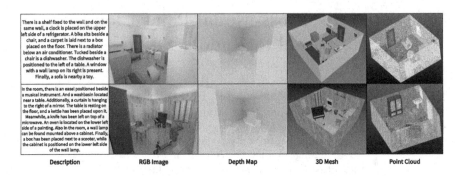

**Fig. 1.** Two samples in CRISP.

studied cross-modal retrieval is the retrieval between text and images [13,16]. With the development of cross-modal methods, BLIP [9] has even achieved close to 100% @1 accuracy on the Flickr30k [16] dataset. Compared to 2D data, 3D data at scene level can provide more spatial information and is less susceptible to occlusion. Despite the development of 3D technology driven by deep learning, there is still a lack of exploration of 3D data in cross modal tasks. 3D data retrieval can establish a bridge between 3D data and other modal data, will be a critical tool of the effective management and utilization of massive 3D data in the future. However, this fundamental task is constrained by the lack of datasets and has developed slowly. Although some previous work [4,20,22] has explored real indoor scenes with RGB-D or point cloud, the existing indoor scene point cloud datasets still have the following issues. a) Small size, the most widely used Scannet [4] dataset only contains around 1,500 scans of different rooms, whereas the text-image cross-modal retrieval dataset Flickr30k has a test set with 1k samples. b) Suffering from long tail distribution, the unbalanced distribution of object categories in dataset makes the models easy to learn the bias.

To address these issues, inspired by some 2D image synthetic datasets such as CLEVR [8] and CLEVRER [29], we designed a novel pipeline to automatically compose a **balanced** large number of realistic 3D indoor scenes with scanned objects and collect point clouds based on them. We use synthetic technology to build our dataset because it is low cost and can be easily scaled to larger numbers while being more balanced without bias [8]. To collect textual description and 2D photographs for the query, we recorded the generated scene information and created a scene graph for each point cloud, with which we constructed the *textual description* of the scene. We also took a photograph from a random corner of a room to obtain a *RGB image* along with its *depth map*. With these three modalities, we finally constructed the **CRISP** (**C**ross-modal **R**etrieval on **I**ndoor **S**cenes **P**oint-cloud ) dataset. Two examples of CRISP are shown in Fig. 1.

To our knowledge, CRISP stands as the first and largest balanced indoor dataset that facilitates point cloud data retrieval, either through scene descriptions or rendered images. This dataset serves as a robust benchmark, offering researchers a valuable tool to assess the efficacy of their models. To demonstrate the usefulness and reasonability of our dataset, we leveraged some state-of-the-art methods to establish strong baselines for CRISP, including Text-PointCloud retrieval and Image-PointCloud retrieval. Specifically, we proposed a new framework called ModalBlender, which brings cross-modal attention and intermediate-modal alignment into Text-PointCloud retrieval and greatly improves the overall performance while maintaining retrieval efficiency, providing a novel and useful approach to enhance text retrieval of point clouds. Our experiments showed that CRISP is challenging and, together with our baselines, provides a strong starting point for future research in indoor scene understanding.

Our main contributions in this study can be summarized as follows: (1) We proposed a general and flexible pipeline that utilizes existing 3D object models to generate a large quantity of photorealistic indoor scenes together with the scene graph; (2) Based on this pipeline, we constructed a dataset called CRISP, which contains massive of text, images, and point cloud that makes it now possible to research cross-modal point cloud retrieval; (3) We evaluated existing SOTA methods and proposed a new framework ModalBlender to further improve the retrieval performance and proved its validity through detailed experiments.

## 2   Related Work

### 2.1   Indoor Scene Datasets

➤ **Real Scanned Datasets:** One of the earliest indoor scene datasets is NYU Depth Dataset V2 [20], which contains RGB-D data of indoor scenes with labeled objects and semantic segmentation masks. The SUN RGB-D [22] dataset is another popular dataset, which includes RGB-D data of indoor scenes with semantic annotations, object instances, and scene categories. The Scannet [4] dataset is a large-scale indoor scene dataset that includes both RGB-D and point cloud data of indoor scenes. The Matterport3D [1] dataset provides high-quality RGB-D data and 3D panoramic of large-scale indoor scenes with object instances and semantic annotations.

➤ **Synthetic Scene Datasets:** Compared to scanning scenes, synthetic datasets offer the advantage of easy scalability to larger amounts of data. SceneNet [6] comprises a variety of annotated indoor scenes that have been widely utilized for object detection, semantic segmentation, and depth estimation. SUNCG [23] contains numerous diverse indoor scenes with highly detailed object annotations, making it useful for 3D reconstruction and semantic parsing. InteriorNet [11] is a dataset featuring photo-realistic indoor scenes, which is useful for evaluating methods on real-world data.

### 2.2   3D Retrieval Datasets

Previous 3D retrieval related dataset and research is mostly about retrieving single 3D objects using 2D images. Some of the popular datasets include the Princeton Shape Benchmark (PSB) [19], the ShapeNet [2], ModelNet40 [27], and MI3DOR [21]. The PSB contains a large collection of 3D models with varying

**Table 1.** Comparison of CRISP with other indoor scene datasets. Our dataset is currently the only one that suitable for benchmark cross modal point cloud retrieval.

| Dataset | Sample | Room | Obj Cate. | Type | RGBD | Desc. | Scene Graph | Point Cloud |
|---|---|---|---|---|---|---|---|---|
| Scannet [4] | 1,513 | 707 | 21 | Real | ✓ | ✗ | ✗ | ✓ |
| NYU-Depth [20] | 1,449 | 464 | 13 | Real | ✓ | ✗ | ✗ | ✗ |
| SUN-RGBD [22] | 10,335 | 10,335 | 800 | Real | ✓ | ✗ | ✗ | ✗ |
| 3DSSG [25] | 1,482 | 478 | 160 | Real | ✓ | ✗ | ✓ | ✓ |
| SUNCG [23] | 45,622 | 45,622 | 84 | Syn | ✓ | ✗ | ✗ | ✗ |
| InteriorNet [11] | 5M | 1.7M | 158 | Syn | ✓ | ✗ | ✗ | ✗ |
| SceneNet [6] | 10,030 | 57 | 13 | Syn | ✓ | ✗ | ✗ | ✗ |
| CRISP (Ours) | 10,000 | 10,000 | 62 | Syn | ✓ | ✓ | ✓ | ✓ |

levels of complexity, while ShapeNet and ModelNet40 aim to provide large-scale objects with different annotated categories. The MI3DOR dataset offers monocular image-based 3D object retrieval. These datasets have been used extensively in the literature to test and compare different 3D object retrieval algorithms and have led to significant advancements in the field. The only related 3D *scene* retrieval dataset is SHREC'19 [30], which uses one 2D sketch-based image to retrieve 3D scenes with 30 categories such as library and supermarket.

## 3 CRISP Dataset

As shown in Table 1, existing indoor scene datasets are divided into real and synthetic types. Some commonly used real-scene datasets like Scannet [4] have a small number of room scans, limited by the difficulty of data collection, while synthetic datasets can have larger numbers of scenes. Additionally, most of these datasets do not consider data distribution and often exhibit long-tail effects. The scarcity of examples for rare categories hinders models from learning robust 3D features. To our knowledge, CRISP is the **earliest** and **largest balanced** indoor retrieval dataset that facilitates exploration of the interaction between 3D data with text or image data. Figure 1 presents examples from the dataset. The creation of CRISP began with the collection of objects and hierarchical scene generation. Throughout the generation phase, our utmost priority was to maintain dataset equilibrium, thereby minimizing potential learning biases within models trained on the dataset. Using the generated scenes, we collected distinct sets of point cloud data, image data, and textual description data, all amalgamated to form the comprehensive CRISP dataset as shown in Fig. 4.

### 3.1 Object Collection

We first selected 62 common seen indoor object categories into our dataset which can encompass the majority of indoor objects encountered in daily life and meanwhile guarantee the scene diversity. Each object category contains 1~5 different

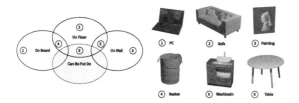

**Fig. 2.** This is the Venn diagram we used to classify object properties. The light-colored ovals represent the positional attributes of the objects, namely board, floor, and wall. Board refers to any flat surface other than the floor. The gray ovals indicate whether an object can have something placed on top of it. (Color figure online)

instances. Totally we collected 171 different instances. We aimed to create a well-balanced dataset while also ensuring that the scenes composed of these objects are more realistic. To achieve the second goal, we analyzed the layout of indoor scenes and categorized the objects accordingly. Inspired by some automated facilities layout technologies [12], we determined the common locations of each type of object in the scene and sorted them into the following categories: "Object On Board", "Objects On Floor", "Objects on Wall", as illustrated in Fig. 2. Those categories were established based on the spatial positional relationships of objects. Following the classification of an object's spatial position, we further categorized it according to the presence of a support surface that can hold other objects, as attribute "Can Be Put On". By combining these attributes, we can ensure that the generated scenes are reasonably plausible.

### 3.2   Hierarchical Generation

We used Kubric [5] library to load and organize objects. During the generation process, we employed a hierarchical generation method that incorporates the object category information obtained in Sect. 3.1 to achieve a higher degree of realism, i.e., one indoor scene should contain various categories of objects, as well as rational horizontal and vertical spatial object relationships. Detailed steps are shown in Fig. 3(a) and an example is shown in Fig. 3(b).

Hierarchical generation can combine single simple objects into a complex indoor scene and ensure rich spatial relationship between objects, clearly display the dependency relationships between objects, which facilitates the generation of scene graph. We maintained a global 3D occupancy map that records the space occupied by objects to prevent collisions during generation. Meanwhile, we maintained the randomness of object categories and positions, making the dataset more balanced and free from long-tailed distributions, providing a more fair and robust benchmark for further research. The generated scene was collected in the format of **3D mesh** for subsequent generation.

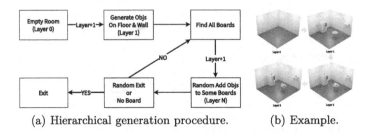

(a) Hierarchical generation procedure.          (b) Example.

**Fig. 3.** Our hierarchical generation procedure begined with layer 0, which represents an empty room. We then proceed to layer 1, where we added objects to the floor and walls. Next, we identified all available boards, and randomly place some objects on them, which became the next layer. This step was repeated until a random exit was triggered or there are no more boards available.

### 3.3   Point Cloud Collection

We converted the 3D scene mesh generated in Sect. 3.2 to ".obj" format and used CloudCompare[1], a 3D point cloud processing software, to sample point cloud. We sampled 200,000 points on surface which ensuring the collected point cloud much closer to the actual collected, and adjusted the rendering effect to make the collected point cloud as similar as possible to the object model.

### 3.4   Scene Image Collection

We used the bpy library provided by Blender[2] to generate photorealistic images from pre-built indoor scenes. For each room, we randomly selected a corner and simulated the breadth and height of human vision using a camera to capture the corresponding RGB image. We took one photograph for each room. To achieve more realism, we adjusted the rendering effect of Blender and added point light sources of different energy to simulate changes in lighting that occur in real scenes. We also extracted the depth image corresponding to the RGB image collected above to provide more information and support further research.

### 3.5   Scene Graph Generation

With the help of controllable generation procedure, we were able to record lots of useful information during scene generation such as the precise coordinate position of each object and where they were put on. We use this information to generate the scene graph.

In our designed scene graph, the nodes represent different objects, while the edges represent the spatial relationships between the nodes. Several types of relationships are defined, including "Next To", "Support" and "Positional Realation".

---

[1] http://www.cloudcompare.org/.
[2] https://www.blender.org/.

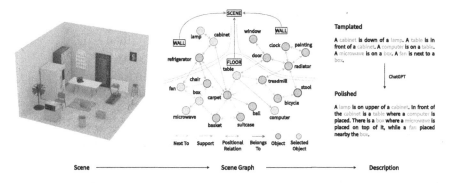

**Fig. 4.** Pipeline of the generation of CRISP, including scene generation (Sect. 3.2), scene graph generation (Sect. 3.5) and discription generation (Sect. 3.6). (Color figure online)

When two object are placed on the same surface and close enough, they form a "Next To" relationship. When one object is placed on the other object, these two object form a "Support" relationship. When for two objects, there are at least one object is on a wall, then they compose a "Positional Relation". The generation of scene graph transformed the layout of the scene into a form that the computer can directly process. An example can be found in Fig. 4. Compared to human annotation, our scene graph generation is fast while ensuring the correctness.

### 3.6 Textural Description Collection

After scene graph generation, as shown in Fig. 4, a random subset of nodes and edges was selected to form a description. The template to generate the descriptions is shown in the following prompt "<Object_A> is <R> <Object_B>", where "<Object_A>" and "<Object_B>" represent the categories of different nodes, and "<R>" denotes a spatial relationship between the two nodes. After generating the templated descriptions, we utilized the OpenAI ChatGPT API[3] to optimize them and make them more consistent with human language conventions. Once the rewriting was finished, we employed manual check by human to proofread the description meticulously, ensuring that it will adhere to the original meanings without mistakes and obtained the final textual descriptions.

### 3.7 Dataset Statistics

CRISP comprises 10,000 scenes, each including one or three descriptions, a set of paired RGB and depth images, and a point cloud of the scene. The detailed statistic is shown in Table 2. On average, a scene in CRISP contains 27.36 objects. We then analyzed the frequency of object category occurrences and Fig. 5 illustrates the top 20 most commonly found categories in point cloud or description.

---

[3] https://platform.openai.com/docs/models/gpt-3-5.

Table 2. Numerical Statistics of CRISP.

| Modal | # Train | # Val | # Test | # Total |
|---|---|---|---|---|
| Text | 24,000 | 1,000 | 1,000 | 26,000 |
| Image | 8,000 | 1,000 | 1,000 | 10,000 |
| Point Cloud | 8,000 | 1,000 | 1,000 | 10,000 |

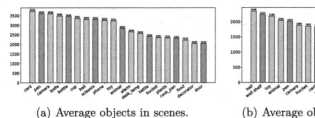

(a) Average objects in scenes.    (b) Average objects in descriptions.

Fig. 5. Distribution of top-20 object categories in CRISP.

It can be seen that both the occurrence frequency of objects in the generated scenes and in the descriptions is almost the same, indicating a well-balanced dataset compared to other indoor datasets like Scannet [4]. The characteristic of CRISP enables models trained on it to focus more on extracting different modal features rather than learning biases.

### 3.8  Unique Features of CRISP

We summarize the highlights of CRISP as follows: **(1) Easy to expand.** We developed an automatic and highly efficient pipeline for generating synthetic indoor scenes and collecting data in various modals, which allows us to easily expand the size of CRISP, and can be extended to construct datasets for other tasks. **(2) Large-Scale.** CRISP is currently the largest point cloud dataset for cross-modal exploration with 10,000 scenes. The extensive data in CRISP ensure a more diverse and representative set of multi-modal data, enabling more accurate and robust models. **(3) Without small data bias.** As shown in Sect. 3.7, our dataset exhibits a strong balance in terms of object category distribution, scene variability, and textual description diversity , ensuring that there is no small data bias. **(4) Novelty and broader applicability.** CRISP is a novel and versatile cross-modal retrieval dataset, making it a cutting-edge and highly promising development in cross-modal retrieval. Its multimodal nature enables various applications such as scene understanding, and more, allowing for future research not only in retrieval, but also other multimodal point cloud tasks.

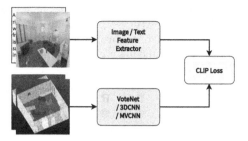

**Fig. 6.** The Pipeline of we used two-stream cross-modal retrieval baseline model.

## 4   Baseline Methods

### 4.1   Model Architecture

Considering the exceptional performance of models based on the CLIP architecture [9,10,18] on the MSCOCO [13] and Flickr [16] datasets, to create a strong baseline, in this paper, we adopted a similar dual-stream model with CLIP loss, see Formula (1). The pipeline is shown in Fig. 6. One stream of the model consists of queries, which in this case are either images or textual descriptions of rooms, while the other stream comprises queried values, specifically point cloud of the rooms. State-of-the-art methods were utilized to extract features from the inputs of both streams, and the CLIP loss, was then applied to align the features of the two different modalities by calculating the loss.

$$L_C = \frac{1}{2}\left[\frac{1}{B}\sum_{k=1}^{B}\frac{\exp\left(S_{v_k,t_k}/\tau\right)}{\sum_{l}^{B}\exp\left(S_{v_k,t_l}/\tau\right)} + \frac{1}{B}\sum_{k=1}^{B}\frac{\exp\left(S_{v_k,t_k}/\tau\right)}{\sum_{l}^{B}\exp\left(S_{v_l,t_k}/\tau\right)}\right], \qquad (1)$$

Where $B$ is the batch size and $\tau$ is the temperature hyper-parameter. $S_{v,t}$ is the total similarity score, defined as Formula (2):

$$S_{v,t} = \frac{1}{2}\left(\sum_{i=1}^{N_v} w_v^i \max_{j} a_{ij} + \sum_{j=1}^{N_t} w_t^i \max_{i} a_{ij}\right). \qquad (2)$$

The variables $u$ and $v$ represents two different modalities, such as text and point clouds, while $a_{ij}$ is a feature similarity matrix obtained by multiplying features from different modalities. The weights of the modal features are represented by $w_v^i$ and $w_t^j$, obtained by $\left[w_v^0, w_v^1, ..., w_v^{N_v}\right] = \text{Softmax}\left(\text{MLP}_v\left(V_f\right)\right)$, $\text{MLP}_v$ represents the fully connected layers used to encode modal $v$.

After training and before testing, for each kind of data in test set, we used its corresponding feature extractor to precalculate the feature of each sample, and then when testing, we would directly use these features to calculate the cosine similarity for retrieving to ensure the retrieval efficiency.

**Table 3.** Experimental Result on CRISP. The parentheses under the "InferTime" indicate the time taken for preprocessing MVCNN, which involves generating multi-view images for 1000 point cloud scenes. "w/o" means "without". Recall is used as an evaluation matrix. One out of a thousand candidates is chosen as the result, and its correctness is noted as R@1, and R@5, R@10 and R@100 are defined in the same way.

| Query | PC Model | R@1 | R@5 | R@10 | R@100 | # Param | | InferTime |
|---|---|---|---|---|---|---|---|---|
| | | | | | | Train | Test | |
| Rand | - | 0.1 | 0.5 | 1.0 | 10.0 | – | – | – |
| **Cross-Modal Retrieval** | | | | | | | | |
| Image | VoteNet | 43.8 | 90.5 | 97.4 | 99.8 | 89M | 89M | 77 s |
| Image | 3DCNN | 17.4 | 46.8 | 62.5 | 99.1 | 136M | 136M | 179 s |
| Image | MVCNN | 92.5 | 99.4 | 99.7 | 100 | 188M | 188M | 144 s (+86m) |
| Text | VoteNet | 0.1 | 0.5 | 1.0 | 10.0 | 134M | 134M | 73 s |
| Text | 3DCNN | 1.5 | 4.7 | 7.9 | 37.7 | 172M | 172M | 159 s |
| Text | MVCNN | 0.5 | 2.9 | 5.6 | 42.7 | 213M | 213M | 144 s (+86m) |
| **ModalBlender** | | | | | | | | |
| Text | VoteNet | 1.1 | 3.0 | 6.2 | 37.8 | 222M | 134M | 73 s |
| Text | 3DCNN | 5.6 | 13.3 | 20.7 | 55.1 | 261M | 172M | 159 s |
| Text | MVCNN | 8.7 | 28.5 | 40.9 | 89.5 | 302M | 213M | 144 s (+86m) |
| w/o CMM | MVCNN | 6.9 | 20.3 | 31.4 | 72.3 | – | – | – |
| w/o IMA | MVCNN | 2.1 | 5.7 | 10.2 | 58.7 | – | – | – |

### 4.2  Implementation Details

We utilized pretrained Swin Transformer [15] and RoBERTa [14] to extract features from images and scene descriptions separately. To extract features from point clouds and compare pros and cons of different 3D approaches, we employed three different methods. The first method, VoteNet [17], which extracted features directly from the point cloud data. The second method, 3DCNN [28], first voxelized the point cloud and then used sparse convolution to extract features, here we used our own designed sparse convolution network with Spconv [3] which had a model architecture similar to ResNet-50 [7]. The third method, MVCNN [24], captured point cloud information from different angles by taking snapshots, which were then used to extract features. For MVCNN, we also used pretrained Swin Transformer as the backbone.

### 4.3  Experimental Result

The experimental results of the cross-modal retrieval are shown in Table 3.

For **Image-PointCloud retrieval**, MVCNN performed the best, far surpassing VoteNet and 3DCNN. This was because MVCNN converts 3D point cloud features to 2D pixel features by rendering the point cloud. Retrieving 2D

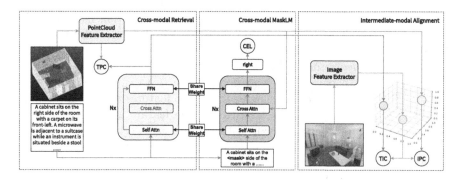

**Fig. 7.** Pipeline of our proposed ModalBlender.

images using 2D image search benefits from high-performance pretrained models. However, compared to VoteNet and 3DCNN, which can process point cloud data directly, MVCNN must first render and photograph point cloud data, making it much slower. It took around 86 min to preprocess all the CRISP test set for MVCNN before testing. Additionally, because MVCNN typically used a fixed camera angle to photograph the rendered result, its performance was greatly affected when there were obstructions, such as a ceiling in a room. On the contrary, VoteNet and 3DCNN were more robust. How to improve model accuracy and maintaining efficiency remains a worthy research question.

While for **Text-PointCloud retrieval**, all three models performed poorly. VoteNet performed the worst and resulted in an accuracy similar to random selection. 3DCNN, which was based on 3D data and 3D convolution kernels, performed the best in Text-PointCloud retrieval, because it had better spatial perception than MVCNN and VoteNet. Overall, the accuracy of all three backbones was very low because of the huge feature gap between text and point cloud. Meanwhile, currently there was no widely-used pretrained model to help Text-PointCloud alignment, making the situation worse. CRISP as the first cross-modal retrieval dataset for text and point cloud, provides a convenient and useful benchmark for studying the alignment of text and point cloud modalities.

## 5  Text Point-Cloud Alignment

### 5.1  Architecture and Implementation

To address the challenge of aligning textual and point cloud features, we then proposed a model called **ModalBlender** which comprises three submodules, as shown in Fig. 7. The first submodule, **CMR** (Cross-modal Retrieval), was the same as the one described in Sect. 4.1, and its output loss is Text-PointCloud CLIP loss (TPC). The second submodule, **CMM** (Cross-modal MaskLM), was a model similar to RoBERTa, but with cross-attention layers that take text feature (Q) and point cloud feature (K and V) as inputs. The self-attention layers in CMR and CMM share the same weights. We randomly masked 15% of the original text query and use it as input to CMM, and the output loss of CMM is cross-entropy loss (CEL). We adopt cross attention mechanism on text and point cloud features to guide the understanding of each other. The third submodule, **IMA** (Intermediate-modal Alignment), used two additional CLIP losses that involve image features as the intermediate modal to align text and point cloud. We paired text and image features to calculate Text-Image CLIP loss (TIC) and paired image and point cloud features to calculate Image-PointCloud CLIP loss (IPC). We leveraged image feature as the intermediate modal because there are well-developed text and image pretrained models and also image has RGB information that can be directly mapped into point cloud data, which made image a perfect bridge modal to align text and point cloud features. Finally, we used the weighted sum of these four losses as the final loss of ModalBlender. In particular, when testing, only the CMR module was used.

### 5.2 Quantitative Analysis

➤ **Performance Comparison:** In Table 3, we present the experimental results of ModalBlender. The performance of all three different backbone models had been significantly improved, indicating the effectiveness of the CMM and IMA modules. MVCNN achieved the highest accuracy and the greatest improvement, followed by 3DCNN. Although VoteNet still had the lowest accuracy among the three models, it was able to show some effective accuracy. The experimental results strongly demonstrated that cross-modal attention and the use of image features as an intermediate modal could greatly facilitate alignment between text and point cloud modalities.

➤ **Ablation Study:** We then conducted ablation experiments on ModalBlender with MVCNN. The results in the last two rows of Table 3 show that the removal of the CMM or IMA module degraded the performance. This demonstrated that both CMM and IMA are important in improving the model's accuracy. And compared to CMM, IMA has a greater influence on overall performance.

## 6    Conclusion

In this paper, we introduced CRISP, the first 3D indoor balanced scene cross-modal retrieval dataset that focuses on retrieving point cloud using text or images. Given the difficulty of obtaining 3D point cloud data, we proposed an automated pipeline that can generate a vast number of realistic indoor 3D point cloud scenes and then formed a dataset called CRISP cantains point clouds, RGBD-images and textual descriptions of the scenes. We conducted comprehensive experiments based on CRISP using now SOTA methods, and observed a huge performance gap between CRISP tasks and previous Text-Image retrieval tasks especially when using text. As one step towards better Text-PointCloud retrieval, we proposed a novel architecture named ModalBlender, and experimental evidence demonstrated that ModalBlender could significantly improve the accuracy of retrieval and provide a useful approach for aligning text and point cloud features in the absence of pretraining.

# References

1. Chang, A., et al.: Matterport3D: learning from RGB-D data in indoor environments. arXiv preprint arXiv:1709.06158 (2017)
2. Chang, A.X., et al.: ShapeNet: an information-rich 3D model repository. arXiv preprint arXiv:1512.03012 (2015)
3. Contributors, S.: SpConv: spatially sparse convolution library. https://github.com/traveller59/spconv (2022)
4. Dai, A., Chang, A.X., Savva, M., Halber, M., Funkhouser, T., Nießner, M.: ScanNet: richly-annotated 3D reconstructions of indoor scenes. In: Proceedings of the IEEE Conference on Computer Vision and Pattern Recognition, pp. 5828–5839 (2017)
5. Greff, K., et al.: Kubric: a scalable dataset generator. In: Proceedings of the IEEE/CVF Conference on Computer Vision and Pattern Recognition, pp. 3749–3761 (2022)
6. Handa, A., Pătrăucean, V., Stent, S., Cipolla, R.: SceneNet: an annotated model generator for indoor scene understanding. In: 2016 IEEE International Conference on Robotics and Automation (ICRA), pp. 5737–5743. IEEE (2016)
7. He, K., Zhang, X., Ren, S., Sun, J.: Deep residual learning for image recognition. In: Proceedings of the IEEE Conference on Computer Vision and Pattern Recognition, pp. 770–778 (2016)
8. Johnson, J., Hariharan, B., Van Der Maaten, L., Fei-Fei, L., Lawrence Zitnick, C., Girshick, R.: CLEVR: a diagnostic dataset for compositional language and elementary visual reasoning. In: Proceedings of the IEEE Conference on Computer Vision and Pattern Recognition, pp. 2901–2910 (2017)
9. Li, J., Li, D., Savarese, S., Hoi, S.: BLIP-2: bootstrapping language-image pre-training with frozen image encoders and large language models. arXiv preprint arXiv:2301.12597 (2023)
10. Li, J., Li, D., Xiong, C., Hoi, S.: BLIP: bootstrapping language-image pre-training for unified vision-language understanding and generation. In: International Conference on Machine Learning, pp. 12888–12900. PMLR (2022)
11. Li, W., et al.: InteriorNet: mega-scale multi-sensor photo-realistic indoor scenes dataset. arXiv preprint arXiv:1809.00716 (2018)
12. Liggett, R.S.: Automated facilities layout: past, present and future. Autom. Constr. 9(2), 197–215 (2000)
13. Lin, T.-Y., et al.: Microsoft COCO: common objects in context. In: Fleet, D., Pajdla, T., Schiele, B., Tuytelaars, T. (eds.) ECCV 2014. LNCS, vol. 8693, pp. 740–755. Springer, Cham (2014). https://doi.org/10.1007/978-3-319-10602-1_48
14. Liu, Y., et al.: RoBERTa: a robustly optimized BERT pretraining approach. arXiv preprint arXiv:1907.11692 (2019)
15. Liu, Z., et al.: Swin transformer: hierarchical vision transformer using shifted windows. In: Proceedings of the IEEE/CVF International Conference on Computer Vision, pp. 10012–10022 (2021)
16. Plummer, B.A., Wang, L., Cervantes, C.M., Caicedo, J.C., Hockenmaier, J., Lazebnik, S.: Flickr30k entities: collecting region-to-phrase correspondences for richer image-to-sentence models. In: Proceedings of the IEEE International Conference on Computer Vision, pp. 2641–2649 (2015)
17. Qi, C.R., Litany, O., He, K., Guibas, L.J.: Deep hough voting for 3D object detection in point clouds. In: Proceedings of the IEEE/CVF International Conference on Computer Vision, pp. 9277–9286 (2019)

18. Radford, A., et al.: Learning transferable visual models from natural language supervision. In: International Conference on Machine Learning, pp. 8748–8763. PMLR (2021)

19. Shilane, P., Min, P., Kazhdan, M., Funkhouser, T.: The Princeton shape benchmark. In: Proceedings Shape Modeling Applications, 2004, pp. 167–178. IEEE (2004)

20. Silberman, N., Hoiem, D., Kohli, P., Fergus, R.: Indoor segmentation and support inference from RGBD images. ECCV 5(7576), 746–760 (2012). https://doi.org/10.1007/978-3-642-33715-4_54

21. Song, D., Nie, W.Z., Li, W.H., Kankanhalli, M., Liu, A.A.: Monocular image-based 3-D model retrieval: a benchmark. IEEE Trans. Cybern. 52(8), 8114–8127 (2021)

22. Song, S., Lichtenberg, S.P., Xiao, J.: Sun RGB-D: a RGB-D scene understanding benchmark suite. In: Proceedings of the IEEE Conference on Computer Vision and Pattern Recognition, pp. 567–576 (2015)

23. Song, S., Yu, F., Zeng, A., Chang, A.X., Savva, M., Funkhouser, T.: Semantic scene completion from a single depth image. In: Proceedings of the IEEE Conference on Computer Vision and Pattern Recognition, pp. 1746–1754 (2017)

24. Su, H., Maji, S., Kalogerakis, E., Learned-Miller, E.: Multi-view convolutional neural networks for 3D shape recognition. In: Proceedings of the IEEE International Conference on Computer Vision, pp. 945–953 (2015)

25. Wald, J., Dhamo, H., Navab, N., Tombari, F.: Learning 3D semantic scene graphs from 3D indoor reconstructions. In: Proceedings of the IEEE/CVF Conference on Computer Vision and Pattern Recognition, pp. 3961–3970 (2020)

26. Wang, K., Yin, Q., Wang, W., Wu, S., Wang, L.: A comprehensive survey on cross-modal retrieval. arXiv preprint arXiv:1607.06215 (2016)

27. Wu, Z., et al.: 3D ShapeNets: a deep representation for volumetric shapes. In: Proceedings of the IEEE Conference on Computer Vision and Pattern Recognition, pp. 1912–1920 (2015)

28. Xu, Y., Tong, X., Stilla, U.: Voxel-based representation of 3D point clouds: methods, applications, and its potential use in the construction industry. Autom. Constr. 126, 103675 (2021)

29. Yi, K., et al.: CLEVRER: collision events for video representation and reasoning. arXiv preprint arXiv:1910.01442 (2019)

30. Yuan, J., et al.: SHREC'19 Track: extended 2D scene sketch-based 3D scene retrieval. Training 18, 70 (2019)

# Correlation Visualization Under Missing Values: A Comparison Between Imputation and Direct Parameter Estimation Methods

Nhat-Hao Pham[1,2], Khanh-Linh Vo[1,2], Mai Anh Vu[1,2], Thu Nguyen[3],
Michael A. Riegler[3], Pål Halvorsen[3], and Binh T. Nguyen[1,2(✉)]

[1] Faculty of Mathematics and Computer Science, University of Science,
Ho Chi Minh City, Vietnam
[2] Vietnam National University Ho Chi Minh City, Ho Chi Minh City, Vietnam
ngtbinh@hcmus.edu.vn
[3] SimulaMet, Oslo, Norway

**Abstract.** Correlation matrix visualization is essential for understanding the relationships between variables in a dataset, but missing data can seriously affect this important data visualization tool. In this paper, we compare the effects of various missing data methods on the correlation plot, focusing on two randomly missing data and monotone missing data. We aim to provide practical strategies and recommendations for researchers and practitioners in creating and analyzing the correlation plot under missing data. Our experimental results suggest that while imputation is commonly used for missing data, using imputed data for plotting the correlation matrix may lead to a significantly misleading inference of the relation between the features. In addition, the most accurate technique for computing a correlation matrix (in terms of RMSE) does not always give the correlation plots that most resemble the one based on complete data (the ground truth). We recommend using ImputePCA [1] for small datasets and DPER [2] for moderate and large datasets when plotting the correlation matrix based on their performance in the experiments.

**Keywords:** missing data · correlation plot · data visualization

## 1 Introduction

A correlation plot is a powerful and important visualization tool to summarise datasets, reveal relationships between variables, and identify patterns in the given dataset. However, missing data is a common situation in practice. There are two main approaches to computing correlation when the data contains missing entries. The first is to impute missing data and then estimate the correlation from

---

N.-H. Pham, K.-L. Vo and M. A. Vu—The three authors have equal contribution.

S. Rudinac et al. (Eds.): MMM 2024, LNCS 14557, pp. 103–116, 2024.
https://doi.org/10.1007/978-3-031-53302-0_8

imputed data. The second way is to estimate the correlation matrix directly from missing data.

Various imputation methods have been proposed to handle missing data [3–5]. For example, some traditional methods that have been used for missing data include mean imputation and mode imputation. In addition, multivariate imputation by chained equations (MICE) [6] is an iterative approach that can provide the uncertainty of the estimates. More recently, DIMV [7] is a method that relies on the conditional Gaussian formula with regularization to provide explainable imputation for missing values. Furthermore, machine learning techniques train on the available data to learn patterns and relationships and then are used to predict the missing values. There are many typical approaches, including tree-based models [8], matrix completion [9], or clustering-based methods such as KNN Imputation (KNNI). Finally, deep learning is an emerging trend for handling missing data due to the excellent prediction of deep learning architectures. Notable deep learning methods for imputation can be Generative Adversarial Imputation Nets (GAIN) [10] and Graph Imputer Neural Network (GINN) [11]. These methods have achieved high accuracy but require significant computational resources and are best suited for data-rich cases.

For the direct parameter estimate approach, a simple way to estimate the covariance matrix from the data is to use case deletion. However, the accuracy of such an approach may be low. Recently, direct parameter estimation approaches were proposed to estimate the parameters from the data directly. For example, the EPEM algorithm [12] estimates these parameters from the data by maximum likelihood estimation for monotone missing data. Next, the DPER algorithm [2] is based on maximum likelihood estimation for randomly missing data [2]. The computation for the terms in the covariance matrix is based on each pair of features. Next, the PMF algorithm [13] for missing dataset with many observed features by using principle component analysis on the portion of observed features and then using maximum likelihood estimation to estimate the parameters. Such algorithms are scalable and have promising implementation speed.

Since the correlation plot is an important tool in data analysis, it is necessary to understand the impact of missing data on the correlation plot [14]. This motivates us to study the effects of various imputation and parameter estimation techniques on the correlation plot. In summary, our contributions are as follows. First, we examine the effect of different imputation and direct parameter estimation methods on the correlation matrix and the correlation matrix heat maps for randomly missing data and monotone missing data. Second, we show that imputation techniques that yield the lowest RMSE may not produce the correlation plots that most resemble the one based on completed data (the ground truth). Third, we illustrate that relying solely on RMSE to choose an imputation method and then plotting the correlation heat map based on the imputed data from that imputation method can be misleading. Last but not least, our analysis process using the Local RMSE differences heatmaps and Local difference (matrix subtraction) for correlation provides practitioners a strategy to draw inferences from correlation plots under missing data.

## 2    Methods Under Comparison

This section details techniques under comparison. These include some state-of-the-art techniques (DPER, GAIN, GINN, SoftImpute, ImputePCA, missForest), and some traditional but remaining widely used methods (KNNI, MICE, Mean imputation). More details are as follows (Except for DPER [2], which directly estimates the covariance matrix from the data; the remaining methods involve imputing missing values before calculating the covariance matrix.) (Table 1).

(a) **Mean imputation** imputes the missing entries for each feature by the mean of the observed entries in the corresponding feature.

(b) **Multiple Imputation by Chained Equations (MICE)** [15] works by using a series of regression models that fit the observed data. In each iteration of the process, the missing values for a specific feature are imputed based on the other features in the dataset.

(c) **MissForest** [8] is an iterative imputation method based on a random forest. It averages over many unpruned regression or classification trees, allowing a multiple imputation scheme and allowing estimation of imputation error.

(d) **K-nearest neighbor imputation (KNNI)** is based on the K-nearest neighbor algorithm. For each sample to be imputed, KNNI finds a set of K-nearest neighbors and imputes the missing values with a mean of the neighbors for continuous data or the mode for categorical data.

(e) **SoftImpute** [9] is an efficient algorithm for large matrix factorization and completion. It uses convex relaxation to produce a sequence of regularized low-rank solutions for completing a large matrix.

(f) **ImputePCA (iterative PCA)** [1] minimizes the criterion, and the imputation of missing values is achieved during the estimation process. Since this algorithm is based on the PCA model, it takes into account the similarities between individuals and the relationships between variables.

(g) **Expectation Maximization algorithm (EM)** [16] is a maximum likelihood approach from incomplete data. It iteratively imputes missing values and updates the model parameters until convergence. The algorithm consists of two steps: E-step estimates missing data given observed data, and M-step estimates model parameters given complete data.

(h) **Generative Adversarial Imputation Networks (GAIN)** [10] adapts the GAN architecture where the generator imputes the missing data while the discriminator distinguishes between observed and imputed data. The generator maximizes the discriminator's error rate while the discriminator minimizes classification loss.

(i) **Graph Imputation Neural Networks (GINN)** [11] is an autoencoder architecture that completes graphs with missing values. It consists of a graph encoder and a graph decoder, which utilize the graph structure to propagate information between connected nodes for effective imputation.

(j) **Direct Parameter Estimation for Randomly missing data (DPER algorithm)** [2] is a recently introduced algorithm that directly finds the maximum likelihood estimates for an arbitrary one-class/multiple-class randomly missing dataset based on some mild assumptions of the data. Since

**Table 1.** The description of datasets used in the experiments.

| Dataset | # Classes | # Features | # Samples |
|---------|-----------|------------|-----------|
| Iris | 3 | 4 | 150 |
| Digits | 10 | 64 | 1797 |
| MNIST [17] | 10 | 784 | 70000 |

the correlation plot does not consider the label, the DPER algorithm used for the experiments is the DPER algorithm for one class dataset.

## 3  Experiments

### 3.1  Experiment Settings

This section details the experimental setup for comparing the effects of the methods presented in Sect. 2 on correlation plots with missing data. For the experiments on randomly missing data, we used the Iris and Digits datasets and generated randomly missing data with rates from 0.1 to 0.5. Here, the missing rate is the ratio between the number of missing entries and the total number of entries. For the monotone missing pattern, we removed a $0, 5$ piece of the image in the right corner with $0.4, 0.5$, and $0.6$ height and width in the MNIST dataset [17]. For MNIST, as KNNI and GINN [11] cannot produce the results within 3 h of running, we exclude them from the plot of results. For each dataset, we normalize the data and run the experiments on a Linux machine with 16 GB of RAM and four cores, Intel i5-7200U, 3.100GHz.

To throughout investigate the effects of missing data on the correlation heatmap, we employ three types of heatmap plots: (1) **Correlation Heatmaps** illustrates the correlation matrix and employs a blue-to-white-to-red color gradient to illustrate correlation coefficients. The colormap utilizes distinct colors: blue, white, and red, which are represented by the values $[(0, 0, 1), (1, 1, 1), (1, 0, 0)]$ respectively, as defined by the color map of the *matplotlib* library. Blue indicates negative values, red denotes positive values, and white indicates median (0) values. The colormap's range is bounded within $vmin = -1$ and $vmax = 1$, which are parameters of *matplotlib* that linearly map the colors in the colormap from data values $vmin$ to $vmax$; (2) **Local RMSE Difference Heatmaps for Correlation** portray the RMSE differences on a cell-by-cell basis, comparing correlation matrices produced by each estimation technique with the ground truth correlation matrix. We adopt a different color gradient, which is a green-to-white color map with values $[(1, 1, 1), (0, 0.5, 0)]$, further enhancing the clarity of these visualizations. The intensity of the green colors corresponds to the magnitude of the difference between the correlation matrix and the ground truth (depicted as white); (3) **Local Difference (Matrix Subtraction) Heatmaps for Correlation** show the local differences for a cell-by-cell comparison (matrix subtraction) between correlation matrices generated from each estimation technique and the ground truth correlation matrix. This plot shows whether the difference between the two plots is positive or negative, providing a clear indication of the direction of the difference.

For each dataset, we provide five heatmap plots. The first two plots display the method outcomes across varying levels of missing value, with the first showing the Correlation Heatmaps and the second showing the Local RMSE Difference Heatmaps. The two subsequent plots provide a more detailed examination, specifically highlighting the highest missing rate of the Correlation Heatmaps and Local RMSE difference heatmaps. This enables a more thorough analysis of color variations that might have been challenging to detect in the first two plots due to spatial limitations. The final plot, also focusing on the highest missing rate, shows the Local Difference (Matrix Subtraction) Heatmaps for Correlation, clearly indicating the direction of the difference.

We use gray to differentiate null values, especially when all cells in a feature share the same values, often near the image edges. This gray shade distinguishes null values from 0 values, which are depicted in white. Furthermore, the plot arrangement follows a descending order of RMSE, focusing on the highest missing rate's RMSE due to its significant impact on color intensity across rates. To maintain consistency with the correlation plot, we represent the ground truth heatmaps difference in white. In addition, in the Local RMSE Difference Heatmaps for Correlation charts, we integrate the calculation of the RMSE, which represents the difference between the two correlation matrices while excluding null positions. Moreover, we include the dense rank alongside it in ascending order. Dense ranking is a method for giving ranks to a group of numbers. The ranks are given so that there are no gaps between them. The next rank isn't skipped if some numbers share the same rank. For instance, let's take this set of values: [89, 72, 72, 65, 94, 89, 72]. The resulting dense ranks would be: [1, 2, 2, 2, 3, 3, 4]. Below each Local RMSE Difference Heatmaps plot, we denote the rank and the RMSE value using the following structure: ($<$ rank $>$)RMSE: $<$ RMSE value $>$.

### 3.2   Result and Analysis

**Randomly Missing Data.** A clear trend emerges in the RMSE values across the Iris and Digits datasets as depicted in the initial two line plots in Fig. 1(a).

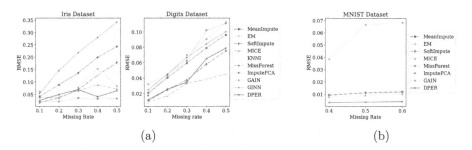

(a)                                                                 (b)

**Fig. 1.** The RMSE difference of the correlation matrix between each method and the ground truth is depicted across missing rates ranging from 0.1 to 0.5 for (a) Iris and Digits dataset with randomly missing patterns and (b) MNIST dataset with monotone missing patterns.

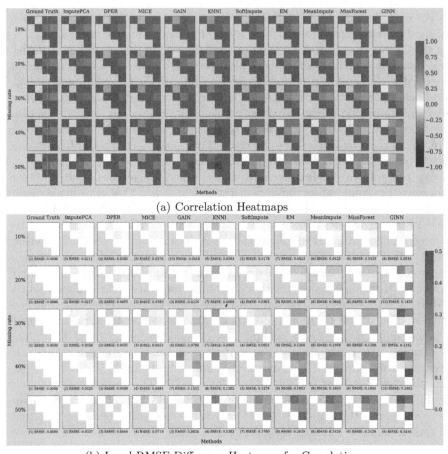

(a) Correlation Heatmaps

(b) Local RMSE Difference Heatmaps for Correlation

**Fig. 2.** Heatmaps for the Iris dataset across missing rates from 0.1 to 0.5

As the missing rate increases, there is a corresponding rise in the RMSE difference. This trend is also observed in the Local RMSE Difference Heatmaps for Correlation at Fig. 3(b) and Fig. 5(b), where the color intensity of the green shades increases in both Fig. 2(b) and Fig. 4(b). The charts also show that the RMSE rankings remain consistent by comparing the changing ranks within the Local RMSE difference heatmaps in the last plot of Figs. 2 and 4. This indicates that the findings about different correlation matrix estimation methods stay the same across different missing rates.

*For the Iris Dataset*, at its highest missing rate in the Local RMSE Difference Heatmaps for Correlation plot in Fig. 3(b), with Mean Imputation as the baseline, it's worth noting that EM and MissForest also exhibit RMSE values that are close to the baseline. Seven of these methods perform better than the baseline, while MissForest and GINN demonstrate higher RMSE values (0.2429 and 0.3416, respectively). Notably, DPER, MICE, and GAIN stand out, as they exhibit RMSE values lower than the median (RMSE = 0.1383, represented by KNNI). These methods are highly recommended for estimating the correlation matrix in this dataset.

The visualization of the Local Difference (Matrix Subtraction) Heatmaps for Correlation in Fig. 3(c) indicates that the discrepancies are both positive and negative, represented by a mixed color between blue and white. However, we observe that the cells with an intense green color for GINN, MissForest, Mean Impute, and EM in Fig. 3(b) appear blue in Fig. 3(c). This implies that features with greater divergence compared to the ground truth tend to have a negative tendency in their differences, suggesting an underestimation of the correlation between these features.

*For the Digits dataset*, in the first row of the plot layout, several methods outperform Mean Impute and have RMSE better than the median value of 0.097, namely KNN, MICE, SoftImpute, and DPER. Interestingly, while EM has a better RMSE than Mean Impute, their RMSE values are quite similar. Like the Iris dataset, EM, Mean Impute, and MissForest exhibit nearly identical RMSE values, while GINN still displays higher RMSE values. Both ImputePCA and GAIN show larger RMSE values compared to the baseline.

Visualizing the Local Difference (Matrix Subtraction), Heatmaps for Correlation in Fig. 5(c) reveals both positive and negative disparities. An intriguing pattern emerges: certain lines running parallel to the diagonal consistently display intense green colors in Fig. 5(b), suggesting significant differences. These areas of intense green indicate that these features tend to underestimate correlation, as evidenced by the corresponding positions of these cells in Fig. 5(c) appearing as blue. These positions correspond to cells with very high correlation values, as visible in Fig. 5(a). This phenomenon is consistent across SoftImpute, EM, Mean Impute, MissForest, GINN, and GAIN. Regarding correlation estimation, methods like KNNI, MICE, SoftImpute, and DPER perform better than the median RMSE value and are thus recommended. However, practitioners should exercise caution when using SoftImpute, as it may underestimate the correlation of some highly correlated features. Throughout the analysis above, techniques including KNN, MICE, SoftImpute, and DPER consistently demonstrate their effectiveness in estimating correlation matrices for the Digits dataset. Within the context of the Iris dataset, DPER, MICE, and GAIN emerge as reliable methods. Moreover, our visual observations highlight certain methods' inclination to underestimate highly correlated features in the correlation matrix within the Digits dataset.

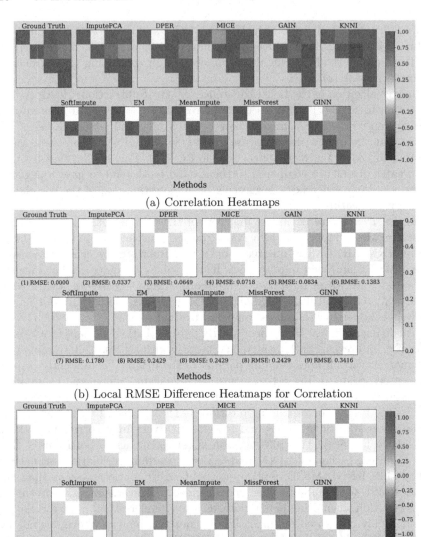

(a) Correlation Heatmaps

(b) Local RMSE Difference Heatmaps for Correlation

(c) Local Difference (Matrix Subtraction) Heatmaps for Correlation

**Fig. 3.** Heatmaps for the Iris dataset at a missing rate of 0.5

**Monotone Missing Data.** In contrast to the Iris and Digits datasets, the results for the plots in Fig. 6(b) across varying missing rates show minimal variation among methods. Most methods, except for GAIN, exhibit closely aligned RMSE values, which demonstrate the highest RMSE and are visually apparent in the RMSE plot. All of these methods, except for GAIN, are recommended for estimating the correlation matrix for this dataset.

**Analysis.** In summary, from the observations for all the datasets, we see that DPER and MICE consistently provide plots that most resemble the ground truth.

From the experiments, one can see that a low RMSE may not imply a correlation plot reflecting the true relation between the features. For example, at a 0.5 missing rate on the Digits dataset, even though softImpute and Missforest have lower RMSE than DPER, DPER still clearly resembles the

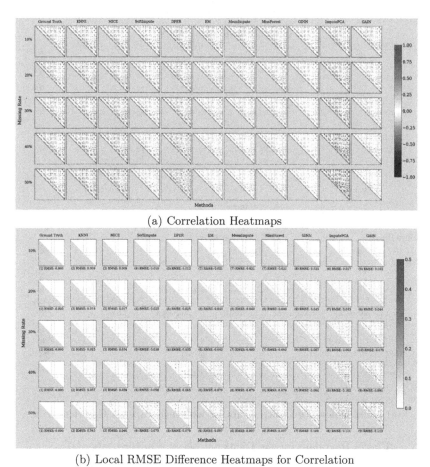

(a) Correlation Heatmaps

(b) Local RMSE Difference Heatmaps for Correlation

**Fig. 4.** Heatmaps for the Digits dataset across missing rates from 0.1 to 0.5

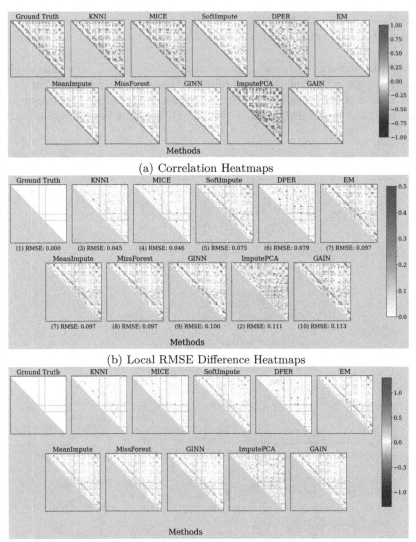

(a) Correlation Heatmaps

(b) Local RMSE Difference Heatmaps

(c) Local Difference (Matrix Subtraction) Heatmaps for Correlation

**Fig. 5.** Heatmaps for the Digits dataset at a missing rate of 0.5

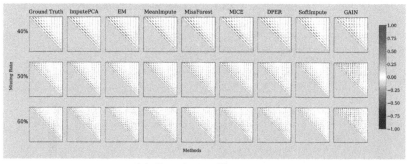

(a) Correlation Heatmaps

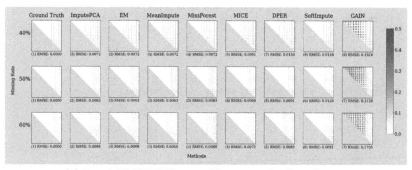

(b) Local RMSE Difference Heatmaps for Correlation

**Fig. 6.** Heatmaps for the MNIST dataset across missing rates from 0.4 to 0.5

ground truth plot. To understand why, suppose that we have a dataset of two features where the covariance matrix is $\Sigma = \begin{pmatrix} \sigma_1 & \sigma_{12} \\ \sigma_{12} & \sigma_2 \end{pmatrix}$. Then $RMSE = \sqrt{(\hat{\sigma}_1 - \sigma_1)^2 + (\hat{\sigma}_2 - \sigma_2)^2 + 2(\hat{\sigma}_{12} - \sigma_{12})^2}$, and whether $((\hat{\sigma}_1 - \sigma_1)^2, (\hat{\sigma}_2 - \sigma_2)^2, (\hat{\sigma}_{12} - \sigma_{12})^2)$ receives the values $(0.1, 0.5, 0.1)$ or $(0.2, 0.2, 0.2)$ will result in the same values of RMSE. In summary, relying solely on RMSE values for selecting an imputation method and then using the imputed data based on that method to create correlation plots and draw inferences can lead to misleading conclusions. Considering Local RMSE differences in heatmaps and Local difference (matrix subtraction), heatmaps are vital to ensure precise method selection (Fig. 7).

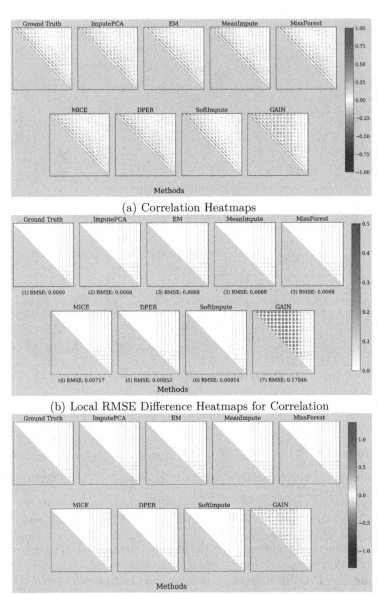

(a) Correlation Heatmaps

(b) Local RMSE Difference Heatmaps for Correlation

(c) Local Difference (Matrix Subtraction) Heatmaps for Correlation

**Fig. 7.** Heatmaps for the MNIST dataset at a missing rate of 0.5

## 4    Conclusions

In this work, we examined the effects of missing data on correlation plots. Cru-
cially, we found that the techniques yielding the lowest RMSE do not consistently produce correlation plots that closely resemble those based on complete

data (the ground truth). This could be because the RMSE is an average measure. Meanwhile, human eyes capture local differences in the matrix, while visual perception is better at capturing local differences, which could be better visualized by representing the local square difference on each cell of the heatmap (employed by the Local RMSE Difference Heatmaps for Correlation in our experiments). In addition, the ground truth data itself may contain noise. Therefore, relying solely on RMSE values for selecting an imputation method and then using the imputed data based on that method to create a correlation plot and draw inferences can lead to misleading conclusions. Considering Local RMSE differences in heatmaps and Local difference (matrix subtraction), heatmaps are vital to ensure proper inferences from correlation plots when there is missing data.

Moreover, our analysis consistently highlights the efficiency of correlation estimation methods such as MICE and DPER across all three datasets. Notably, DPER is scalable and consistently produces the correlation plot that best resembles the ground truth for both small and large datasets. We hypothesize that such a performance is related to the asymptotic consistency property of DPER and will investigate this theoretically in the future. In the future, we want to study whether the methods that help speed up the imputation process, such as PCAI [18] or BPI [19], affect the correlation heatmaps.

**Acknowledgments.** We would love to thank AISIA Research Lab, SimulaMet (Oslo, Norway), the University of Science, and Vietnam National University Ho Chi Minh City (VNU-HCM) for supporting us under the grant number DS2023-18-01 during this project.

# References

1. Josse, J., Husson, F.: missMDA: a package for handling missing values in multivariate data analysis. J. Stat. Softw. **70**, 1–31 (2016)
2. Nguyen, T., Nguyen-Duy, K.M., Nguyen, D.H.M., Nguyen, B.T., Wade, B.A.: DPER: direct parameter estimation for randomly missing data. Knowl.-Based Syst. **240**, 108082 (2022)
3. Nguyen, P., et al.: Faster imputation using singular value decomposition for sparse data. In: Asian Conference on Intelligent Information and Database Systems, pp. 135–146. Springer, Singapore (2023). https://doi.org/10.1007/978-981-99-5834-4_11
4. Lien, P.L., Do, T.T., Nguyen, T.: Data imputation for multivariate time-series data. In: 2023 15th International Conference on Knowledge and Systems Engineering (KSE), pp. 1–6. IEEE (2023)
5. Nguyen, T., Storås, A.M., Thambawita, V., Hicks, S.A., Halvorsen, P., Riegler, M.A.: Multimedia datasets: challenges and future possibilities. In: International Conference on Multimedia Modeling, pp. 711–717. Springer, Cham (2023). https://doi.org/10.1007/978-3-031-27818-1_58
6. van Buuren, S., Groothuis-Oudshoorn, K.: mice: multivariate imputation by chained equations in R. J. Stat. Softw., pp. 1–68 (2010)
7. Vu, M.A., et al.: Conditional expectation for missing data imputation. arXiv preprint arXiv:2302.00911 (2023)

8. Stekhoven, D.J., Bühlmann, P.: Missforest-non-parametric missing value imputation for mixed-type data. Bioinformatics **28**(1), 112–118 (2012)

9. Mazumder, R., Hastie, T., Tibshirani, R.: Spectral regularization algorithms for learning large incomplete matrices. J. Mach. Learn. Res. **11**, 2287–2322 (2010)

10. Yoon, J., Jordon, J., van der Schaar, M.:GAIN: missing data imputation using generative adversarial nets. CoRR, abs/1806.02920 (2018)

11. Spinelli, I., Scardapane, S., Uncini, A.: Missing data imputation with adversarially-trained graph convolutional networks. Neural Netw. **129**, 249–260 (2020)

12. Nguyen, T., Nguyen, D.H.M., Nguyen, H., Nguyen, B.T., Wade, B.A.: EPEM: efficient parameter estimation for multiple class monotone missing data. Inf. Sci. **567**, 1–22 (2021)

13. Nguyen, T., Phan, T.N., Hoang, V.H., Halvorsen, P., Riegler, M., Nguyen, B.: Efficient parameter estimation for missing data when many features are fully observed (2023)

14. Kraus, M., et al.: Assessing 2D and 3D heatmaps for comparative analysis: an empirical study. In: Proceedings of the 2020 CHI Conference on Human Factors in Computing Systems, pp. 1–14 (2020)

15. Pedregosa, F., et al.: Scikit-learn: machine learning in Python. J. Mach. Learn. Res. **12**, 2825–2830 (2011)

16. Dempster, A.P., Laird, N.M., Rubin, D.B.: Maximum likelihood from incomplete data via the EM algorithm. J. Royal Stat. Soc.: Ser. B (Methodol.) **39**(1), 1–38 (1977)

17. LeCun, Y.: The MNIST database of handwritten digits. http://yann.lecun.com/exdb/mnist/ (1998)

18. Nguyen, T., Ly, H.T., Riegler, M.A., Halvorsen, P., Hammer, H.L.: Principal components analysis based frameworks for efficient missing data imputation algorithms. In Asian Conference on Intelligent Information and Database Systems, pp. 254–266. Springer Nature Switzerland, Cham (2023). https://doi.org/10.1007/978-3-031-42430-4_21

19. Do, T.T., et al.: Blockwise principal component analysis for monotone missing data imputation and dimensionality reduction. arXiv preprint arXiv:2305.06042 (2023)

# IFI: Interpreting for Improving: A Multimodal Transformer with an Interpretability Technique for Recognition of Risk Events

Rupayan Mallick$^{(\boxtimes)}$ [ID], Jenny Benois-Pineau [ID], and Akka Zemmari [ID]

LaBRI UMR CNRS 5800, Universite de Bordeaux, 33400 Talence, France
`rupayan.mallick@u-bordeaux.fr`

**Abstract.** Methods of Explainable AI (XAI) are popular for understanding the features and decisions of neural networks. Transformers used for single modalities such as videos, texts, or signals as well as multimodal data can be considered as a state-of-the-art model for various tasks such as classification, detection, segmentation, etc. as they generalize better than conventional CNNs. The use of feature selection methods using interpretability techniques can be exciting to train the transformer models. This work proposes the use of an interpretability method based on attention gradients to highlight important attention weights along the training iterations. This guides the transformer parameters to evolve in a more optimal direction. This work considers a multimodal transformer on multimodal data: video and sensors. First studied on the video part of multimodal data, this strategy is applied to the sensor data in the proposed multimodal transformer architecture before fusion. We show that the late fusion via a combined loss from both modalities outperforms single-modality results. The target application of this approach is Multimedia in Health for the detection of risk situations for frail adults in the @home environment from the wearable video and sensor data (BIRDS dataset). We also benchmark our approach on the publicly available single-video Kinetics-400 dataset to assess the performance, which is indeed better than the state-of-the-art.

**Keywords:** Multimodal Networks · Interpretability · Classification · Risk Events · Transformers

## 1 Introduction

Deep Models are successful in almost all sectors ranging from vision to language, healthcare, etc. With the advent of transformers, their number of parameters ranges from millions to billions. A very large amount of data is needed for the model to generalize well. The interpretation of these models becomes more complex. This research is an attempt to use interpretability techniques for the transformers training to obtain better generalizing models. Before transformer-based

S. Rudinac et al. (Eds.): MMM 2024, LNCS 14557, pp. 117–131, 2024.
https://doi.org/10.1007/978-3-031-53302-0_9

models CNNs dominated the literature for vision, thus a number of explanation and interpretable models were proposed for CNNs [27,40]. One of the categories is *gradient-based models* [27] which highlights the regions for decisions given by the saliency map. This category includes methods of e.g. gradient backpropagation [31] and deconvolution [38]. Other methods such as CAMs [40] are class-specific explainers. They use various feature maps with different weighing schemes. Similarly, in language and time-series models other forms of explainers were used which included masking, LIME [26], shapely values [17], and others.

In our proposed work, we are using a real-world dataset encompassing two modalities, a) an egocentric video modality and b) a combination of physiological and motion sensors. The target application is the detection of risk events amongst old and frail individuals. Risk events are not very common in daily life and thus are rare in the life-log data, making this a complex problem.

In this work, the intuition is to focus on the parts of attention that are relevant to the particular class/label, thus we are weighing the gradient of attention to the attention. Our video models are based on the fine-tuning of models trained on a larger dataset. During the fine-tuning process, our model employs loss by combining two terms. The first one is to compute the classification accuracy, and the second one encourages more focus on the relevant attention weights. To better understand the risk events, an additional modality comprising signals from sensors is added to the visual modality. To the best of our knowledge, this is a novel work using interpretability to train a multimodal network.

Our contributions are as follows.

- We propose a gradient-based interpretable method that can be used at training time to improve classification scores at the generalization step in both video and signal transformers;
- We are using a multimodal architecture, compounding sensor data (signals) to the visual data (videos) for a better understanding of the context in the application of recognition of risks.

The paper is organized as follows. In Sect. 2 we present related work on transformers, interpretability techniques, and recent multimodal methods. In Sect. 3 we present our interpretability technique applicable to both video and signal data and the multimodal architecture that we propose for the recognition of risk events from wearable video and signal data. In Sect. 4, we describe our datasets, multimodal data organization and report on training using a single modality as well as multiple modalities. In Sect. 5, we give the conclusion and discuss our future work.

## 2    Related Works

Transformer models have recently become the most popular tool for data analysis for various classification problems [8,9]. The success of architectures such as BERT [8] and its variants on a number of Natural Language Processing (NLP) tasks has paved the way for presenting the effectiveness of pre-trained trans-formers. Recently, transformers have been used for computer vision tasks such as video understanding [33], object detection [4], action recognition [3], etc. A transformer is a combination of various self-attention modules [9,19,34] but it can also be formed as a combination of convolutional networks and self-attention [30]. Due to the lower inductive bias compared to CNNs, the use of transformers is based on a large amount of data, as the evaluative performance of image trans-formers (ViTs) is shown to decline significantly with limited training data. Video understanding and recognition have been long studied in the literature, with the use of LSTMs on top of convolutional features [12] or with the advent of 3D CNN models [32]. For tasks such as video classification and object detection, the use of self-attention with convolution operation has been studied [11]. Therefore, it is natural to use CNNs for the first layer of feature embedding in video data. In multimedia, data are multi-modal: video, text, audio [25], the use of visual and signal sensor information has shown increased efficiency even without Deep Learning techniques [21]. Hence, in the following two subsections, we briefly review the multimodal models working on multi-modal data and interpretability techniques.

## 2.1   Multimodal Architectures

Multimodal architectures have existed in the literature since the inception of multimodal data. Speech Recognition, Natural Language Processing, Text Gen-eration, etc., used multimodal models to understand the underlying feature rep-resentation. One of the works [23] of audio-visual bimodal fusion uses shared hidden layer representation to understand the higher-level correlation between the audio and visual cues. Other works such as CLIP [25] jointly train the image and the text encoder to get the correct parings in a batch to create the training examples. In the [24] work, the authors predict the temporal synchronization of audio and visual features. These examples are numerous; however, multimodal architectures on visual information and sensor information remain rare. Com-bining various modalities has been well studied in [14], using different fusion methods.

## 2.2   Interpretability Techniques

To circumvent the black-box nature of deep neural networks, recent studies have tried to explain their decisions [2,6,27,28]. In visual recognition tasks, decisions can be explained by the importance of pixels in the input image or features in a feature map. The interpretation of the decisions can be class-dependent or class-agnostic (i.e., for the overall model). In the work [28], partial derivatives of the class scores (using first-order Taylor's expansion) are computed w.r.t the image pixel. Another saliency-based method, guided backpropagation [29] uses higher-level feature activations that are propagated back to the image to find the pixels most responsible for activation. The gradients of the class scores are computed w.r.t the input image. This approach restricts the backpropagation of the negative gradients to have sharper visualizations. Gradient-based methods, including, e.g., Grad-CAM [27] and its variations [31] compute the linear combination of class-specific weights (computed as the gradient of class scores w.r.t. feature maps) in the last convolutional layer. Grad-CAM-like methods are class-specific, as they consider class-dependent gradients.

Another family of methods includes the attribution propagation methods [22], one such method is Layerwise Relevance Propagation (LRP) [2] which uses Deep Taylor Decomposition (DTD) to propagate backward the relevance scores from the last layer to the input. The LRP method is based on the principle of conservation of relevance scores in a layer throughout the network. Another family of methods, such as LIME [26] proceeds by masking the parts in input data and tracking the scores, thus defining important elements in the input. For a detailed review of the explanation methods, we refer the reader to [1]. In Transformers, attention already gives insight into the importance of the input parts. Therefore, it will be an inherent part of our interpretability technique. Furthermore, the gradient-based interpretability approach on the basis of attention will be used to understand the stability and the local interaction of features while using pre-trained models. Indeed, gradient gives the rate of change of a value. When multiplied with a value itself, it stresses the change. In [18], self-attention in a vision transformer is multiplied by its gradient. This also serves to stabilize the training of attention weights.

Explainability techniques have been used for feature selection in [39] to improve primarily the explanation by combining the gradient and masked approaches. Liang et al. [13] use interpretability to decrease the 'many-to-many' correspondences between the CNN filters and the classes. Another work by Chefer et al. [7] uses relevancy maps to increase the robustness of the ViT by using it with foreground and background masks.

## 3 Methodology

Our objective is to identify attention weights in a transformer responsible for classifying the data point to a particular class, therefore added supervision is provided. This added supervision is based on an interpretable method which in itself is a gradient-based method. To cater to the application of recognizing risk events, we apply the same methodology on the video transformer and on the sensor signal transformer.

### 3.1 Training Transformers with Interpretable Methods

The interpretable method for a ViT was proposed in [18]. It uses the gradient of attention. We are similarly using the same gradient of attention to train our video transformer. Our video transformer is based on the Swin-3D model [16]. The intuition to use the gradient of attention is to direct the computation of the attention weights toward the obtained class score as the gradient of attention is computed with respect to the class score.

A vanilla image transformer (ViT) uses the decomposition of the input image into rectangular patches and considers each patch as a vector [9]. The Query $(Q)$, Key$(K)$, and Value $(V)$ matrix is an embedding in the $\mathbb{R}^{(P^2 \times d)}$ space where $(P, P)$ is the patch size and $d$ is the dimension for the linear projection of patches-vectors into the representation space. Self-Attention is computed as the inner product of the *Query* and the *Key* as given in Eq. 1. The final attention is given in Eq. 2

$$A = Q.K^T \tag{1}$$

Thus, the attention as proposed in [34].

$$\text{Attention}(Q, K, V) = \text{Softmax}(QK^T/\sqrt{d})V \tag{2}$$

If we now consider temporal data such as a bunch of video frames or a bunch of sequential signal recordings, then the computation of attention can be performed as in the Video Swin Transformer [16]. They consider blocks of video frames as a 3D structure with time as the $3^{rd}$ axis. And instead of taking patches in images, a 3D block is taken in the video as a data point. Each of these data points is embedded to $Q$, $K$, and $V$ with the dimension of $(MP^2 \times d)$ and $M$ is the dimension of the temporal patch. It is given by introducing a 3D relative position bias and shifted windows. The 3D relative position bias $B$ gives the geometric relationship between the visual tokens. It encodes the relative configurations in the spatial or temporal dimension. $B$ thus encodes the distance between the two tokens. The latter are limited by $[-M + 1, M - 1]$ for the temporal axis and $[-P + 1, P - 1]$ for the spatial axes, that is height and width. The attention for the swin-transformer [15] is given in the Eq. 3 with $Q, K, V \in \mathbb{R}^{(MP^2 \times d)}$ and $B \in \mathbb{R}^{M^2 \times P^2 \times P^2}$. The input video clip, which is our token, is a tensor of $T \times H \times W$ dimension. We divide it into $\lfloor \frac{T}{M} \rfloor, \lfloor \frac{H}{P} \rfloor, \lfloor \frac{W}{P} \rfloor$ non-overlapping windows with the window size of $M \times P \times P$ to get our data points.

$$\text{Attention}(Q, K, V) = \text{Softmax}\left(\frac{QK^T}{\sqrt{d}} + B\right)V \tag{3}$$

For our work, we are using interpretable methods to train models in an optimal direction. The core idea of using the gradient of attention $\nabla A$ helps in the understanding of change in the attention weights w.r.t classes. Thus, we are multiplying the attention weights $A_n$ at each $n^{th}$ epoch with the gradient of attention w.r.t. the class as given in Eq. 4, $\nabla A_{n-1}^c$ is the gradient of attention w.r.t class $c$ and $A_n \in \mathbb{R}^{MP^2 \times MP^2}$.

$$A_{n_{interpret}} = A_n \cdot \nabla A_{n-1}^c \tag{4}$$

We use this to compute the loss for interpretability, as given in Eq. 5:

$$\mathcal{L}_{interpret} = \text{CrossEntropy}(A_n, A_{n_{interpret}}) \tag{5}$$

The final loss of the training algorithm is given by the weighted sum of the classification loss $\mathcal{L}_{class}$ and the interpretability loss $\mathcal{L}_{interpret}$. The final loss function is in Eq. 6 where $\alpha$ and $\beta$ are hyperparameters with $\beta = 1 - \alpha$ and $\alpha >= 0, \beta >= 0$ .

$$\mathcal{L}_{total} = \alpha\mathcal{L}_{class} + \beta\mathcal{L}_{interpret} \tag{6}$$

The algorithm is given in Algorithm 1. The training schemes for single modalities (video and signal) are illustrated in Figs. 1 and 3 respectively.

---

**Algorithm 1.** Training Algorithm for the Transformer model

---

**if** *epoch* is 0 **then**
    $A_n, model_0 \leftarrow$ train *(model_{pt})*          ▷ *model_{pt}* is the pre-trained model
    $\nabla A_0^c \leftarrow$ evaluate *(model_0)* (where $c \in 1, ...C$)
**else if** $1 \leq epoch \leq N$ **then** (where $epoch \in 1, ...N$)
    **if** $c == Label :$ **then**
        $A_n, index, model_n \leftarrow$ train *(model_{n-1}, \nabla A_{n-1}^c)*
        $\nabla A_n^c \leftarrow$ evaluate *(model_n)*
        $\mathcal{L} = \alpha\mathcal{L}_{class} + \beta\mathcal{L}_{interpret}$
        $\arg\min_{\theta}\mathcal{L} \leftarrow \arg\min_{\theta} [\alpha\mathcal{L}_{class} + \beta\mathcal{L}_{interpret}]$
    **else if** $c \neq Label$ **then**
        $A_n, index, model_n \leftarrow$ train *(model_{n-1})*
    **end if**
**end if**

---

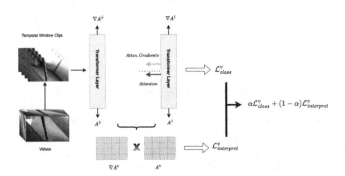

**Fig. 1.** The training scheme for the video modality, using the gradient of the attention for additional supervision. $A$ denotes the attention and $\nabla A$ denotes the gradient of attention. $\mathcal{L}_{class}$ is the classification loss, $\mathcal{L}_{interpret}$ interpretation loss.

## 3.2   Multimodal Architecture

Our multimodal architecture is designed with two streams: a) the video stream which uses Swin-3D [16] architecture, and b) the sensor transformer which uses a LinFormer [35] as the architecture to generalize on the sensor signal data. We have used a late fusion technique via a combined loss on signal and video modalities. Thus, the optimisation for the combination of the losses is given by:

$$\arg\min_{\theta_v,\theta_s}\Big[\lambda_v\mathbb{E}_{(x_i,\hat{x}_i,y_i)\sim\mathbb{V}}\left[L(\theta_v,x_v,y_v)\right]$$
$$+\lambda_s\mathbb{E}_{(m_i,y_i)\sim\mathbb{S}}\left[L(\theta_s,m_i,y_i)\right]\Big] \tag{7}$$

The $\lambda_v$ and $\lambda_s$ are hyperparameters to weight the loss functions ($\lambda_s = 1 - \lambda_v$ where $\lambda_v \geq 0, \lambda_s \geq 0$). $(x_v, y_v)$ pair belongs to the distribution $\mathbb{V}$ which is for the video data, where $x_v$ is the input, $y_v$ is the ground-truth labels. Similarly, $(x_s, y_s)$ is the input and ground-truth label pair of the distribution $\mathbb{S}$, that is, for the sensor signal data. The combination of the two different modalities is illustrated in Fig. 2.

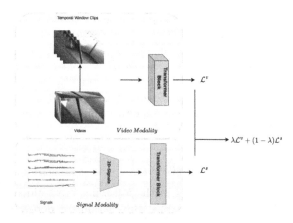

**Fig. 2.** The combination of the two modalities and training using the combined loss as in Eq. (7).

## 4   Experiments and Results

**Kinetics Dataset.** Video (Visual Information) is the most important component in the detection of semantic events. Therefore, we first report on our experiments with only a video transformer on this dataset introduced below.

Kinetics-400 [5] is the dataset for human action classification. The videos are taken from YouTube with real-world scenarios and are about 10 s long.

The dataset contains around 306,000 videos, about 240,000 videos constitute a training set, 20,000 are selected for validation, and 40,000 for the test. It comprises 400 classes of actions. For our experiments, we have sampled video clips of temporal dimension $T = 8$ from the original videos reduced by temporal sampling as in [5]. Thus, each video clip is a tensor in $RGB - T$ space with dimensions $(\zeta \times T \times H \times W)$. Here $\zeta = 3$ is the number of color channels, $T = 8$ is the temporal length, and $H = W = 224$ is the spatial dimension of the cropped video frames.

**Multimodal Dataset.** For this work, a multimodal risk detection dataset named BIRDS (Bio-Immersive Risks Detection System)[1] is used. This dataset contains visual and signal information extracted from physiological and motion sensors. The dataset is recorded according to the "single-person" scenario for the classification of risk situations of people living alone in their homes [20]. The data were recorded by a healthy volunteer during 21 days at home. The two different modalities, i.e. the signals and the videos, are weakly synchronized because of the difference in the frequency rates. The BIRDS dataset is made up of a total of 19,500 videos, each approximately 3 s long. We are working on making it public. We classify 6 situations: 5 risks and No-Risk as mentioned below. The No-Risk class was sub-sampled to balance the dataset and reduce bias in the models. Only 5% of the No-Risk videos were randomly selected.

A taxonomy of five classes has been defined [37] on immediate and long-term risks. The immediate-risks are *Environmental Risk of Fall*, *Physiological Risk of Fall*, and *Risk of Domestic Accident*. The long-term risks are *Risks Associated with Dehydration*, and *Risks Associated with Medication Intake*. This gave a total number of clips as 9600 for all classes of the taxonomy. The training, validation, and test contain 6713, 923, and 1924 clips, respectively.

### 4.1   Multimodal Data Organization

The multimodal data are organized in the following way: Videos are split into temporal clips of duration $T$. Note that this parameter depends on the video frame rate and has to be set experimentally. In our case, we propose $T = 8$ for our wearable video setting and variable frame rate in the BIRDS corpus, as well as for the Kinetics-400 dataset. We have optimized this parameter by a sparse grid search with respect to the global accuracy objective. These clips are sampled with a sliding window approach with a covering of $\Delta_T$ frames. We have set this parameter $\Delta_T = 4$ frames on BIRDS and $\Delta_T = 5$ on Kinetics-400 datasets, respectively.

The signals and videos are referenced using timestamps. Similarly, we take the corresponding windows from the signals using the timestamp of the videos. A fixed length window of around $\eta = 20$ with a sliding stride of $\Delta_\eta = 4$ samples partitions signal along the time. The organization of signal data is illustrated in Fig. 3. The temporal windows of $\eta$ samples taken from the bunch of $s$ sensors

---

[1] BIRDS will be publicly available upon GDPR clearance.

represent a tensor of dimension $\eta \times s$. Note that the time intervals between different samples are not stable because of the sensors' imprecision. If there is imprecision on the labels of these windows in the sensors, then we take the most represented label in the particular window.

## 4.2 Training of 3D-Swin with Interpretability Techniques on Video

**Kinetics-400.** To train on the Kinetics-400 dataset, we are using the pre-trained weights on the ImageNet-1K dataset for the 3D-Swin transformer. For the Kinetics-400 dataset, in our experiments, we are using the 8-frame clip with a resolution of 224 × 224. Our method outperforms all the baseline methods; see Table 1. In [18], the authors use the class-specific form of interpretation by elementwise multiplying the gradient of attention along with the attention. In our transformer, we do the same as in [18], see Sect. 3, to weigh the importance of attention with respect to the ground truth labels for videos.

**BIRDS Dataset.** For training on the BIRDS dataset, we are using the pre-trained model on the Kinetics-400 and fine-tune our training scheme with these pre-trained weights. Our training scheme generalizes better than the state-of-the-art on the video part of the BIRDS corpus. The improvement in the average top-1 accuracy compared to the vanilla training scheme (without additional supervision) of Video Swin Transformer [16] is ∼2.98% as depicted in the last column of Table 2. These results validate our proposed method of additional supervision. 3D-BotNet in the second column of the Table 2 is a model for videos adapted from [30] which was devised for images. The nature of the videos for the BIRDS are different from Kinetics-400 as BIRDS are ego-centric videos.

**Table 1.** Test accuracy scores (top-1 accuracy) on the Kinetics-400 Dataset

| Algorithms | Top-1 Accuracy | Pre-Train |
|---|---|---|
| 3D-ConvNet [32] | 56.1% | ✓ |
| I3D [5] | 71.1% | ✓ |
| I3D NL [36] | 77.7% | ✓ |
| X3D-M [10] | 76.0% | ✓ |
| TimesFormer [3] | 75.1% | ✓ |
| Video Swin Transformer [16] | 78.8% | ✓ |
| **Video Swin T-In (VS-T-In, ours)** | **79.1%** | ✓ |

**Table 2.** Test accuracy scores (top-1) on the BIRDS dataset for the video modality.

| Algorithms | Top-1 Acc. | Pre-Train |
|---|---|---|
| 3D-ConvNet [32] | 69.27% | ✓ |
| 3D-BotNet | 70.61% | ✓ |
| TimesFormer [3] | 74.11% | ✓ |
| Video Swin Transformer (Swin-T) [16] | 73.39% | ✓ |
| Video Transformer (with pooling) [19] | 75.19% | ✓ |
| **Video Swin T-In (VS-T-In, ours)** | **76.37%** | ✓ |

As illustrated in Fig. 4, the per-class accuracy for the video modality (orange bars) and video modality with interpretability (violet bars) differs. In four classes out of six, the accuracy of the transformer with supervision using interpretability is higher. In two classes: *Environmental Risk of Fall* and *Dehydration* (which means detection of drinking action) the baseline *Video Swin Transformer* performs better. We can explain this by the discrepancy of the data in these two classes corresponding to a very different viewpoint on the environment (class 2) and on the difference of close views (bottles, mugs, and glasses). However, the overall accuracy with the additional supervision by the attention gradient is higher ∼3% (average top-1 accuracy improvement).

### 4.3   Training of Signal Transformer

The training scheme of the signal transformer is similar to that of the video transformer. The challenge with signal data is the non-recording of approximately 95% of the data across various features due to sensor failures. The data is imputed by replacing the missing values with random values from the normal distribution computed on the particular signal (feature) then the data are linearly scaled along that particular signal. In signal transformers, additional supervision using interpretation gives an improvement of ∼4.7%. We are not pre-training the signal transformer due to the unavailability of similar datasets, which would comprise

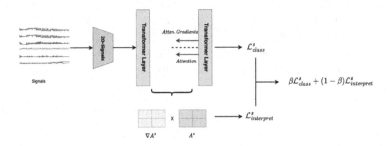

**Fig. 3.** The training scheme for generalization in the signal part in the multimodal transformer.

both physiological and motion sensors. The training scheme is presented in Fig. 3 and the comparative study is given in Table 3.

**Table 3.** Test accuracy (top-1 accuracy) on the signal sensor data with balanced BIRDS dataset.

| Algorithms | Top-1 Acc. | Pre-Train |
|---|---|---|
| Vanilla Transformer [34] | 32.68 | × |
| Vanilla Transformer-In(ours) | 34.07 | × |
| LinFormer [35] | 35.55 | × |
| **LinFormer-In (ours)** | **40.26** | × |

### 4.4   Multimodal Transformer with Interpretability Results

The accuracy score using the multimodal approach, i.e. using the videos and signals together is **78.26%** in this BIRDS dataset, which is an improvement of ~1.9% while using the singular video modality of the network. The overall accuracy score using both modalities together **without** interpretability assisted additional supervision is 76.51%. The improvement of ~1.8% of our method also validates our training scheme. Figure 5 shows the accuracy scores of various architecture models on single sensor modality, the single video modality, and the combined video and signal modalities. It illustrates the better performance of multimodal models and shows that our model with interpretability is the best (~1.8%)

### 4.5   About Ablation

The ablation in a multimodal architecture means the use of only one modality, such as video or signals in our case. We have implicitly done it in the previous section when we were talking about training videos and signal transformers. In Sect. 4.4 we have reported, that our multimodal scheme brings improvement of ~1.8% compared to the best modality (video).

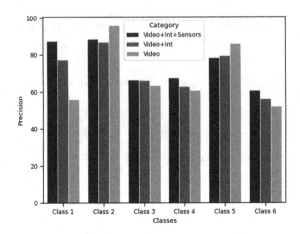

**Fig. 4.** Precision scores for each class: 1-No Risk, 2-Environmental Risk of Fall, 3-Physiological Risk of Fall, 4-Risk of Domestic Accident, 5-Risk Associated with Dehydration, 6-Risk Associated with Medication Intake

### 4.6  Training Settings

For the video transformers, all models use $224 \times 224$ resolution frames, with a patch size of $16 \times 16$ for transformers. For all the video transformers we have used tiny models for the fine-tuning, due to resource constraints. We use pre-trained weights. For the training, we have used a single Tesla A40 GPU. For the BIRDS dataset, the fine-tuning process is for a period of 25 epochs with a batch size of 16. A grid search is used to obtain the learning rate between $1e-3$ and $1e-5$. Small changes in learning rate do not have a strong impact on the classification accuracy of the models. For our approach, we have used

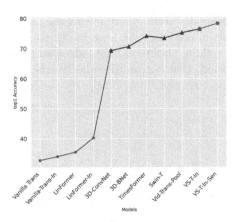

**Fig. 5.** Top-1 accuracy scores. In Green: Signal Modality. In blue: video modality only. In red: video and sensor modalities together for BIRDS corpus (Color figure online)

an SGD optimizer with Nesterov momentum and weight decay. To weigh the loss function of various modalities, the weighting constants are also obtained using a grid search approach where $\alpha = 0.7$, $\beta = 0.65$, and $\lambda_v = 0.7$ and $\lambda_s = 1 - \lambda_v = 0.3$; see Sect. 3.

## 5    Conclusion and Future Works

The key element of our work is a scheme to use additional interpretability-based supervision which improves the overall accuracy of any self-attention-based network. To validate our approach, we took a multimodal dataset comprising videos and signals from different sensors. First, our approach is validated on a single modality data i.e., video or signals. This approach gives a tangible improvement in every single modality tested. In signals, it outperforms the best baseline of ~4.8%, on video ~1.2%. In the video modality of BIRDS, the proposed method outperforms the vanilla version of the Video Swin Transformer by ~3%. For the publicly available Kinetics-400 dataset, we also achieved better results on video than the best-performing baseline Video Swin Transformer. Computing on multimodal architecture on the multimodal BIRDS corpus, we achieved 1.8% top-1 accuracy improvement using interpretation-supervised training. The BIRDS dataset for our target application of recognition of risk events is challenging in both modalities, as the videos are egocentric, while the signal modality is weakly synchronized, and values are missing in each sensor. Therefore, we can conclude, that our proposed interpretability technique for training of transformers helps in both single-modality and multimodal model training.

For our future work, we will focus more on using this approach in other modalities (e.g.: sound, speech, and text) and additional data and also apply data visualization techniques for human interpretation of the results.

## References

1. Ayyar, M.P., Benois-Pineau, J., Zemmari, A.: Review of white box methods for explanations of convolutional neural networks in image classification tasks. J. Electron. Imaging **30**(5), 050901 (2021)
2. Bach, S., Binder, A., Montavon, G., Klauschen, F., Müller, K.R., Samek, W.: On pixel-wise explanations for non-linear classifier decisions by layer-wise relevance propagation. PLoS ONE **10**(7), e0130140 (2015)
3. Bertasius, G., Wang, H., Torresani, L.: Is space-time attention all you need for video understanding? In: ICML (2021)
4. Carion, N., Massa, F., Synnaeve, G., Usunier, N., Kirillov, A., Zagoruyko, S.: End-to-end object detection with transformers. In: Vedaldi, A., Bischof, H., Brox, T., Frahm, J.-M. (eds.) ECCV 2020. LNCS, vol. 12346, pp. 213–229. Springer, Cham (2020). https://doi.org/10.1007/978-3-030-58452-8_13
5. Carreira, J., Zisserman, A.: Quo vadis, action recognition? A new model and the kinetics dataset. In: Proceedings of the IEEE Conference on CVPR 2017
6. Chefer, H., Gur, S., Wolf, L.: Transformer interpretability beyond attention visualization. In: Proceedings of the IEEE/CVF Conference on Computer Vision and Pattern Recognition (CVPR), pp. 782–791, June 2021

7. Chefer, H., Schwartz, I., Wolf, L.: Optimizing relevance maps of vision transformers improves robustness. In: NeuRIPS 2022
8. Devlin, J., Chang, M.W., Lee, K., Toutanova, K.: BERT: pre-training of deep bidirectional transformers for language understanding. In: NAACL (2019)
9. Dosovitskiy, A., et al.: An image is worth 16x16 words: transformers for image recognition at scale. In: ICLR 2021
10. Feichtenhofer, C.: X3d: expanding architectures for efficient video recognition. In: 2020 IEEE/CVF Conference on CVPR (2020)
11. Guo, X., Guo, X., Lu, Y.: SSAN: separable self-attention network for video representation learning. In: Proceedings of the IEEE/CVF CVPR 2021
12. Li, Q., Qiu, Z., Yao, T., Mei, T., Rui, Y., Luo, J.: Action recognition by learning deep multi-granular spatio-temporal video representation. In: Proceedings of the 2016 ACM on International Conference on Multimedia Retrieval. ACM (2016)
13. Liang, H., et al.: Training interpretable convolutional neural networks by differentiating class-specific filters. In: Vedaldi, A., Bischof, H., Brox, T., Frahm, J.-M. (eds.) ECCV 2020. LNCS, vol. 12347, pp. 622–638. Springer, Cham (2020). https://doi.org/10.1007/978-3-030-58536-5_37
14. Liu, K., Li, Y., Xu, N., Natarajan, P.: Learn to combine modalities in multimodal deep learning. ArXiv abs/1805.11730 (2018)
15. Liu, Z., et al.: Swin transformer: hierarchical vision transformer using shifted windows. In: Proceedings of the IEEE/CVF International Conference on Computer Vision (ICCV), pp. 10012–10022, October 2021
16. Liu, Z., et al.: Video swin transformer. arXiv preprint arXiv:2106.13230 (2021)
17. Lundberg, S.M., Lee, S.I.: A unified approach to interpreting model predictions. In: Guyon, I., et al. (eds.) Advances in NeuRIPS 2017
18. Mallick, R., Benois-Pineau, J., Zemmari, A.: I saw: a self-attention weighted method for explanation of visual transformers. In: IEEE ICIP 2022
19. Mallick, R., et al.: Pooling transformer for detection of risk events in in-the-wild video ego data. In: 26th ICPR 2022
20. Mallick, R., Yebda, T., Benois-Pineau, J., Zemmari, A., Pech, M., Amieva, H.: Detection of risky situations for frail adults with hybrid neural networks on multimodal health data. IEEE Multim. **29**(1), 7–17 (2022)
21. Meditskos, G., Plans, P., Stavropoulos, T.G., Benois-Pineau, J., Buso, V., Kompatsiaris, I.: Multi-modal activity recognition from egocentric vision, semantic enrichment and lifelogging applications for the care of dementia. J. Vis. Commun. Image Represent. **51**, 169–190 (2018)
22. Montavon, G., Lapuschkin, S., Binder, A., Samek, W., Müller, K.R.: Explaining nonlinear classification decisions with deep Taylor decomposition. Pattern Recogn. **65**, 211–222 (2017)
23. Ngiam, J., Khosla, A., Kim, M., Nam, J., Lee, H., Ng, A.Y.: Multimodal deep learning. In: Proceedings of the 28th ICML 2011
24. Owens, A., Efros, A.A.: Audio-visual scene analysis with self-supervised multisensory features. In: Ferrari, V., Hebert, M., Sminchisescu, C., Weiss, Y. (eds.) ECCV 2018. LNCS, vol. 11210, pp. 639–658. Springer, Cham (2018). https://doi.org/10.1007/978-3-030-01231-1_39
25. Radford, A., et al.: Learning transferable visual models from natural language supervision (2021)
26. Ribeiro, M.T., Singh, S., Guestrin, C.: "Why should I trust you?": Explaining the predictions of any classifier. In: KDD, pp. 1135–1144. ACM (2016)

27. Selvaraju, R.R., Cogswell, M., Das, A., Vedantam, R., Parikh, D., Batra, D.: Grad-CAM: visual explanations from deep networks via gradient-based localization. In: Proceedings of the IEEE/CVF ICCV (2017)
28. Simonyan, K., Zisserman, A.: Two-stream convolutional networks for action recognition in videos. In: NIPS (2014)
29. Springenberg, J., Dosovitskiy, A., Brox, T., Riedmiller, M.: Striving for simplicity: the all convolutional net, December 2014
30. Srinivas, A., Lin, T.Y., Parmar, N., Shlens, J., Abbeel, P., Vaswani, A.: Bottleneck transformers for visual recognition. In: CVPR (2021)
31. Srinivas, S., Fleuret, F.: Full-gradient representation for neural network visualization. In: Advances in Neural Information Processing Systems (2019)
32. Tran, D., Bourdev, L., Fergus, R., Torresani, L., Paluri, M.: Learning spatiotemporal features with 3d convolutional networks. In: Proceedings of the IEEE/CVF International Conference on Computer Vision (ICCV), pp. 4489–4497, December 2015
33. Tran, D., Wang, H., Torresani, L., Feiszli, M.: Video classification with channel-separated convolutional networks. In: Proceedings of the IEEE/CVF International Conference on Computer Vision (ICCV), October 2019
34. Vaswani, A., et al.: Attention is all you need. In: Advances in Neural Information Processing Systems, pp. 5998–6008 (2017)
35. Wang, S., Li, B.Z., Khabsa, M., Fang, H., Ma, H.: Linformer: self-attention with linear complexity (2020). https://arxiv.org/abs/2006.04768
36. Wang, X., Girshick, R., Gupta, A., He, K.: Non-local neural networks. In: Proceedings of the IEEE Conference on Computer Vision and Pattern Recognition (CVPR), June 2018
37. Yebda, T., Benois-Pineau, J., Pech, M., Amieva, H., Middleton, L., Bergelt, M.: Multimodal sensor data analysis for detection of risk situations of fragile people in @home environments. In: Lokoč, J., et al. (eds.) MMM 2021. LNCS, vol. 12573, pp. 342–353. Springer, Cham (2021). https://doi.org/10.1007/978-3-030-67835-7_29
38. Zeiler, M.D., Fergus, R.: Visualizing and understanding convolutional networks. In: Fleet, D., Pajdla, T., Schiele, B., Tuytelaars, T. (eds.) ECCV 2014. LNCS, vol. 8689, pp. 818–833. Springer, Cham (2014). https://doi.org/10.1007/978-3-319-10590-1_53
39. Zhang, H., Torres, F., Sicre, R., Avrithis, Y., Ayache, S.: Opti-CAM: optimizing saliency maps for interpretability. CoRR abs/2301.07002 (2023)
40. Zhou, B., Khosla, A., Lapedriza, A., Oliva, A., Torralba, A.: Learning deep features for discriminative localization. In: 2016 IEEE Conference on Computer Vision and Pattern Recognition (CVPR), pp. 2921–2929 (2016)

# OOKPIK- A Collection of Out-of-Context Image-Caption Pairs

Kha-Luan Pham[1,2], Minh-Khoi Nguyen-Nhat[1,2], Anh-Huy Dinh[1,2],
Quang-Tri Le[1,2], Manh-Thien Nguyen[1,2], Anh-Duy Tran[3], Minh-Triet Tran[1,2],
and Duc-Tien Dang-Nguyen[4(✉)]

[1] University of Science, Ho Chi Minh City, Vietnam
[2] Vietnam National University, Ho Chi Minh City, Vietnam
[3] imec-DistriNet, KU Leuven, Belgium
[4] University of Bergen, Bergen, Norway
pkluan18@apcs.fitus.edu.vn

**Abstract.** The development of AI-based Cheapfakes detection models
has been hindered by a significant challenge - the scarcity of real-world
datasets. Our work directly tackles this issue by focusing on out-of-
context (OOC) image-caption pairs within the Cheapfakes landscape.
Here, genuine images come with misleading captions, making it tough
to identify them accurately. This study introduces a dataset manually
collected for OOC detection. Previous attempts at solving this problem
often used strategies like automatically making OOC data or providing datasets made by people. However, these approaches had limitations
affecting how practical they were. Our contribution stands out by providing a carefully selected dataset to fill this gap. We tested our dataset
using existing OOC detection models, showing it works well. Additionally, we suggest practical ways to use our work to improve Cheapfakes
detection in real-world situations.

**Keywords:** Cheapfakes · Out-of-context · Dataset · SBERT-WK ·
OOKPIK

## 1 Introduction

Information disorder, including mis/disinformation, is an alarming problem due
to the rapid development of social media platforms. Misinformation can be
spread extraordinarily further and faster than the truth [23]. It also has a substantial menace on all categories of information [16,17,20].

The advancement of deep learning, especially in generative AI, creates a
class of fake news with deceiving manipulated media using complicated techniques such as face swapping, face reenactment, and others, known as Deepfakes.
These manipulation methods are popular, and humans cannot differentiate their
realistic results. As a result, Deepfakes complicates the issue of fake news [22].
However, Paris et al. [5] show another group of media manipulation methods,

Supported by NORDIS.

**Out-of-context**     **In-context**

The picture shows the wreckage of two military aircraft that crashed on September 27, 2018.

Cars wait in line at gas stations in Nairobi

The photo shows the wreckage of a Nigerian government fighter jet shot down by the Eastern Security Network (ESN).

Cars stuck in line outside a gas station in Nairobi.

**Fig. 1.** Out-of-context (left) and in-context (or not out-of-context) (right) image-caption pairs. Two true (green) and/or false (red) captions are accompanied with an image (Color figure online)

which is "cheaper" in terms of technology sophistication, allowing a broader public to produce misleading news, known as Cheapfakes. Cheapfakes only use conventional methods, such as exploiting lookalikes and recontextualizing a piece of media. These techniques do not require extensive knowledge in using complicated deep learning models yet still create persuasively fake news, which is more accessible to a broader range of people. Therefore, these types of fake news grow significantly in size. As a matter of fact, Cheapfakes have proven to be more prevalent than Deepfakes [4,19].

Exploiting lookalikes is one of the common ways to mislead the audience. The technique involves reusing pictures/videos of someone who looks similar to a target and misleading the audience based on the given media. The manipulation technique is known as re-contextualization. On social media, an image often goes with a caption describing its content and/or other related information. Although the image is genuine, the caption can still lead the audience to different, incorrect information. Figure 1 shows out-of-context (OOC) and in-context media examples. As the media's content is not altered, only the caption paired with it creates misleading information, an exceptional challenge that the public can detect. In this paper, we focus our work on detecting out-of-context image-caption pairs.

Recently, challenges for detecting out-of-context image-caption pairs have been developed and extended to address this problem. A particularly prominent challenge in this area is the ACM Multimedia Grand Challenge on Detecting Cheapfakes [3]. Several solutions [1,2,8,21] were proposed to address the challenge. Due to the lack of training data, most work reuses pre-trained modules in computer vision and natural language processing without tuning specifically for OOC detection.

The common drawback of the proposed methods is their use of multiple pre-trained modules as components in their classification pipeline. Since each component is only trained on a separate task, the proposed methods must ensemble different models to handle various cases of the Cheapfakes problem. As a result, the overall computational cost of the whole pipeline increases. The main reason for not having a single model working on the Cheapfakes problem is insufficient data for training. Therefore, the dataset is a crucial determinant for working with this idea.

Synthetic data has recently been considered a solution for expanding data without compromising privacy. Furthermore, this type of data generation is cheaper than real data collection and is usually used in scenarios where data is short. However, synthetic dataset comes with a few remarkable limitations. Generating output resembling real-world scenarios remains a difficult challenge, especially in the news domain, where data is diverse and constantly changing. As a result, there is a massive demand for high-quality and practical data, which is also the motivation for our work.

In this paper, we introduce our OOKPIK dataset comprising of **545** images and **1,090** real captions. These image-caption pairs are collected from social networks and carefully check using a reliable Fack-checking online website. Our dataset, in general, is similar to COSMOS [2]: an image is accompanied by two captions (from two distinct news sources) in which one caption describes the image correctly, and the other can be true or false. The significant difference between the two datasets is that we collect the photos and captions from various sources. The collected caption is kept raw since the data, such as hashtags, emotion icons, etc., might provide additional information.

In this work, our contributions are:

- Providing a practical dataset collected from several sources and carefully verified using a fact-checking website. The dataset is publicly available[1].
- Evaluating the proposed dataset with various baseline models.
- Proposing applications as well as challenges corresponding to the dataset.

## 2  Related Work

### 2.1  Related Datasets

Some of the previously proposed datasets for multimedia misinformation detection are the Multimodal Information Manipulation Dataset (MAIM) [6] and the Multimodal Entity Image Repurposing (MEIR) [18]. Jaiswal et al. [6] presented the MAIM dataset with multimedia packages, where each package contains an image-caption pair. To mimic image repurposing, the associated image captions were replaced with randomly selected captions, which may result in semantically inconsistent. Sabir et al. [18] created the MEIR dataset by randomly swapping named entities (person, location, or organization) occurring in Flickr

[1] https://anonymous.4open.science/r/OOKPIK-Dataset.

to create manipulated packets that contained manipulated GPS metadata where appropriate. One assumption was that for each caption package, there is a non-manipulated related package (geographically and semantically similar) in the reference set. In this way, the integrity of the query packet can be verified by retrieving the related package and then comparing the two. A closer study proposed TamperedNews [11], a dataset that performed similar manipulations on entities and used Wikidata to apply more rules on entity substitution. The rules include replacing a person with another person of the same gender or same country of citizenship, changing a location with another location in the same country or city, etc.

In recent works, Aneja et al. [2] have collected the dataset COSMOS, a large dataset of about 200,000 images tagged with 450,000 captions, from two primary sources: News outlets and fact-checking websites. First, they used publicly available APIs from news agencies to collect the images with the corresponding captions. When searching for these images using Google's Cloud Vision API, the authors found different contexts on the Internet where the same image was shared. This gave them multiple captions with different contexts for each image. Finally, they aimed to detect conflicts in English-written captions of the same image. In another work, Luo et al. [10] created the large-scale dataset NewsCLIP-pings, which contains both pristine and falsified (OOC) images. It is based on the VisualNews [9] corpus of news articles from 4 news outlets: The Guardian, BBC, USA Today, and The Washington Post. Matching was done automatically using trained vision and language models (such as SBERT-WK [24], CLIP [15] or scene embedding [25]). The falsified examples may misrepresent the context, location, or person in the image, with inconsistent entities or contexts. The NewsCLIP-pings dataset contains different subsets depending on the method of matching the image to the caption (e.g., text-text similarity, image-image similarity, etc.).

## 2.2   Cheapfakes Detection Methodology

The popular technique used in the topic is to capture the similarity of the two captions and the image to classify whether the triplet is OOC. One of the pioneering works in the field is COSMOS. Shivangi et al. [2] proposed a self-supervised learning framework that aligns objects in an image with their textual description. The algorithm then classifies whether the two captions are OOC by verifying the entities in both captions and comparing their similarity using Semantic Textual Similarity (STS) score. If the two captions mention the same object in the image and their similarity is less than the predefined threshold, they are classified as OOC.

However, the STS score does not capture the similarity of the two captions well. Tankut et al. [1] proposed four components to boost the performance of the COSMOS baseline, including differential sensing and fake-or-fact checking to measure the similarity of two captions. The work also attempts to solve the problem when the caption aligns with multiple objects in the image by combining numerous bounding boxes and decreasing the IoU threshold.

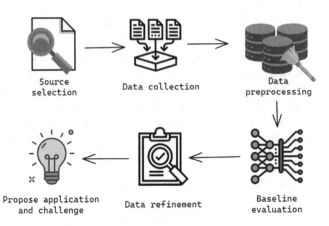

Source
selection                Data collection                    Data
                                               preprocessing

Propose application      Data refinement                   Baseline
and challenge                                              evaluation

**Fig. 2.** The proposed workflow for constructing the OOKPIK dataset, including 7 steps, starts with source selection and finishes with proposing challenge and application.

La et al. [8] proposed various techniques focusing on utilizing image captioning to extract the correlation between images and captions and using different methods to measure the similarity of the three captions. Tran et al. [21] proposed a multimodal heuristic consisting of four components: natural language inference, fabricated claim detection, visual entailment, and online caption checking. The solution improves the detection accuracy by a significant margin of 7.2%. An attempt by Pham et al. [14] to recover the original context by collecting all related background information of the image via search engines, the method is expected to be adaptive to up-to-date data, hence opening another direction in OOC detection approaches.

Addressing the issue of Cheapfakes is crucial across multiple domains, particularly in improving the quality and reliability of journalism. Extensive research has been conducted on leveraging advancements in machine learning to facilitate the production of high-quality journalism [12] and on proposing standards for verifying Visual User-Generated Content in journalism [7].

## 3    OOKPIK Dataset

OOKPIK dataset focuses on detecting out-of-context captions. Specifically, each entry in the dataset is a triplet $(I, C_1, C_2)$, where $I$ is an original image and $(C_1, C_2)$ is a pair of captions describing the content of the image $I$. The caption could either correctly describe the image with its original context or mislead viewers by distorting facts.

To construct the OOKPIK dataset, we first filter out appropriate sources containing both authentic and misleading descriptions of an image. As the problem requires careful information verification, we manually check and collect the photos and the captions. Finally, the data is processed to be compatible with existing out-of-context detection methods. Our workflow is demonstrated in Fig. 2.

## 3.1   Data Sources and Data Collection

Our data are mostly collected from online websites. Images with true labels are obtained from *news websites*, while those with false captions are from *fact-checking websites*. Ensuring the reliability of our collected data is a priority. Hence, all chosen fact-checking websites have transparent processes, allowing readers to retrace the steps of news verification. Some fact-checking sites have gained a reputation as a reliable verification source for a long time, such as Snopes[2] and Reuters[3]. Some data sources are affiliated with reputable organizations, such as Factcheck[4] (which belonged to the Annenberg Public Policy Center of the University of Pennsylvania). In the actual captioned image aspect, we gather the data from BBC[5] and Deutsche Welle[6] as they are among the oldest and biggest news international channels.

Among various types of fake news from fact-checking sites, we only selected news consisting of unmodified images, i.e., images that underwent no visual manipulation. To construct the triple sample $(I, C_1, C_2)$, we need to extract a pair of captions $(C_1, C_2)$ describing the image $I$. The caption pair could be out-of-context or not out-of-context. The first caption is the true caption used to describe the image in the original context correctly. The true captions come from the analysis included in the post.

We employ different approaches to collect the second caption based on the sample type. For out-of-context caption pairs, the second captions come from the main posting attached to the analysis or from the text in the Claim header for Snopes(See footnote 2) or under the FactChecks section of the website for FactChecks[7]. On the other hand, for not-out-of-context caption pairs, the second caption is taken from the public image description and the `alt text` tag. If the public description is unavailable, we perform a reverse image search with the Google Lens tool to find another caption from another data source to construct the caption pair. This caption is guaranteed to be in the same context as the original caption by the manual data collector.

The significant improvement in the data-collecting process is retrieving the original falsified caption of the image instead of the conclusion about the fake news for the second caption. In the COSMOS dataset, multiple samples are used after the verification. As a result, these conclusions usually contain terms such as *fake, hoax, fraud*, etc. However, in a realistic setting, our task is to verify the claim's veracity, and we do not typically have the conclusion as the input for our model. Therefore, eliminating the pattern could shift the attention of the models to more robust textual features.

---

[2] https://www.snopes.com/.
[3] https://www.reuters.com/fact-check.
[4] https://www.factcheck.org/.
[5] https://www.bbc.com/.
[6] https://www.dw.com/.
[7] https://www.factcheck.org/.

## 3.2   Data Organization and Statistics

The dataset is organized primarily in a `json` file with an image folder named `img`. Each line of a file represents one sample and has the following fields:

- `img_local_path`: path to a corresponding image in `img` folder.
- `caption1`: first caption for image
- `caption2`: second caption for image
- `article_url`: link to the article from which we retrieved data.
- `label`: label of the data, `label = 1` means one of two captions are mismatched, otherwise `label = 0`.
- `caption1_entities,caption2_entities`: list of named entities appear in each caption, detected using *spacy* entity recognizer.
- `maskrcnn_bboxes`: list of detected bounding boxes corresponding to the image.

There is a total of 545 images in the dataset. As can be seen from the Fig. 3(*a*), the number of images corresponding to each website from left to right in the figure is 58, 48, 70, 73, 23, 71, 17, 32, 57, and 96, respectively. In Fig. 3(*b*), the number of images with two in-context captions is 153 while the out-of-context one is 392.

Compared with the COSMOS dataset, a widely used benchmark for Cheapfakes detection, the OOKPIK dataset poses more challenging problems. Since the captions are collected from social media platforms, they vary in length and complexity. Some of them might contain emojis, hashtags, and irrelevant information. The number of tokens in a single sentence varies from 14 to 256, with an average of 54 tokens per sentence. At the same time, for the COSMOS dataset, the caption length is limited from 6 to 206 tokens, with an average of 51 tokens per caption.

Another challenge, suggested by Akgul et al. [1], is that the caption could mention multiple subjects in the image. Existing works heavily rely on the COSMOS method, where the author aligns the caption with a single object in the image. As a result, the method would incorrectly associate the caption with the corresponding object, hence misclassifying the caption pair as not out-of-context. To address the problem, we include samples with multiple entities within a single caption. On average, each caption contains 2.5 entities.

## 4   Evaluation

To demonstrate the discussed problems to existing methods, we evaluate the following methods on the OOKPIK dataset: the COSMOS method [2], COSMOS on Steroid [1] and the Textual-Visual-Entailment-based algorithm (TVE) [21]

We also conducted an experiment to show the textual bias in the original COSMOS dataset, increasing the accuracy performance of existing works. The bias is due to the lack of diversity in news sources, causing the consistent use of some specific terms in OOC samples. As a result, multiple existing methods exploit this pattern to leverage their overall performance by a large margin.

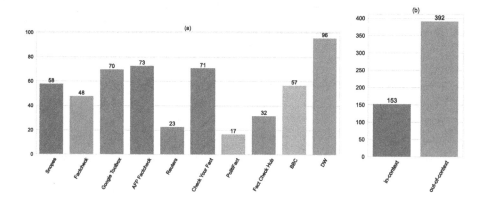

**Fig. 3.** Sample histogram by (a) channel and (b) label

## 4.1 Existing Work Evaluation

For all models using COSMOS as a component, we perform bounding box extraction using **COCO-Keypoints/keypoint_rcnn_X_101_32x8d_FPN_3x** model in the detectron2 framework following the instruction of the original COSMOS work. The image-caption matching model is trained on the COSMOS training set and used for the inference process on the OOKPIK dataset.

**COSMOS on Steroid**: We reuse the best-performed configuration reported in the work. The model consists of four components, including Differential Sensing (FS), Fake-or-Fact (FoF) checking, Object-Caption matching (OCM), and the IoU threshold adjustment to be 0.25

**TVE**: We reuse the model with the best accuracy as reported in the work. The model consists of four components: Natural Language Inference (NLI), Fabrication Claim Detection (FCD), Visual Entailment (VE), and Online Caption Checking(OCC).

We evaluate the performance of the methods in terms of accuracy, precision, recall, and F-1 score.

The dataset also introduces some challenges to the researchers. As mentioned in the previous section, all existing methods are based on the COSMOS baseline as a crucial pipeline component, accounting for most of the predicted outcomes. As a result, these methods do not perform well for images containing multiple entities. The COSMOS on the Steroid Object Caption Matching component attempts to solve the problem using a bounding box for the entire image in such cases, increasing 0.9%. The Object Caption Maching shows only slight improvement because it does not correctly identify and associate the caption with the entities in the image.

**Table 1.** Result on the dataset

| Model | COSMOS | | | OOKPIK | | |
|---|---|---|---|---|---|---|
| | Accuracy | Precision | Recall | Accuracy | Precision | Recall |
| COSMOS | | | | | | |
|    Textual model | 76.80 | 73.51 | 83.80 | 77.25 | 88.07 | 79.08 |
|    COSMOS (Base) | 81.90 | 85.20 | 77.20 | 71.56 | 89.37 | 68.62 |
| COSMOS on Steroid | | | | | | |
|    Base + DS | 85.60 | 85.60 | 85.60 | 71.93 | 89.19 | 69.39 |
|    Base + FoF | 84.90 | 85.83 | 83.60 | 72.11 | 88.96 | 69.90 |
|    Base + OCM | 82.00 | 85.56 | 77.00 | 72.29 | 89.37 | 68.62 |
|    COSMOS on Steroid | 88.30 | 85.79 | 91.80 | 73.03 | 88.40 | 71.94 |
| TVE | | | | | | |
|    Base + NLI | 87.10 | 87.00 | 87.20 | 79.82 | 90.52 | 80.36 |
|    Base + FCD | 86.00 | 83.80 | 87.20 | 73.39 | 88.96 | 71.94 |
|    Base + OCC + VE | 84.90 | 88.50 | 80.20 | 72.29 | 89.00 | 70.15 |
| **TVE** | **89.10** | **87.80** | **90.80** | **80.73** | **90.42** | **81.89** |

Another challenging problem for associating between the images and the captions is their semantic gaps. Some images contain no known detectable objects, images that consist of multiple smaller images, images of events that require external knowledge to identify, etc. There are also new challenges with the captions. Since some of the data is crawled from the news channels, their images are too abstract, making it difficult for the machine to relate them with the captions-for example, an image of cracking soil with captions about drought conditions.

## 4.2   Textual Bias

We conjecture that the COSMOS dataset contains textual bias, especially for out-of-context samples. The pattern is typically used by most existing works in different forms to increase their performance. Specifically, the data in the COSMOS dataset is collected from fact-checking sites such as Snopes, which is also a part of the data source for the OOKPIK dataset. However, the captions collected for the samples show significant differences. There are caption pairs for the COSMOS dataset, the first being the caption used in fake news, while the second is usually the conclusion from the fact-checking sites. For instance, the first caption is shocking news about a celebrity. In contrast, the second caption is 'A shocking report about the former child actor Kirk Cameron was just another hoax'. As the second claim is usually a conclusion from fact-checking sites, it is sufficient to detect words such as hoax to know that the caption pair is out-of-context despite the content of the two captions and the images. On the other hand, for the OOKPIK dataset, we select the genuine caption as the original caption of the image. The genuine caption could be found in the analysis log on fact-checking sites since the analysis process is transparent. This would help eliminate the mentioned bias.

To verify our conjecture, we create a simple classifier to check if there are specific terms in the captions. It would mark the caption pair as out-of-context. The set of words is selected from the Fabricated Claim Detection module from the Textual and Visual Entailment (TVE) method. Specifically, the list of words are *fake, hoax, fabrication, supposedly, falsification, propaganda, deflection, deception, contradicted, defamation, lie, misleading, deceive, fraud, concocted, bluffing, made up, double meaning, alternative facts, trick, half-truth, untruth, falsehoods, inaccurate, disinformation, misconception*

As a result, the classifier found 53 pairs containing the pattern for the COSMOS dataset, and 51/53 pairs are actually out-of-context. This means the specified pattern accounts for nearly (5%) of the dataset, and only using a simple classifier could achieve a precision of 96.15%. Based on the observation, several components are proposed, exploiting the pattern to create another classifier that works well for these edge cases. Finally, the proposed component is combined with the COSMOS baseline using a heuristic prediction rule to boost the performance.

In contrast to the COSMOS dataset, the classifier only found four pairs containing the specified words, which account for less than 1 % of samples in the dataset. As a result, Table 1 shows a slight improvement for methods combining the COSMOS baseline with components using the mentioned patterns compared to those on the COSMOS dataset.

## 5    Applications

Prior work showed that the larger the available training data corpus is, the higher the accuracy a supervised training model could achieve. To obtain more data, synthetic data can be taken into consideration. Even though synthetic data can be created with different methods, it is essential to ensure that the generated dataset is as realistic as the original one. Luo et al. [10] introduced several strategies for producing persuasive image-caption pairs automatically using multiple text and image embedding models. Another way to gain more data (including noisy data) is to use reinforcement learning. Also, since these kinds of data are created from a real dataset, their diversity relies significantly on the primary one. We believe that our diverse and practical data exceptionally contribute to producing better synthetic data.

Cheapfake's techniques primarily relied on reusing existing media in a misleading context, such as using an inaccurate caption. Therefore, combining OOC detection and multimedia retrieval techniques is a promising solution to automate the verification process. Detecting OOC captions could be integrated into media verification systems with powerful event retrieval abilities, such as TrulyMedia, Invid, etc., as an attempt to create an AI-assisted framework for fact-checking organizations, journalists, and activists in the fight against fake news.

Misinformation is a prominent societal issue that lacks sufficient training data. We are providing a realistic dataset containing a variety of categories of image-caption pairs. Our contribution would be the core to expanding the current community's out-of-context dataset. From this manually curated dataset, not only can we generate more synthetic data, but we also open more research challenges, providing more control over information manipulation issues on social media.

The dataset is made available to everyone. This paper is an open-access paper distributed under the terms of the MIT license.

## 6    Discussions

News patterns continue to be updated every day, accompanied by misleading information. Therefore, we will continue to collect and continuously update the current dataset. We will not only collect older sources but also collect more misleading captions from a scene in the videos.

In addition, we also discussed ethical concerns regarding our dataset. Firstly, our most concern is data privacy and the personally identifiable information in the data (in both captions and images). Our data is collected from news and fact-checking sites whose policies have ensured the ability to use these data publicly. Besides, they are reliable sources that verify the truth of the information, and we kept the collected data as pure as possible, retaining the accuracy of our collected data. Since they are pristine, the captions might have additional information not describing the images, such as the creator's stories, comments, and hashtags. This redundant information makes the dataset more practical and challenging for current AI models. However, extraneous information in captions is unsuitable for training supervised learning models.

Secondly, we tried to avoid categorical bias. We did not subjectively try to narrow our data towards any topics but manually collected information from the latest posts/news, which can be objectively affected by current real-world concerns to a small extent. Finally, our final concern is the population we expect our dataset to work with. As our data are mainly from US-based news and fact-checking sites, they largely focused on events, people, and places from Western and European countries such as the US, the UK, and Russia. Also, a small piece of data containing information from Southern Asia is gathered from an Indian fact-checking site.

## 7    Conclusion

In this paper, we curated a dataset through meticulous manual collection. Through a baseline model, the dataset has shown its advantage as a benchmark tool rather than a dataset for training purposes. We also provide a baseline model for evaluation and point out some benefits and limitations of the dataset. In the end, we propose applications and also challenges of the dataset. We aim to encourage researchers to create innovative strategies for dealing with these issues and assess several suggested models for detecting OOC abuse of genuine photos in news articles.

**Acknowledgements.** This study results from the collaborative partnership established in the Memorandum of Understanding (MoU) between Infomedia, DCU, and VNUHCM-US.

We acknowledge the use of ChatGPT [13] for checking and correcting the grammar of this paper. It's important to note that we did not use ChatGPT to generate totally new text; instead, we use it to check/correct our provided text.

# References

1. Akgul, T., Civelek, T.E., Ugur, D., Begen, A.C.: Cosmos on steroids: a cheap detector for cheapfakes. In: Proceedings of the 12th ACM Multimedia Systems Conference, MMSys 2021, pp. 327–331. Association for Computing Machinery, New York, NY, USA (2021). https://doi.org/10.1145/3458305.3479968
2. Aneja, S., Bregler, C., Nießner, M.: COSMOS: catching out-of-context misinformation with self-supervised learning. In: ArXiv preprint arXiv:2101.06278 (2021), https://arxiv.org/pdf/2101.06278.pdf
3. Aneja, S., et al.: ACM multimedia grand challenge on detecting cheapfakes. arXiv preprint arXiv:2207.14534 (2022)
4. Brennen, J.S., Simon, F.M., Howard, P.N., Nielsen, R.K.: Types, sources, and claims of COVID-19 misinformation. Ph.D. thesis, University of Oxford (2020)
5. Britt Paris, J.D.: Deepfakes and cheap fakes the manipulation of audio and visual evidence (2019). https://datasociety.net/wp-content/uploads/2019/09/DS_Deepfakes_Cheap_FakesFinal-1-1.pdf
6. Jaiswal, A., Sabir, E., AbdAlmageed, W., Natarajan, P.: Multimedia semantic integrity assessment using joint embedding of images and text. In: Proceedings of the 25th ACM International Conference on Multimedia, pp. 1465–1471 (2017)
7. Khan, S.A., et al.: Visual user-generated content verification in journalism: an overview. IEEE Access **11**, 6748–6769 (2023)
8. La, T.V., Tran, Q.T., Tran, T.P., Tran, A.D., Dang-Nguyen, D.T., Dao, M.S.: Multimodal cheapfakes detection by utilizing image captioning for global context. In: Proceedings of the 3rd ACM Workshop on Intelligent Cross-Data Analysis and Retrieval, ICDAR 2022, pp. 9–16. Association for Computing Machinery, New York, NY, USA (2022). https://doi.org/10.1145/3512731.3534210
9. Liu, F., Wang, Y., Wang, T., Ordonez, V.: Visual news: benchmark and challenges in news image captioning. arXiv preprint arXiv:2010.03743 (2020)
10. Luo, G., Darrell, T., Rohrbach, A.: Newsclippings: automatic generation of out-of-context multimodal media (2021). https://doi.org/10.48550/ARXIV.2104.05893, https://arxiv.org/abs/2104.05893
11. Müller-Budack, E., Theiner, J., Diering, S., Idahl, M., Ewerth, R.: Multimodal analytics for real-world news using measures of cross-modal entity consistency. In: Proceedings of the 2020 International Conference on Multimedia Retrieval, pp. 16–25 (2020)
12. Opdahl, A.L., et al.: Trustworthy journalism through AI. Data Knowl. Eng. **146**, 102182 (2023)
13. OpenAI: Introducing ChatGPT. https://openai.com/blog/chatgpt (2021). Accessed 08 Aug 2023

14. Pham, K.L., Nguyen, M.T., Tran, A.D., Dao, M.S., Dang-Nguyen, D.T.: Detecting cheapfakes using self-query adaptive-context learning. In: Proceedings of the 4th ACM Workshop on Intelligent Cross-Data Analysis and Retrieval, ICDAR 2023, pp. 60–63. Association for Computing Machinery, New York, NY, USA (2023). https://doi.org/10.1145/3592571.3592972
15. Radford, A., et al.: Learning transferable visual models from natural language supervision. In: International Conference on Machine Learning, pp. 8748–8763. PMLR (2021)
16. Rocha, Y.M., de Moura, G.A., Desidério, G.A., de Oliveira, C.H., Lourenço, F.D., de Figueiredo Nicolete, L.D.: The impact of fake news on social media and its influence on health during the COVID-19 pandemic: a systematic review. J. Public Health, pp. 1–10 (2021). https://doi.org/10.1007/s10389-021-01658-z
17. Roozenbeek, J., et al.: Susceptibility to misinformation about covid-19 around the world. R. Soc. Open Sci. **7**(10), 201199 (2020). https://doi.org/10.1098/rsos.201199
18. Sabir, E., AbdAlmageed, W., Wu, Y., Natarajan, P.: Deep multimodal image-repurposing detection. In: Proceedings of the 26th ACM international conference on Multimedia, pp. 1337–1345 (2018)
19. Schick, N.: Don't underestimate the cheapfake (2020). https://www.technologyreview.com/2020/12/22/1015442/cheapfakes-more-political-damage-2020-election-than-deepfakes/
20. Tandoc Jr., E.C.: The facts of fake news: a research review. Soc. Compass **13**(9), e12724 (2019). https://doi.org/10.1111/soc4.12724
21. Tran, Q.T., Nguyen, T.P., Dao, M., La, T.V., Tran, A.D., Dang Nguyen, D.T.: A textual-visual-entailment-based unsupervised algorithm for cheapfake detection, August 2022. https://doi.org/10.1145/3503161.3551596
22. Vo, N.H., Phan, K.D., Tran, A.-D., Dang-Nguyen, D.-T.: Adversarial attacks on deepfake detectors: a practical analysis. In: Þór Jónsson, B., et al. (eds.) MMM 2022. LNCS, vol. 13142, pp. 318–330. Springer, Cham (2022). https://doi.org/10.1007/978-3-030-98355-0_27
23. Vosoughi, S., Roy, D., Aral, S.: The spread of true and false news online. Science **359**(6380), 1146–1151 (2018). https://doi.org/10.1126/science.aap9559
24. Wang, B., Kuo, C.C.J.: Sbert-wk: a sentence embedding method by dissecting bert-based word models. IEEE/ACM Trans. Audio Speech Lang. Process. **28**, 2146–2157 (2020)
25. Zhou, B., Lapedriza, A., Khosla, A., Oliva, A., Torralba, A.: Places: a 10 million image database for scene recognition. IEEE Trans. Pattern Anal. Mach. Intell. **40**(6), 1452–1464 (2017)

# LUMOS-DM: Landscape-Based Multimodal Scene Retrieval Enhanced by Diffusion Model

Viet-Tham Huynh[1,2], Trong-Thuan Nguyen[1,2], Quang-Thuc Nguyen[1,2], Mai-Khiem Tran[1,2], Tam V. Nguyen[3], and Minh-Triet Tran[1,2(✉)]

[1] Software Engineering Laboratory and Faculty of Information Technology, University of Science, VNU-HCM, Ho Chi Minh City, Vietnam
{hvtham,ntthuan,nqthuc,tmkhiem}@selab.hcmus.edu.vn
[2] Vietnam National University, Ho Chi Minh City, Vietnam
tmtriet@hcmus.edu.vn
[3] Department of Computer Science, University of Dayton, Dayton, U.S.A.
tamnguyen@udayton.edu

**Abstract.** Information retrieval is vital in our daily lives, with applications ranging from job searches to academic research. In today's data-driven world, efficient and accurate retrieval systems are crucial. Our research focuses on video data, using a system called LUMOS-DM: Landscape-based Multimodal Scene Retrieval Enhanced by Diffusion Model. This system leverages Vision Transformer and Diffusion Models, taking user-generated sketch images and text queries as input to generate images for video retrieval. Initial testing on a dataset of 100 h of global landscape videos achieved an 18.78% at Top-20 accuracy rate and 36.45% at Top-100 accuracy rate. Additionally, video retrieval has various applications, including generating data for advertising and marketing. We use a multi-modal approach, combining sketch and text descriptions to enhance video content retrieval, catering to a wide range of user needs.

**Keywords:** Information Retrieval · Stable Diffusion · ControlNet · Vision Transformer · Sketch · Text

## 1 Introduction

Information retrieval [6,8,12,17] is far from being a novelty in today's world. It plays a pivotal role in our daily lives, with numerous practical applications such as job hunting, academic research, and accessing work-related documents. In an era where the daily influx of information is rapidly expanding, one of the paramount concerns in information retrieval is ensuring the continuous and timely updating of data. In addition, the Diffusion Model (DM) [7], notably the Stable Diffusion [13] variant, is a cutting-edge generative modeling framework in machine learning. It excels at progressively enhancing random noise to generate high-quality and diverse data samples. Fine-tuning diffusion steps, provides precise control over the trade-off between sample quality and diversity, making it a crucial tool for applications like image synthesis and text generation, driving innovation in AI and data science.

Video retrieval offers a multitude of possibilities, including the synthesis of new data for various purposes, such as advertising campaigns and marketing efforts. It is employed to retrieve appropriate content that can be used to create new media with the potential to go viral. Therefore, retrieval is essential. In this paper, our focus lies on the cross-modal of video retrieval, particularly for natural and outdoor scenes. We adopt a multi-modal approach, utilizing both sketch and text descriptions to depict queries for various scenes. Users have the ability to quickly sketch or draw the main features of a scene, such as a waterfall or a house with a windmill, and provide textual descriptions. By combining these two different sources, we synthesize a scene that closely mimics reality, which serves as input for our retrieval process. We employ this approach to enhance the retrieval of video content.

In an era of unprecedented data expansion, creating a swift and efficient information retrieval system is vital. Our research focuses on extracting concise and contextually relevant video clips from the dynamic landscape of video data. We introduce a straightforward, yet remarkably effective solution in this paper, emphasizing video content retrieval. Leveraging the power of the diffusion model, specifically the Stable Diffusion variant, empowers us to generate diverse, high-quality video clips. By integrating this generative model, our information retrieval system identifies relevant video segments and enhances search results with visually captivating and contextually pertinent content. This synergy of generative modeling and information retrieval pioneers a new approach to navigating the data deluge, ensuring that the extracted video clips align seamlessly with user expectations and preferences. Our proposed solution leverages the Vision Transformer (ViT) and Stable Diffusion Model to enhance information retrieval. Using user-generated sketch images and queries as input, our system generates images for video retrieval. The outcome is a video frame closely resembling the sketch image, accompanied by a relevant description, revolutionizing content retrieval.

To evaluate our system, we also carefully collected a dataset of 100 h of YouTube videos featuring various landscapes from different regions around the world. We conducted extensive testing on this dataset, and the results demonstrated that our proposed baseline method achieved a high accuracy rate (above 18.78%) in information retrieval and search.

The key results achieved in this paper are as follows:

- We successfully collected 100 h of landscape videos from various countries around the world. In this paper, we provide a list of YouTube links to the 100 h of collected videos.
- We propose a LUMOS-DM: Landscape-based Multimodal Scene Retrieval Enhanced by Diffusion Model method with input consisting of both sketches and text queries. This method has demonstrated its effectiveness with an 18.78% accuracy (Top-20) on the dataset we collected. Furthermore, we also tested on Top-50 and Top-100, respectively, resulting in 27.8% and 36.45%, respectively.

The structure of this paper is as follows. In Sect. 2, we briefly review recent work on information retrieval and the types of input data used in information retrieval. We present our process to collect data in Sect. 3. Then we propose in Sect. 4 our method using Vision Transformer, namely LUMOS-DM: Landscape-based Multimodal Scene Retrieval Enhanced by Diffusion Model, to conduct information retrieval. The experimental results in the landscape dataset using our proposed method are in Sect. 5. Finally, Sect. 6 discusses the conclusion and future work.

## 2  Related Work

### 2.1  Benchmarks

**Information Retrieval Using Image.** A novel CBIR method based on the transfer learning-based visual geometry group (VGG-19) approach, genetic algorithm (GA), and extreme learning machine (ELM) classifier was proposed by Bibi et al. [2] in 2022. The problem of semantic-based image retrieval of natural scenes was addressed by Alqasrawi [1]. It is acknowledged that the convolutional part of the network contains many neurons, with the majority having minimal impact on the final classification decision. In this context, a novel algorithm was proposed by Staszewski et al. [16], enabling the extraction of the most significant neuron activations and their utilization in constructing effective descriptors. In 2023, Chang et al. [3] introduced a Deep Supervision and Feature Retrieval network (Dsfer-Net) designed for bitemporal change detection. MindEye, an innovative fMRI-to-image approach for retrieving and reconstructing viewed images from brain activity, was presented by Scotti et al. [14]. Gong et al. [6] introduced Boon, a novel cross-modal search engine that combines the GPT-3.5-turbo large language model with the VSE network VITR (Vision Transformers with Relation-focused learning) to enhance the engine's capability in extracting and reasoning with regional relationships in images. This approach also provides textual descriptions alongside corresponding images when using image queries. Other influential contributions in this field include the work by Pradhan et al. [12].

**Information Retrieval Using Text.** A method for retrieving patents relevant to an initial set of patents, utilizing state-of-the-art techniques in natural language processing and knowledge graph embedding, was proposed by Siddharth et al. [15]. These methods struggle with mitigating ambiguity in video-text correspondence when describing a video using only one feature, necessitating multiple simultaneous matches with various text features. Lin et al. [10] introduced a Text-Adaptive Multiple Visual Prototype Matching Model to tackle this challenge in 2022. In 2023, Lin et al. [11] explored a principled model design space along two axes: representing videos and fusing video and text information. Chen et al. [5] utilized tags as anchors to enhance video-text alignment. However, the primary objective of the text-to-video retrieval task is to capture complementary audio and video information relevant to the text query, rather than solely

**Fig. 1.** Some illustrations of data collected from different countries

improving audio and video alignment. To address this concern, Ibrahimi et al. [8]. Introduced TEFAL in 2023, a method that produces both audio and video representations conditioned on the text query. To facilitate effective retrieval with aspect strings, Sun et al. [17] introduced mutual prediction objectives between the item aspect and content text.

## 2.2 Discussion

To capture more ranking information, Chen et al. [4] proposed a novel ranking-aware uncertainty approach for modeling many-to-many correspondences using provided triplets. In light of the existing research progress in text-pedestrian image retrieval, Li et al. [9] proposed a progressive feature mining and external knowledge-assisted feature purification method to address the associated challenges comprehensively. These two methods have achieved some quite outstanding results. However, Chen et al.'s method focuses on human subjects, while Li's method focuses on fashion. With this research, we want to focus on the subject of natural landscapes. The highlight of LUMOS-DM is that we will try to combine text and sketch input data instead of combining text and natural images like the two methods above.

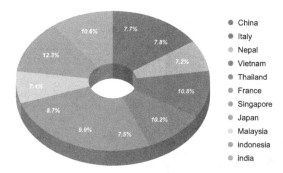

**Fig. 2.** The percentage of video hours by country in our collection

# 3   Landscape Dataset

## 3.1   Dataset Overview

**Dataset Collection.** We embarked on an extensive data collection endeavor, curating a rich repository of videos that showcased the unique landscapes of several diverse countries across the globe. Among the nations included in our comprehensive dataset are France, Vietnam, Japan, Thailand, Singapore, Malaysia, and more. Figure 1 illustrates the dataset we collected through 100h of video.

**Dataset Statistics.** Our collective efforts culminated in acquiring an impressive 100 h of video content, with the corresponding breakdown of video hours for each country thoughtfully illustrated in Fig. 2.

Our data collection strategy was meticulously designed to achieve two distinct objectives. First and foremost, we aimed to imbue our dataset with diversity, drawing from the rich tapestry of natural and urban environments found across these nations. This diversity ensures that our dataset encapsulates a broad spectrum of terrains, climates, and cultural influences, enabling a more comprehensive analysis.

In parallel, our data collection approach was strategically directed towards capturing the essence of each location. To achieve this, we meticulously selected videos featuring iconic landmarks, renowned architectural marvels, and the breathtaking natural wonders that define each region. By doing so, we have diversified our dataset and accentuated the distinctive characteristics that make each country unique.

In summary, our data collection efforts spanned multiple countries, focusing on both diversity and the representation of notable locations. This carefully cultivated dataset serves as the foundation for our research, enabling a robust exploration of landscape imagery and its connection to information retrieval.

## 3.2   Dataset Construction Pipeline

We initially converted the data collected from videos into images to prepare for the experimental phase. We opted to create image data by extracting frames

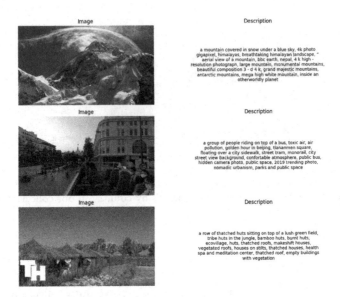

Image                          Description

a mountain covered in snow under a blue sky, 4k photo
gigapixel, himalayas, breathtaking himalayan landscape, "
aerial view of a mountain, bbc earth, nepal, 4 k high -
resolution photograph, large mountain, monumental mountains,
beautiful composition 3 - d 4 k, grand majestic mountains,
antarctic mountains, mega high white mountain, inside an
otherworldly planet

Image                          Description

a group of people riding on top of a bus, toxic air, air
pollution, golden hour in beijing, tiananmen square,
floating over a city sidewalk, street tram, monorail, city
street view background, confortable atmosphere, public bus,
hidden camera photo, public space, 2019 trending photo,
nomadic urbanism, parks and public space

Image                          Description

a row of thatched huts sitting on top of a lush green field,
tribe huts in the jungle, bamboo huts, burnt huts,
ecovillage, huts, thatched roofs, makeshift houses,
vegetated roofs, houses on stilts, thatched houses, health
spa and meditation center, thatched roof, empty buildings
with vegetation

**Fig. 3.** Description of the image created from CLIP-Interrogator tool

from videos at a rate of 0.5 frames per second. In total, we obtained 360,000 images. We will refer to this dataset as the *LandscapeDataset*.

We further transformed the *LandscapeDataset* into sketch-like representations using the Holistically-Nested Edge Detection technique to facilitate experimentation involving the fusion of sketches and text for information retrieval.

Additionally, from the *LandscapeDataset*, we generated descriptive sentences for the images using the CLIP-Interrogator tool (Fig. 3).

With these preparatory steps completed, we are fully equipped for information retrieval through the fusion of text and sketch to generate images using Diffusion Models.

## 4 Methodology

### 4.1 Stable Diffusion Model

**Diffusion probabilistic models (DM)** [7] represents a novel and potent paradigm in the realm of probabilistic generative modeling. These models manifest as parameterized Markov chains, fine-tuned via variational inference, with the primary objective of generating samples that closely align with the underlying data distribution at a finite temporal horizon. Intriguingly, the transitions within this chain, denoted as $p_\theta(x_{t-1}|x_t)$, are systematically trained to reverse a diffusion process in a Markov chain that progressively injects noise into the data, counter to the sampling direction represented by $q_\theta(x_t|x_{t-1})$, ultimately leading to the gradual erosion of the original signal. Notably, when the diffusion mechanism involves a subtle infusion of Gaussian noise, the sampling chain transitions can be elegantly streamlined to conditional Gaussians, simplifying the neural network parameterization. This simplification is encapsulated in the following objective function:

$$L_{DM} = E_{x,\epsilon_\theta \sim N(0,1),t}[||\epsilon - \epsilon_\theta(x_t, t)||_2^2] \tag{1}$$

Furthermore, the framework employs denoising autoencoders $(\theta(x_t, t); t = 1, \ldots, T)$ trained meticulously to predict denoised variants of input data $(x_t)$ derived from their noisy counterparts $(x)$. This synthesis of principles from physics, variational inference, and neural networks underpins a powerful approach that holds significant promise for high-quality image synthesis and generative modeling, notwithstanding the substantial computational demands it entails.

**Stable Diffusion.** [13] advances the Diffusion model's foundational principles, significantly enhancing generative capabilities. Central to this innovation is integrating a Variational Autoencoder (VAE) with an encoder and decoder, elevating image fidelity. The VAE's encoder transforms data into a lower-dimensional latent form, efficiently reconstructed by the decoder. Stable Diffusion also employs a ResNet-enhanced U-Net architecture, preserving image integrity while reducing noise. A text encoder with CLIP text embeddings and cross-attention layers also enhances contextual awareness, conditioning image generation on textual cues. This pioneering approach merges diffusion principles, VAEs, U-Net structures, and text embeddings, yielding high-quality, contextually informed images with unmatched precision.

## 4.2  Architecture

**ControlNet.** [19], a vital component of Stable Diffusion, enhances image synthesis by providing precise control over content, style, and context. It employs advanced neural network structures, including attention mechanisms and disentangled representations, to focus on specific image regions and manipulate factors like pose and lighting. During training, ControlNet's loss function, similar to Stable Diffusion's but with added text $(c_T)$ and latent conditions $(c_F)$, ensures output consistency with specified criteria, enabling precise control and superior image synthesis:

$$L_{ControlNet} = E_{x,\epsilon_\theta \sim N(0,1),t,c_T,c_F}[||\epsilon - \epsilon_\theta(x_t, t, c_T, c_F)||_2^2] \tag{2}$$

This loss function is meticulously designed to optimize the model's performance by minimizing the discrepancy between the generated images and the target images, while also considering the alignment with textual prompts and control parameters. The loss function serves as a crucial metric for monitoring the model's progress and convergence behavior, ensuring that the generated images exhibit high fidelity and align closely with the intended context and narrative.

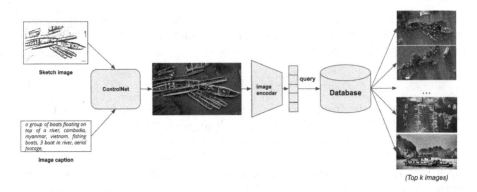

**Fig. 4.** System overview

**Landscape-Based Multimodal Scene Retrieval Enhanced by Diffusion Model:** We propose a novel system with three integral components, each contributing to the seamless synthesis and retrieval of landscape images from user inputs. We begin the process with the input sketch image and the user's descriptive sentence. We perform feature extraction from a sketch, utilizing techniques that capture salient visual attributes, contours, and structural elements. Simultaneously, we extract features from the user's textual input, harnessing natural language processing methods to identify keywords and context within the sentence. The significance of this dual feature extraction is twofold:

- Empower us to distill essential visual and semantic information from both modalities.
- Ensure that the extracted data have the same dimensionality, enabling seamless integration. The result is a pair of feature vectors, one representing the sketch and the other encapsulating the textual input, quantifying the input data's essence.

Building on these feature vectors, we leverage the ControlNet architecture, a cutting-edge generative model, to expedite the image generation process. ControlNet expertly synthesizes an image that aligns with the user's input, bridging the gap between the sketch and the textual description. Figure 4 comprehensively illustrates our system's workflow, highlighting the intricate synergy between these components. Furthermore, armed with the newly generated images, we embark on a subsequent phase dedicated to meticulous feature extraction. This process involves a detailed analysis of the visual attributes and characteristics embedded within the synthesized images. These extracted image features form the foundation for our querying procedure, which revolves around our Landscape Dataset - a meticulously curated repository housing an extensive assortment of landscape images. Implementing robust techniques, including similarity measures and content-based image retrieval methodologies, facilitates the querying process. These techniques are instrumental in identifying and ranking the top $k$ images that closely align with the user's input, with the flexibility to adjust the value of $k$ according to specific preferences and requirements.

# 5   Experiments

## 5.1   Experimental Settings

In pursuit of a comprehensive investigation into the matter, we embarked on an exhaustive exploration utilizing various tools at our disposal. To ensure a rigorous examination, we undertook a systematic approach by experimenting with various combinations of components. For the vital aspect of information representation, we diligently employed no fewer than three distinct models to generate the essential feature maps from both the source data and query data. These models included ViT-B-16, which was employed exclusively for images, ViT-B-32, which was utilized for both images and text, and EfficientNet-B4 [18], specially chosen for its efficacy in processing image data. This judicious selection of models was undertaken to ensure a comprehensive analysis yielding insightful results. Using these models as components, we have created six experiments reported in Table 1:

- **ViT ControlNet:** In this experiment, we use ViT B-16 to compute the feature map from ControlNet-generated query image. Thus, this feature map should contain text and sketch data information. ViT B-16 is also used to compute the feature maps of the whole source dataset. Thus, these computed feature maps exist in the same feature space, all having the same dimension of (1000) and can be compared with each other.
- **ViT Edge:** This experiment is similar to ViT ControlNet but differs at the input stage. Instead of using fused ControlNet-generated images, only the edge map is used, and the ControlNet image generation step is skipped altogether. Because both the dataset feature map and the edge map features are generated with ViT B-16, they, therefore, are also in the same feature space. This makes it possible to make comparisons of the edge maps against the image dataset.
- **EfficientNetB4-ControlNet:** This experiment is also similar to ViT ControlNet. However, we replace ViT-B16 with EfficientNet-B4. EfficientNet-B4 is used to compute the feature maps of ControlNet-generated images and the source dataset. The feature space is thusly the same with dimensions of (1792, 7, 7), and we make comparisons in this space.
- **ViT-CLIP-ControlNet:** We use CLIP-ViT B32, a cross-modal model capable of fusing both textual and visual inputs into a common feature space. ControlNet-generated images and the source dataset are used to compute feature maps in this experiment. The resultant feature maps have a dimension of (1, 512) and are comparable across the query and the source dataset.
- **ViT-CLIP-Prompt:** This method is similar to ViT-CLIP-ControlNet. However, instead of using visual inputs, we only use textual inputs from the ViT L-14 generated prompts to generate feature maps. We can perform the same comparisons using the source dataset feature maps from ViT-CLIP-ControlNet.

– **Vit-CLIP-Edge:** Similar to ViT-CLIP-ControlNet, we use CLIP-ViT B32
as feature map generator. However, in this experiment, we only use the edge
maps instead of generating images of ControlNet. Thus, the information about
the prompt is absent in the generated feature map.

**Table 1.** Overview of the experiments

| Experiment name | Textual input | Visual input | Feature space (feature size) |
|---|---|---|---|
| ViT-Edge | (none) | Edge maps | ViT-B-16 (1000) |
| ViT-ControlNet | (ViT-L-14) generated prompts | ControlNet-generated images | ViT-B-16 (1000) |
| EfficientNetB4-ControlNet | ViT-L-14 generated prompts | ControlNet-generated images | EfficientNet-B4 (1792, 7, 7) |
| ViT-CLIP-ControlNet | (none) | ControlNet-generated images | ViT-B-32 (1, 512) |
| ViT-CLIP-Prompt | ViT-L-14 generated prompts | (none) | ViT-B-32 (1, 512) |
| ViT-CLIP-Edge | (none) | Edge maps | ViT-B-32 (1, 512) |

### 5.2   Implementation Details

After obtaining the data described in Sect. 3.2, we conducted experiments on
one-third of the dataset. The images selected for experimentation were chosen randomly. We conducted the experiments and extracted feature vectors for
both the images and their corresponding descriptive sentences. With the outputs
obtained during this phase, we evaluated the performance using metrics such as
Top-1, Top-5, and Top-20, as explained in the evaluation methodology outlined
in Sect. 5.1. These experimental results are presented in Sect. 5.3.

### 5.3   Experimental Results

**Qualitative Observations.** We investigate and assess the performance of our
innovative method, leveraging a carefully curated testing dataset as depicted in
Fig. 5. Our evaluation process is thoughtfully designed to encompass a wide spectrum of real-world scenarios, ensuring our approach's robustness and versatility.
To elaborate on LUMOS-DM, we begin with the original image, positioned on
the left side, serving as the primary target for retrieval. Within our extensive
image repository, we initiate the process by subjecting the image to holistic edge
detection as a pivotal initial step. The outcome of this stage, illustrated in the
middle, is an edge map that accentuates the structural contours and distinctive

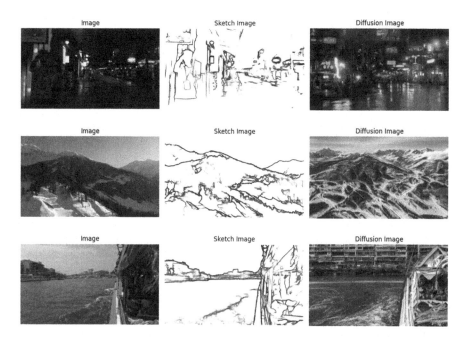

**Fig. 5.** Illustration images in order: original image (left), sketch image (middle) and image created from diffusion model (right)

features within the image. However, our innovation doesn't halt here; we further enhance the interpretability and predictive prowess of LUMOS-DM by generating a descriptive textual prompt using the ViT model. This prompt is seamlessly integrated with the previously derived edge map, culminating in creating the prediction image, elegantly showcased on the right side of the visual spectrum. This intricate sequence of transformations represents the core of LUMOS-DM, where both visual and semantic information harmoniously converge to yield accurate and meaningful predictions. Subsequently, the ensuing evaluation delves into the performance of this method across diverse scenarios, shedding light on its inherent strengths and capabilities within the domain of image retrieval and prediction.

**Quantitative Observations.** Figure 6 visually represents the initial effectiveness of various methods and models in interacting with a query image. This preliminary observation sets the stage for a detailed exploration of specific accuracy metrics. Notably, our VIT-CLIP-ControlNet achieves a remarkable Top-1 accuracy of 3.22%, underscoring its capacity to consistently rank the most pertinent image as the top prediction. Extending our analysis to Top-5 accuracy, we maintain strong performance at 9.27%, affirming the model's ability to encompass the correct result within the top quintet of predictions. Notably, Top-20 accuracy soars to 18.79%, accentuating its proficiency in capturing pertinent

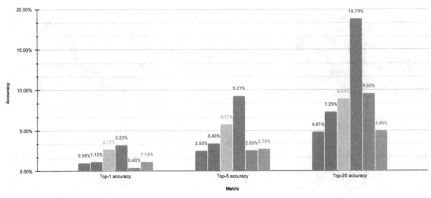

**Fig. 6.** Accuracy of different retrieval methods across different metrics on a dataset of 21,000 pictures.

images across a broader spectrum of predictions-particularly valuable in contexts featuring expansive image repositories. Our meticulous comparative analysis substantiates VIT-CLIP-ControlNet's supremacy over alternative methods; for instance, ViT-Edge attains a modest 0.99% Top-5 accuracy, while ViT-CLIP-Prompt fares slightly better at 2.53%, yet both fall significantly short of our model's performance. Furthermore, statistical significance tests substantiate the reliability and significance of these performance disparities, solidifying the excellence of VIT-CLIP-ControlNet in image retrieval and prediction tasks.

## 6   Conclusion

In this paper, we introduce the Landscape Dataset, which was collected from YouTube and comprises a total duration of 100 h of video content. This dataset encompasses various landscapes from different countries around the world. We propose a LUMOS-DM: Landscape-based Multimodal Scene Retrieval Enhanced by Diffusion Model method using Vision Transformer and Stable Diffusion. This method enhances information retrieval efficiency by expanding the input modalities to include descriptive sentences and sketch-like images. This facilitates more user-friendly information retrieval from videos.

LUMOS-DM opens up numerous potential applications in our daily lives. One of the simplest examples of this can be seen in how LUMOS-MD can significantly support individuals involved in affiliate marketing. Content creators will no longer face the challenges of searching for videos. Instead, they only need to provide a description and a basic sketch of the scene they desire, and LUMOS-MD will suggest corresponding video clips. This streamlines the content creation process and facilitates reaching customers more efficiently.

# References

1. Alqasrawi, Y.: Bridging the gap between local semantic concepts and bag of visual words for natural scene image retrieval (2022)
2. Bibi, R., Mehmood, Z., Munshi, A., Yousaf, R.M., Ahmed, S.S.: Deep features optimization based on a transfer learning, genetic algorithm, and extreme learning machine for robust content-based image retrieval. PLoS ONE **17**(10), e0274764 (2022)
3. Chang, S., Kopp, M., Ghamisi, P.: Dsfer-Net: a deep supervision and feature retrieval network for bitemporal change detection using modern hopfield networks (2023)
4. Chen, J., Lai, H.: Ranking-aware uncertainty for text-guided image retrieval. ArXiv abs/2308.08131 (2023). https://api.semanticscholar.org/CorpusID:260926537
5. Chen, Y., Wang, J., Lin, L., Qi, Z., Ma, J., Shan, Y.: Tagging before alignment: integrating multi-modal tags for video-text retrieval. In: Proceedings of the AAAI Conference on Artificial Intelligence, vol. 37, no. 1, pp. 396–404 (2023). https://doi.org/10.1609/aaai.v37i1.25113, https://ojs.aaai.org/index.php/AAAI/article/view/25113
6. Gong, Y., Cosma, G.: Boon: a neural search engine for cross-modal information retrieval (2023)
7. Ho, J., Jain, A., Abbeel, P.: Denoising diffusion probabilistic models. Adv. Neural. Inf. Process. Syst. **33**, 6840–6851 (2020)
8. Ibrahimi, S., Sun, X., Wang, P., Garg, A., Sanan, A., Omar, M.: Audio-enhanced text-to-video retrieval using text-conditioned feature alignment (2023)
9. Li, H., Yang, S., Zhang, Y., Tao, D., Yu, Z.: Progressive feature mining and external knowledge-assisted text-pedestrian image retrieval (2023)
10. Lin, C., et al.: Text-adaptive multiple visual prototype matching for video-text retrieval. In: Oh, A.H., Agarwal, A., Belgrave, D., Cho, K. (eds.) Advances in Neural Information Processing Systems (2022). https://openreview.net/forum?id=XevwsaZ-4z
11. Lin, X., et al.Towards fast adaptation of pretrained contrastive models for multi-channel video-language retrieval (2023)
12. Pradhan, J., Pal, A.K., Hafizul Islam, S.K., Bhaya, C.: DNA encoding-based nucleotide pattern and deep features for instance and class-based image retrieval. IEEE Trans. Nanobiosc. **23**, 190–201 (2023)
13. Rombach, R., Blattmann, A., Lorenz, D., Esser, P., Ommer, B.: High-resolution image synthesis with latent diffusion models. In: Proceedings of the IEEE/CVF Conference on Computer Vision and Pattern Recognition, pp. 10684–10695 (2022)
14. Scotti, P.S., et al.: Reconstructing the mind's eye: fMRI-to-image with contrastive learning and diffusion priors (2023)
15. Siddharth, L., Li, G., Luo, J.: Enhancing patent retrieval using text and knowledge graph embeddings: a technical note (2022)
16. Staszewski, P., Jaworski, M., Cao, J., Rutkowski, L.: A new approach to descriptors generation for image retrieval by analyzing activations of deep neural network layers. IEEE Trans. Neural Netw. Learn. Syst. **33**(12), 7913–7920 (2022)
17. Sun, X., et al.: Pre-training with aspect-content text mutual prediction for multi-aspect dense retrieval (2023)

18. Tan, M., Le, Q.: EfficientNet: rethinking model scaling for convolutional neural networks. In: International Conference on Machine Learning, pp. 6105–6114. PMLR (2019)
19. Zhang, L., Agrawala, M.: Adding conditional control to text-to-image diffusion models. arXiv preprint arXiv:2302.05543 (2023)

# XR-MACCI: Special Session on eXtended Reality and Multimedia - Advancing Content Creation and Interaction

# Mining Landmark Images for Scene Reconstruction from Weakly Annotated Video Collections

Helmut Neuschmied$^{(\boxtimes)}$ and Werner Bailer

JOANNEUM RESEARCH – DIGITAL, Graz, Austria
`helmut.neuschmied@joanneum.at`

**Abstract.** Many XR productions require reconstructions of landmarks such as buildings or public spaces. Shooting content on demand is often not feasible, thus tapping into audiovisual archives for images and videos as input for reconstruction is a promising way. However, if annotated at all, videos in (broadcast) archives are annotated on item level, so that it is not known which frames contain the landmark of interest. We propose an approach to mine frames containing relevant content in order to train a fine-grained classifier that can then be applied to unlabeled data. To ensure the reproducibility of our results, we construct a weakly labelled video landmark dataset (WAVL) based on Google Landmarks v2. We show that our approach outperforms a state-of-the-art landmark recognition method in this weakly labeled input data setting on two large datasets.

**Keywords:** Landmark retrieval · image classification · few-shot learning · fine-grained classification · learning from weakly-labeled data

## 1 Introduction

The creation of XR applications depends on the availability of high-quality 3D models. For many XR use cases in culture and media, models of buildings or public spaces, commonly referred to as landmarks, are of great importance. They can be used as basis for augmenting them with information or educational content, or they can be used in virtual studies to enhance storytelling. Performing a high-quality capture process on demand is often not feasible due to cost reasons or time limitations (e.g., when producing for news). However, the archives of broadcasters and cultural institutions hold a vast amount of image and video data that is a valuable source for creating 3D models. Novel scene representation methods such as NERFs [11] also open new possibilities for creating such assets from 2D imagery.

Video data is a particularly interesting source for this purpose, as a moving camera shot may provide a number of views of the landmark of interest, taken at the same time under the same conditions. There are possibly also multiple shots from the same recording in a video. However, if the landmarks of interest are at

S. Rudinac et al. (Eds.): MMM 2024, LNCS 14557, pp. 161–174, 2024.
https://doi.org/10.1007/978-3-031-53302-0_12

all annotated, the annotation is weak, i.e., metadata exist on the level of a video file, broadcast, or in the best case a story, but do not provide information about the exact temporal location of views of the landmarks in each frame. Despite their limitations, these weakly annotated videos offer a valuable resource, as they can be used to train models that are able to detect the same landmark in entirely unannotated content, replacing the labor-intensive and time-consuming process of manually labeling training samples. However, the success of deep learning models depends on the quality and size of annotated datasets used for training. The video thus contains "noise" from the point of view of this purpose, such as interviews, shots of anchorpersons, close-up interior views etc. In order to effectively use these data, the first task is to isolate the relevant ranges of frames showing the landmark of interest. Apart from very popular landmarks, the number of samples may still be small, thus requiring a method capable of learning from a small set of samples (few-shot learning).

In order to mine images for scene reconstruction, the problem to be solved can be defined as follows. Given is a collection of $N$ videos $\mathcal{V} = \{v_1, \ldots, v_N\}$, where each video $v_k$ is composed of a set of $M_k$ frames $v_k = \{f_1^k, \ldots, f_{M_k}^k\}$, a set of $P$ landmarks $\mathcal{L} = \{L_1, \ldots, L_P\}$, and a set of landmarks contained in a video $A_k = \{L_p, L_q, \ldots\}$ (with $|A_k| \ll P$, and likely $A_k = \{\}$ for some videos). The first step is to mine a training set $\mathcal{T}_p = \{f_j^k | L_p \in A_k \wedge \text{vis}(f, L_p) = 1\}$, where the function $\text{vis}(f, L_p)$ returns 1 if the landmark $L_p$ is at least partly visible in frame $f$, and 0 otherwise. Then a landmark classifier can be trained from $\mathcal{L}$ in order to annotate further videos and increase the set of images mined for reconstruction. By training our deep learning models on the training data mined from the weakly annotated video dataset, we aim to develop algorithms capable of recognizing landmarks with high precision, thus being able to collect an image set providing the basis for reliable scene reconstruction.

Although landmark recognition is a widely addressed research topic, there is a challenge of finding a dataset that is appropriate for our task setting. While there are a number of public landmark recognition datasets, they are image based with per image annotations (see discussion in Sect. 2.3). There are relevant weakly annotated video datasets [3], but those are not openly available.

The main contributions of this paper are as follows:

- We construct a novel weakly annotated video landmark dataset (WAVL), based on the widely used Google Landmarks v2 dataset [23] that is publicly available to the research community.
- We propose a pipeline for training data selection and training a fine-grained landmark classifier from a small set of training samples per class.
- In order to study the case of unannotated collections, we propose an alternative approach using data mined from web search to feed the same pipeline.
- We analyze the impact of weakly annotated video data on robustness in landmark recognition on two datasets, comparing against a state-of-the-art method.

The rest of this paper is organized as follows. In Sect. 2 we review related approaches and datasets. We describe the construction of our dataset in Sect. 3

and present the proposed method in Sect. 4. Section 5 provides evaluation results and Sect. 6 concludes the paper.

## 2 Related Work

### 2.1 Landmark Recognition

Earlier landmark recognition methods were based on the extraction of local image features, often represented as visual words. With the advent of deep learning, convolutional neural networks (CNNs) were introduced to extract features from images, enabling both landmark recognition and the use of similarity measures between pairs of images. Noh et al. [12] introduced DELF (DEep Local Features), a fusion of classical local feature methods with deep learning techniques. DELF leverages features from CNN layers and integrates an attention module to enhance recognition accuracy. Boiarov et al. [1] extended this by utilizing the center loss function to train CNNs, which penalizes the distances between image embedding and their corresponding class centers. In handling variations due to different viewpoints, they employed hierarchical clustering to compute centroids for each landmark, effectively managing the variability inherent in landmarks. Razali et al. [17] propose a lightweight landmark recognition model using a combination of Convolutional Neural Network (CNN) and Linear Discriminant Analysis (LDA) for feature size reduction. They compared different CCNs and showed that the EfficientNet architecture with a CNN classifier outperformed the other models evaluated. Yang et al. [24] train two different variants of the ResNet network architecture (ResNeSt269 and Res2Net200_vd) with an increasing image resolution (step by step) in order to improve the landmark recognition feature vectors obtained from a CNN. They normalize and merge the embedded layer descriptors of the two models above, doubling their size. As a post-processing step, the retrieval results can be significantly improved by re-ranking methods, e.g. by spatial verification [13] which is a method that checks the geometric consistency using local descriptors.

### 2.2 Learning from Weakly-Labeled Data

The challenge of learning from weakly-labeled data has spurred innovative approaches to filter out potential temporal noise in annotations [19]. A different perspective is presented by Li et al. [8], who propose transforming potential noise in weakly-labeled videos into valuable supervision signals. This is achieved through the concept of sub-pseudo labels (SPL), where a new set of pseudo-labels is generated, expanding the original weak label space. This creative approach demonstrates a shift from noise filtering to converting noise into useful information, harnessing the power of weakly-labeled data for improved learning.

The rise of the internet and social media has simplified the acquisition of data relevant to specific classification tasks. In cases where supervision is incomplete and only a portion of training data carries labels, harnessing the abundance of online data becomes crucial. The substantial information available from online sources contributes significantly to robust model development by augmenting labeled training data with additional instances from diverse contexts. But also images from web search engines like Google or DuckDuckGo tend to be biased toward images where a single object is centered with a clean background and a canonical viewpoint [10] and depending on the search term, there might also be pictures included that correspond to a completely different context than intended. Chen et al. [4] propose a two-stage CNN training approach for leveraging noisy web data. They initially employ simple images to train a baseline visual representation using a CNN. Subsequently, they adapt this representation to more challenging and realistic images by capitalizing on the inherent structure of the data and category.

## 2.3   Weakly Annotated Landmark Datasets

While there exist many datasets relevant to our task, most of them are fully labeled image datasets. Oxford [15] and Paris [16] buildings are early datasets used in landmark recognition and visual place recognition tasks. Over the years datasets grew in size, such as Pittsburg250k [22] and data diversity, e.g. including different captures times as in Tokyo 24/7 [21]. The Google Landmarks datasets v1 [12] and v2 [23] have become a widely adopted benchmark for this task. Datasets created for autonomous driving research such as BDD100k [26] provide location indexed videos and are thus sometimes used for visual geolocation. However, these datasets show landmarks always from a vehicle point of view, and in contrast to weakly annotated content from media archives the remaining content contains other street views.

The Italian public broadcaster RAI assembled a dataset with monuments of Italy [3]. The dataset contains about 2,000 clips depicting about 200 monuments from all regions of Italy, mainly acquired from RAI regional newscasts, collected for assessing similarity search in video. Annotations of the monument are provided on clip level. Each clip contains typically a news story, of which one or more shots contain an exterior view of the relevant monument, and in some cases also interior views. Some of the shots may show the monument occluded or in the background (e.g., as backdrop of an interview). In addition, the clips often contain other material of the story, e.g., the anchor in the studio introducing the topic (with an image that shows a view of the monument or something else), views of people in the street, close-up shots of people or interior items etc. As such, the dataset is typical for the type of content and the granularity of annotation to be found in a broadcast archive. The dataset is not available for public download but provided by RAI under a custom license agreement.

V3C (Vimeo Creative Commons Collection) [18] is a very large dataset (28,450 videos, about 3,800 h) assembled for benchmarking video retrieval. We have considered amending the existing metadata with landmark annotations for a subset. However, a preliminary experiment found a too small number of clips (using landmark or city names as initial queries).

The lack of a dataset that matches the characteristics of data and annotation granularity found in media archives and that is openly available led to the decision to construct such a dataset by combining commonly used datasets.

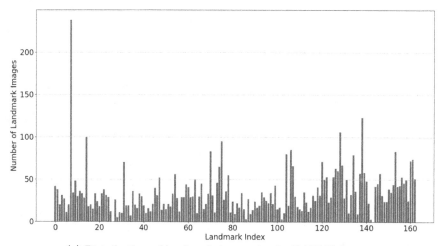

(a) Distribution of landmark images in the RAI-MI dataset.

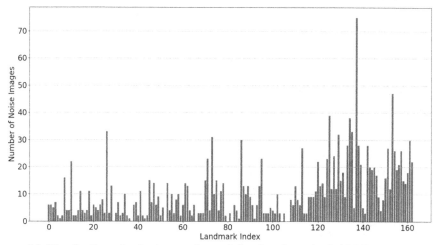

(b) Distribution of noise images for each landmark in the RAI-MI dataset.

**Fig. 1.** Statistic information of the RAI-MI dataset.

## 3  Weakly Annotated Dataset

We propose the Weakly Annotated Video Landmarks (WAVL) dataset. For evaluating landmark recognition, there is no need to use temporal correlations

**Table 1.** Key figures for images per landmark from the RAI-MI and WAVL data sets (LM: landmarks, Imgs.: images, Std Dev.: standard deviation).

| Dataset | LM | LM Imgs | Noise Imgs. | Mean LM Imgs | Std Dev. LM Imgs | Mean Noise Imgs. | Std Dev. Noise Imgs. |
|---------|-----|--------|-------------|--------------|------------------|------------------|----------------------|
| RAI-MI  | 163 | 5,620  | 1,871       | 34.55        | 26.92            | 10.63            | 10.60                |
| WAVL    | 141 | 4,230  | 1,592       | 29.42        | 1.33             | 11.29            | 1.07                 |

between successive video frames. This allowed us to simplify our dataset by focusing only on keyframes, which represent either a single frame per video sequence or frames extracted at defined temporal intervals. In order to mimic a keyframe dataset as extracted from archive video content, with similar characteristics as the RAI monuments of Italy (RAI-MI) dataset, we merged images from two different sources: the Google Landmarks v2 dataset [23] and the V3C [18] video dataset. These combined sources allowed us to create a dataset that contains sets of keyframes as they would be extracted from one video, containing both keyframes of a particular landmark as well as unrelated keyframes (noise). The video is annotated with the landmark visible in the subset of keyframes taken from Google Landmarks.

For each set of keyframes representing a video, we combined on average 30 associated images from the Google Landmarks v2 dataset with on average 11 keyframes (noise images) from one of the videos in the V3C1 dataset. This process has been done for 141 landmarks, resulting in a dataset of 5,770 images. The dataset has been made available at https://github.com/XRecoEU/WAVL-Dataset.

To perform a comparative evaluation on the RAI-MI dataset, we selected a subset of 163 different videos of landmarks representing buildings, and extract keyframes from them. The key figures for both datasets are shown in Table 1. The distributions of the landmark and noise images for the RAI-MI dataset vary more strongly, thus we plot them in Fig. 1. For a number of cases the number of samples is in a range of 10 or less, so that the problem can at least partly be considered a few-shot learning problem.

## 4    Proposed Approach

We propose an approach to landmark recognition that treats the problem as a fine-grained image classification task. To realize this, we build on the Attentive Pairwise Interaction Network (API-Net) [27] with EfficientNet B3 [20] as the backbone architecture. However, the presence of weak annotations introduces a challenge as a substantial portion of training images contains incorrect labels. To address this issue, we have explored two different approaches.

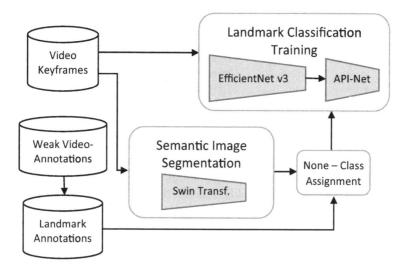

**Fig. 2.** Swin+API-Net: Training of the EfficientNet B3 network with API-Net from a filtered set of keyframes. The corresponding landmark class labels are changed to *None* class if no or only small building areas are detected in the images.

### 4.1 Semantic Segmentation Prefiltering

In the first approach (Swin+API-Net) (see Fig. 2), given our focus on landmarks associated with buildings, we leverage a Swin transformer network for semantic image segmentation [9]. This process enables us to filter out images where structures like buildings or walls are observable, providing insights into the size of these areas. Additionally, we utilize information about the recognition of extensive regions where human figures are present, aiding in identifying anchorpersons or interior views with close-up shots, discarding these images as well. All images filtered in this process are put into a *None* class for the training process. This strategy of region-based filtering contributes to balancing the impact of keyframes unrelated to the landmark on the training process.

For inference, a simplified approach is adopted for landmark classification on images, using only the retrained EfficientNet backbone with the API-Net classification layer. This allows a straightforward evaluation of the performance of the trained model on individual test images, focusing solely on the ability of EfficientNet to recognise the trained landmarks.

### 4.2 Web Image Mining

In the second approach (CDVA+API-Net) (see Fig. 3), we perform a targeted web search for relevant landmark training images using the landmark information embedded in the weak video annotations. To facilitate this, we employ the DuckDuckGo[1] image search engine, retrieving a set of 40 images for each

---

[1] https://duckduckgo.com/.

**Table 2.** Images obtained using the web image mining process for the RAI-MI and the WAVL datasets.

| Dataset | Landmarks | Downloaded images | Rejected images | Landmarks w. $\geq 1$ img. | Landmarks w. $\geq 10$ img. |
|---------|-----------|-------------------|-----------------|----------------------------|-----------------------------|
| RAI-MI  | 163       | 6,130             | 1,177           | 162                        | 153                         |
| WAVL    | 141       | 5,585             | 166             | 140                        | 139                         |

landmark term. Recognizing that web data can introduce considerable noise, our selection process is refined. We only retain images that show similarity to keyframes extracted from the weakly annotated videos. This similarity check is done by computing CDVA [5] descriptors from the images. In particular, we use the learned component of the descriptor, which is binarised to obtain 512 bit binary vector that can be efficiently matching using Hamming distance.

To filter the web images, we compare them to the keyframes and keep only those that have a similarity score above a certain threshold. We use a threshold of 0.6 to select web images in the first step. However, to account for the variability of web images, in a subsequent step we compare the remaining web images with the previously selected ones. If the similarity score exceeds a slightly lower threshold of 0.58, these previously rejected images are also used. These threshold values were determined experimentally with the help of a visualization of the filter results. Table 2 gives an overview of the web images obtained for the RAI-MI and the WAVL landmarks. Following the filtering of web images, we proceed to train an EfficientNet B3 using the API-Net framework, mirroring the process employed in our first approach. This strategy capitalizes on the wealth of online resources while maintaining a rigorous validation process to ensure the quality and relevance of the acquired images. Once the training has been completed, the images are again evaluated using only the EfficientNet B3 backbone and the classification layer of API-Net.

## 5    Evaluation

We evaluate the two proposed training approaches on both the RAI-MI and WAVL datasets. In the case of the WAVL dataset, we also compare our approach with a state of the art landmark recognition method in order to establish a baseline for the dataset. Given the composition of the WAVL dataset, which uses Google Landmarks v2 images, we chose to compare our approaches with a well-performing method on this benchmark dataset.

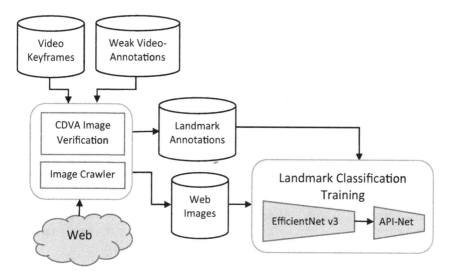

**Fig. 3.** CDVA+API-Net: Training of the EfficientNet B3 network with API-Net by using images which are crawled from the web. Only those web images are used which have sufficient similarity (based on CDVA descriptors) with one of the video keyframes.

In this context, we considered the solution developed by the smlyaka team [25], for which the source code is publicly available[2]. It has demonstrated exceptional performance, winning the first place in the Google Landmark Retrieval 2019 Challenge and the third place in the Google Landmark Recognition 2019 Challenge on Kaggle. This approach involves the initial pre-training of a ResNet-101 backbone on ImageNet and the Google Landmark Dataset v1 (GLD-v1) [12] training dataset. This pre-trained model has not been made available. We thus start from a model pre-trained on ImageNet. To compensate for pre-training on landmarks, we extended the number of training epochs on the WAVL dataset to 14 (instead of 5 as reported in [25]). The authors propose an automated data cleaning process to remove wrong annotations. The cleaning process involves a three-step approach that uses spatial verification to filter images by k-NN search. The authors use RANSAC [6] with affine transformation and deep local attentive features (DELF) [12] for spatial verification. If the count of verified images reaches a certain threshold, the image is added to the cleaned dataset. We use this approach once without (labelled smlyaka) and once with the data cleaning process (labelled smlyaka with data cleaning).

In evaluating our approaches, we employed a comprehensive suite of evaluation metrics. In addition to precision and recall we use balanced accuracy (BA), the mean of true positive and true negative rate [2] and symmetric balanced accuracy (SBA) [7], which aims to eliminate bias by the choice of the positive class. These metrics are related to a specific recognition threshold. In line with

---

[2] https://github.com/lyakaap/Landmark2019-1st-and-3rd-Place-Solution.

**Table 3.** Evaluation results for the RAI-MI and WAVL dataset.

| Method | GAP | Threshold | Precision | Recall | BA | SBA |
|---|---|---|---|---|---|---|
| API-Net | 54.67 | 7.5 | 86.80 | 41.53 | **63.67** | **63.41** |
| Swin+API-Net | **56.09** | 7.5 | 84.88 | **43.54** | 63.39 | 63.05 |
| CDVA+API-Net | 44.50 | 7.5 | **88.26** | 27.43 | 59.69 | 61.02 |

(a) Results on the RAI-MI dataset (best score in bold).

| Method | GAP | Threshold | Precision | Recall | BA | SBA |
|---|---|---|---|---|---|---|
| smlyaka [25] | 47.34 | 2.0 | 53.56 | **66.31** | 54.51 | 54.64 |
| smlyaka with data cleaning. [25] | 47.59 | 2.5 | 54.08 | 57.22 | 54.31 | 54.32 |
| API-Net | 37.43 | 9.0 | 56.27 | 32.22 | 53.59 | 53.99 |
| Swin+API-Net | **53.86** | 6.5 | 69.63 | 62.06 | 68.76 | 68.96 |
| CDVA+API-Net | 53.03 | 6.0 | **79.77** | 52.57 | **70.61** | **72.43** |

(b) Results on the WAVL dataset (best score in bold).

the Google Landmarks benchmark [23], we also use Global Average Precision (GAP, originally proposed as micro-AP in [14]) as a metric. GAP differs from the more commonly used mean average precision (MAP) in that mean of precision at rank for all relevant returned results is determined, not taking the number of ground truth positives into account. Evaluation data are selected from videos which are not used for training data.

Table 3 lists the results of the two datasets. The values for precision, recall, BA and SBA are given for a specific threshold, which is determined with respect to a maximum value of BA. The variation of these metrics as a function of the threshold value can be seen in Figs. 4 and 5 for methods Swin+API-Net and smlyaka with data cleaning, respectively.

Filtering out training images using image region classification (Swin+API-Net) has very different effects on the two datasets. In the RAI-MI dataset, the filtered images sometimes contain information that adds value to landmark recognition. These could be, for example, interior views of buildings or studio backgrounds where a newsreader is visible. On the other hand, selecting images by image region classification from the WAVL dataset significantly improves recognition results. In any case, filtering out images reduces the risk of learning unrelated information.

Using web images for training does not provide good results for the RAI-MI dataset. As can be seen in Table 2, one reason is that the web search for this dataset did not yield a sufficient number of usable images for all landmarks. For three of the 163 landmarks, the annotation is incorrect, which means that no landmark related images could be found on the web. Nevertheless, this approach achieves high precision values. We believe that this is due to the similarity constraint in the selection of web images, which will reduce the risk of false positives but also harm the diversity of samples. Another aspect is that in both datasets there are a relatively large number of images for each landmark. If this were

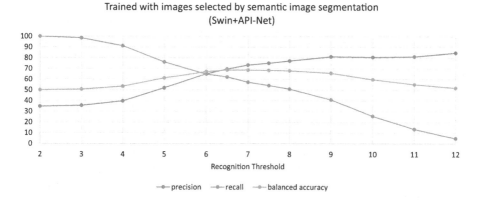

**Fig. 4.** Threshold dependent results on the WAVL dataset for the approach Swin+API-Net.

not the case, the web-based approach would have the advantage that sufficient images for such landmarks could still be found on the web for training.

The comparison with the winning method of the Google Landmark Retrieval 2019 Challenge (smlyaka [25]) shows that while it performs best in terms of recall, it is otherwise outperformed by one of the proposed approaches on the WAVL dataset. It is particularly interesting that the data cleaning proposed in this approach does not provide any improvement under these conditions. This is because the frequency of similar images is not a good criterion for selecting training images in this use case.

For the purpose of mining images for 3D reconstruction the fact that the proposed approaches have higher precision than other methods is beneficial, as it improves the robustness of the reconstruction process, in particular for methods such as NERF, that do not include a feature matching and filtering step.

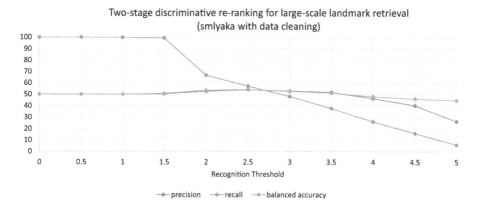

**Fig. 5.** Threshold dependent results on the WAVL dataset for the approach smlyaka with data cleaning.

# 6    Conclusion

XR content creators need to mine existing archive content for landmark recon-struction. In order to address this problem, we have proposed two approaches for training landmark classifiers on weakly annotated video data. The training pipeline assumes weakly labelled input data, which represents the real world scenario of content available from audiovisual media archives. For this purpose we published the Weakly Annotated Video Landmarks (WAVL) dataset, con-structed from the Google Landmarks v2 dataset and V3C1 as noise content. We show that the two proposed methods (either using semantic segmentation for pre-selection or mining web images) perform well, outperforming a state-of-the-art approach proposed for the Google Landmarks V2 dataset on our noisy version of it. A possible direction for future work is the integration of a re-ranking method as a post-processing step. We would also like to investigate approaches for incremental training of new landmarks.

**Acknowledgements.** The authors would like to thank Stefanie Onsori-Wechtitsch for providing the segmenter implementation.

The research leading to these results has been funded partially by the European Union's Horizon 2020 research and innovation programme, under grant agreement n° 951911 AI4Media (https://ai4media.eu), and under Horizon Europe under grant agree-ment n° 101070250 XRECO (https://xreco.eu/).

# References

1. Boiarov, A., Tyantov, E.: Large scale landmark recognition via deep metric learn-ing. In: Proceedings of the 28th ACM International Conference on Information and Knowledge Management, pp. 169–178 (2019)
2. Brodersen, K.H., Ong, C.S., Stephan, K.E., Buhmann, J.M.: The balanced accu-racy and its posterior distribution. In: 2010 20th International Conference on Pat-tern Recognition, pp. 3121–3124. IEEE (2010)
3. Caimotti, E., Montagnuolo, M., Messina, A.: An efficient visual search engine for cultural broadcast archives. In: AI* CH@ AI* IA, pp. 1–8 (2017)
4. Chen, X., Gupta, A.: Webly supervised learning of convolutional networks. In: Proceedings of the IEEE International Conference on Computer Vision (ICCV), December 2015
5. Duan, L.Y., et al.: Compact descriptors for video analysis: the emerging mpeg standard. IEEE Multimedia **26**(2), 44–54 (2019). https://doi.org/10.1109/MMUL.2018.2873844
6. Fischler, M.A., Bolles, R.C.: Random sample consensus: a paradigm for model fitting with applications to image analysis and automated cartography. Commun. ACM **24**(6), 381–395 (1981). https://doi.org/10.1145/358669.358692

7. Gösgens, M., Zhiyanov, A., Tikhonov, A., Prokhorenkova, L.: Good classification measures and how to find them. Adv. Neural. Inf. Process. Syst. **34**, 17136–17147 (2021)
8. Li, K., et al.: Learning from weakly-labeled web videos via exploring sub-concepts. CoRR abs/2101.03713 (2021). https://arxiv.org/abs/2101.03713
9. Liu, Z., et al.: Swin transformer: hierarchical vision transformer using shifted windows. In: Proceedings of the IEEE/CVF International Conference on Computer Vision, pp. 10012–10022 (2021)
10. Mezuman, E., Weiss, Y.: Learning about canonical views from internet image collections. In: Pereira, F., Burges, C., Bottou, L., Weinberger, K. (eds.) Advances in Neural Information Processing Systems, vol. 25. Curran Associates, Inc. (2012)
11. Mildenhall, B., Srinivasan, P.P., Tancik, M., Barron, J.T., Ramamoorthi, R., Ng, R.: NeRF: representing scenes as neural radiance fields for view synthesis. In: Vedaldi, A., Bischof, H., Brox, T., Frahm, J.-M. (eds.) ECCV 2020. LNCS, vol. 12346, pp. 405–421. Springer, Cham (2020). https://doi.org/10.1007/978-3-030-58452-8_24
12. Noh, H., Araujo, A., Sim, J., Weyand, T., Han, B.: Large-scale image retrieval with attentive deep local features. In: Proceedings of the IEEE International Conference on Computer Vision, pp. 3456–3465 (2017)
13. Perd'och, M., Chum, O., Matas, J.: Efficient representation of local geometry for large scale object retrieval. In: 2009 IEEE Conference on Computer Vision and Pattern Recognition, pp. 9–16 (2009). https://doi.org/10.1109/CVPR.2009.5206529
14. Perronnin, F., Liu, Y., Renders, J.M.: A family of contextual measures of similarity between distributions with application to image retrieval. In: 2009 IEEE Conference on Computer Vision and Pattern Recognition, pp. 2358–2365. IEEE (2009)
15. Philbin, J., Chum, O., Isard, M., Sivic, J., Zisserman, A.: Object retrieval with large vocabularies and fast spatial matching. In: 2007 IEEE Conference on Computer Vision and Pattern Recognition, pp. 1–8. IEEE (2007)
16. Philbin, J., Chum, O., Isard, M., Sivic, J., Zisserman, A.: Lost in quantization: improving particular object retrieval in large scale image databases. In: 2008 IEEE Conference on Computer Vision and Pattern Recognition, pp. 1–8. IEEE (2008)
17. Razali, M.N.B., Tony, E.O.N., Ibrahim, A.A.A., Hanapi, R., Iswandono, Z.: Landmark recognition model for smart tourism using lightweight deep learning and linear discriminant analysis. Int. J. Adv. Comput. Sci. Appl. (2023). https://api.semanticscholar.org/CorpusID:257386803
18. Rossetto, L., Schuldt, H., Awad, G., Butt, A.A.: V3C – a research video collection. In: Kompatsiaris, I., Huet, B., Mezaris, V., Gurrin, C., Cheng, W.-H., Vrochidis, S. (eds.) MMM 2019. LNCS, vol. 11295, pp. 349–360. Springer, Cham (2019). https://doi.org/10.1007/978-3-030-05710-7_29
19. Song, H., Kim, M., Park, D., Shin, Y., Lee, J.G.: Learning from noisy labels with deep neural networks: a survey. IEEE Trans. Neural Netw. Learn. Syst. **34**, 8135–8153 (2022)
20. Tan, M., Le, Q.: EfficientNet: rethinking model scaling for convolutional neural networks. In: International Conference on Machine Learning, pp. 6105–6114. PMLR (2019)
21. Torii, A., Arandjelović, R., Sivic, J., Okutomi, M., Pajdla, T.: 24/7 place recognition by view synthesis. In: CVPR (2015)

22. Torii, A., Sivic, J., Pajdla, T., Okutomi, M.: Visual place recognition with repetitive structures. In: Proceedings / CVPR, IEEE Computer Society Conference on Computer Vision and Pattern Recognition, June 2013, vol. 37, pp. 883–890. IEEE Computer Society Conference on Computer Vision and Pattern Recognition (2013). https://doi.org/10.1109/CVPR.2013.119

23. Weyand, T., Araujo, A., Cao, B., Sim, J.: Google landmarks dataset v2-a large-scale benchmark for instance-level recognition and retrieval. In: Proceedings of the IEEE/CVF Conference on Computer Vision and Pattern Recognition, pp. 2575–2584 (2020)

24. Yang, M., Cui, C., Xue, X., Ren, H., Wei, K.: 2nd place solution to google landmark retrieval 2020 (2022)

25. Yokoo, S., Ozaki, K., Simo-Serra, E., Iizuka, S.: Two-stage discriminative re-ranking for large-scale landmark retrieval. In: Proceedings of the IEEE/CVF Conference on Computer Vision and Pattern Recognition Workshops, pp. 1012–1013 (2020)

26. Yu, F.: BDD100K: a large-scale diverse driving video database. BAIR (Berkeley Artificial Intelligence Research) (2018). https://bair.berkeley.edu/blog/2018/05/30/bdd

27. Zhuang, P., Wang, Y., Qiao, Y.: Learning attentive pairwise interaction for fine-grained classification. CoRR abs/2002.10191 (2020). https://arxiv.org/abs/2002.10191

# A Framework for 3D Modeling of Construction Sites Using Aerial Imagery and Semantic NeRFs

Panagiotis Vrachnos$^{(\boxtimes)}$, Marios Krestenitis, Ilias Koulalis,
Konstantinos Ioannidis, and Stefanos Vrochidis

Information Technologies Institute - CERTH, Thessaloniki, Greece
{panosvrachnos,stefanos}@iti.gr

**Abstract.** The rapid evolution of drone technology has revolutionized data acquisition in the construction industry, offering a cost-effective and efficient method to monitor and map engineering structures. However, a significant challenge remains in transforming the drone-collected data into semantically meaningful 3D models. 3D reconstruction techniques usually lead to raw point clouds that are typically unstructured and lack the semantic and geometric information of objects needed for civil engineering tools. Our solution applies semantic segmentation algorithms to the data produced by NeRF (Neural Radiance Fields), effectively transforming drone-captured 3D volumetric representations into semantically rich 3D models. This approach offers a cost-effective and automated way to digitalize physical objects of construction sites into semantically annotated digital counterparts facilitating the development of digital twins or XR applications in the construction sector.

**Keywords:** Neural Radiance Fields · Digital Twins · Drone Technology · 3D Modeling · Digitalization in Construction

## 1 Introduction

In the construction industry, efficient management is crucial and relies heavily on precise 3D representations of project sites. Traditional methods of acquiring data for these representations often involve expensive hardware and time-consuming tasks, a gap filled by the high-resolution data provided by drone technology and the emerging capabilities of computer vision. Drones have been used in various industries, including agriculture, public safety, and military purposes, and have recently been introduced to the construction industry [2]. They provide invaluable help and cost savings with wide views of inaccessible and otherwise difficult and tough to navigate locations, facilitating aerial surveying of buildings, bridges, roads, and highways, and ultimately saving project time and costs [2,19,23]. Digital twins and XR applications can benefit from computer vision to enable construction progress monitoring, safety inspection, and quality management [12,15]. However, challenges exist in the analysis process.

S. Rudinac et al. (Eds.): MMM 2024, LNCS 14557, pp. 175–187, 2024.
https://doi.org/10.1007/978-3-031-53302-0_13

Transforming drone-collected visual data into semantically meaningful 3D models remains a challenge. Recent implicit neural reconstruction techniques, such as Neural Radiance Fields (NeRF) [21], are appealing as they rapidly generate 3D representations, but these are not directly usable for civil engineering tools, digital twins, or XR applications as they do not include semantic labels.

Semantic scene understanding involves attaching class labels to a geometric model. The tasks of estimating the geometric model from images can benefit from the smoothness, coherence, and self-similarity learned by self-supervision methods [38]. Scene consistency is inherent to the training process and enables the network to produce accurate geometry of the scene. However, unlike scene geometry, semantic classes are human-defined concepts, and it is not possible to semantically label a novel scene in a purely self-supervised manner. Moreover, the majority of existing works pursue offline, passive analysis, in which scene understanding is conducted over already acquired sequences of data [37]. Regrettably, this approach often significantly restricts the effectiveness of scene understanding, as data collection is separate from the corresponding scene understanding algorithms. Moreover, the absence of extensive labeled 3D training datasets, like those available for 2D images (ImageNet, COCO, etc.), prevents 3D methods from benefiting from pre-training on large-scale 2D datasets.

In this paper, we propose an innovative application that leverages NeRF for generating 3D representations from drone data. Specifically, we develop an end-to-end pipeline that combines the benefits of UAV-based data capture, semantic segmentation, and NeRF's continuous scene representation capabilities to transform raw drone footage into semantically-rich 3D models of construction sites. Our key technical contributions include an automated UAV footage collection methodology tailored for construction sites that generates images from multiple viewpoints to enable accurate 3D reconstruction. A data preprocessing stage extracts frames, annotates them using semantic segmentation to identify construction objects, and estimates camera poses. We train a custom Semantic-NeRF model that takes the UAV images, semantic labels of classes of interest, and poses as input to jointly learn the scene's appearance, geometry, and semantics. Marching cubes algorithm is used to extract the final 3D model labeled with semantic classes from the trained model. Compared to existing techniques that produce unstructured point clouds lacking semantics, our approach generates textured 3D models, addressing a key gap in digitizing as-built construction sites. We demonstrate our pipeline on a real highway bridge construction project. The ability to automatically create an as-built 3D model from only drone video footage can enable powerful digital twinning and immersive applications for construction industry.

## 2    Background: The Intersection of NeRF, Semantic Segmentation, and UAV Technology

The integration of advanced computational techniques with UAV technology has led to significant advancements in 3D modeling and monitoring of construction projects. This section provides a comprehensive review of three key components of our proposed end-to-end pipeline for 3D construction site modeling: Neural Radiance Fields (NeRF), Semantic Analysis, and UAV Technology. We explore the foundational knowledge and related works in each area, highlighting their intersection and potential for enhancing construction site modeling.

In recent years, unmanned aerial vehicles (UAVs), also known as drones, have revolutionized data acquisition in various fields, such as agriculture [30], disaster management [1], environmental monitoring [10], extending even to military applications [32] and construction project monitoring [3,18]. Over the past few years, the construction sector has notably evolved, driven by technological advancements. The integration of UAVs as a pioneering technological advancement, holds the promise to boost various construction activities. These enhancements include tasks ranging from observation and inspection to the monitoring of safety practices, resulting in significant savings in terms of time, expenditures, and the mitigation of injuries. Furthermore, it augments the overall quality of work performed. UAVs have transformed the project planning, execution and maintenance in the construction industry. Their agility and accessibility to capture high-resolution images and navigate challenging surfaces makes them ideal platforms for collecting visual data. Several studies have explored the use of UAVs for construction site monitoring and 3D modeling. For instance, Anwar et al. [2] utilized UAVs to capture aerial images for monitoring and progress assessment of construction projects. By leveraging UAV technology, construction tasks can benefit from efficient and cost-effective data collection, enabling real-time monitoring and analysis, 3D modeling and remote inspections with remarkable precision.

Creating realistic renderings of real-world scenes from various viewpoints using traditional computer vision techniques is a complex task, as it involves capturing detailed models of appearance and geometry. However, recent research has shown encouraging progress by employing learning-based approaches that can implicitly encode both geometry and appearance without relying on explicit 3D supervision. These studies have demonstrated promising outcomes in generating photo-realistic renderings with the freedom to view scenes from different angles. Neural Radiance Fields (NeRF) [21] have emerged as a powerful technique for scene reconstruction and rendering, revolutionizing the field of computer vision. Particularly it is a state-of-the-art technique for scene reconstruction that models volumetric scenes as continuous functions, enabling the synthesis of novel views and realistic rendering [34]. In simple terms, NeRF demonstrates its capability to reconstruct complex 3D scenes from 2D images. Since its inception, and given its results, it has witnessed significant progress and inspired numerous extensions and variations. For instance, Zhang et al. [35] introduced improvements to the original NeRF formulation, enhancing its efficiency and robustness. Martin-Brualla et al. [20] extended NeRF to handle unstructured and diverse real-world scenes, enabling reconstruction from internet photo collections. Also, Barron et al. [4] in their approach mitigated undesirable aliasing artifacts and greatly enhanced the NeRF model's capacity to capture intricate details.

Semantic segmentation [22], a fundamental task in computer vision and machine learning, plays a crucial role in understanding and extracting meaningful information from images by classifying each pixel into specific semantic categories. Before the rise of deep learning methods, traditional approaches heavily relied on handcrafted feature extraction and classification design such as K-means algorithm, support vector machines, and decision trees. In recent years, deep neural networks have achieved great performances in semantic segmentation tasks [7,31,36] which provide valuable insights for incorporating contextual information into the 3D modeling process. Semantically annotated point clouds, which constitute a crucial geometric data structure, have attracted numerous researchers. In contrast to two-dimensional data, their sparsity and randomness make them full of challenges. Some of the most common challenges are the scale of the data and the lack of clear structures. However, Landrieu et al. [17] proposed a deep learning-based method for semantic segmentation in large scale point clouds. Also, Hu et al. [13] designed an efficient neural architecture to directly infer per-point semantics for large-scale point clouds, in a single pass, significantly reducing computational cost and memory requirements. Furthermore, Qi et al. [24] designed a deep network architecture that directly consumes point clouds for semantic segmentation in order to minimize data corruption which may be caused by transforming data prior feed them to the network, as traditional methods apply. Later, as an extension of the previous work, a hierarchical neural network architecture was designed [25] to efficiently capture both local and global features from point clouds, enhancing the accuracy of point-wise semantic segmentation.

Integration of semantic segmentation with the Neural Radiance Fields (NeRF) has gained attention due to its potential to enhance scene reconstruction and rendering. Several studies have explored the combination of these two techniques, highlighting their complementary nature and the benefits they offer. One line of research has focused on leveraging semantic segmentation to improve the quality and accuracy of 3D reconstructions with NeRFs. For instance, Fu et al. [11] unified coarse 3D annotations with 2D semantic cues transferred from existing datasets, utilizing NeRF. With this combination both semantic and geometry are optimized. However, this method requires a priori knowledge of 3D objects and their boundary boxes for improvement. By incorporating semantic information, the aim is to guide the reconstruction process and achieve more meaningful and semantically consistent 3D models [14]. Towards this direction, Zhi et al. [38] enhanced NeRF's robust 3D representations with sparse or noisy annotations to produce a scene-specific 3D semantic representation. This work had the biggest influence on the development of our proposal pipeline. In the same direction, Kundu et al. [16] preprocessed 2D images to extract 2D semantic segmentations and camera parameters. Additionally he utilized jointly a set of 3D oriented bounding boxes and category labels to optimize the extracted results. Another direction in the literature has explored the use of NeRF for refining and enhancing semantic segmentation results. By leveraging the continuous scene representation offered by NeRF, researchers strove to improve the precision and robustness of semantic segmentation [33]. In the context of construction site modeling [5], semantic segmentation enables the identification and segmentation of various objects and structures.

## 3 Methodology: Building a Pipeline for Semantic NeRF Analysis of UAV Footage

In this section the deployed end-to-end framework for 3D modeling of construction sites is presented. As illustrated in Fig. 1, the proposed pipeline consists of three distinct stages, each of which will be thoroughly explained in the subsequent sections.

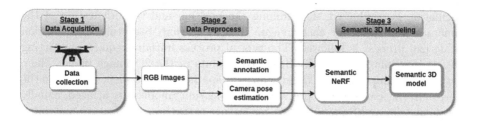

**Fig. 1.** Illustration of the proposed pipeline for 3D modeling of a construction site. The process includes three main stages: UAV data acquisition (Stage 1), data preprocessing to create a scene-specific dataset (Stage 2) and reconstruction of the semantic 3D model via Semantic-NeRF (Stage 3).

### 3.1 UAV Technology and Data Acquisition

In the first stage, data acquisition in the designated target area is conducted. For this purpose, a UAV equipped with a high resolution camera is utilized to capture footage from the scene. UAV utilization enables a seamless process of data collection, as it can access otherwise inaccessible areas and capture the target region from multiple perspectives, while minimizing human effort. The main objective is to create a meaningful dataset, containing images from varying viewing angles of the examined scene, that will allow accurate 3D reconstruction. To this end, the UAV path is carefully designed to scan the whole area and capture the existing essential instances in detail, from diverse distances and heights relative to the target. The UAV speed is also adjusted accordingly, aiming to minimize any distortion or blurring effects and receive high resolution visual data. In this manner, sufficient overlap of imagery is ensured, resulting in a dataset that is usable for 3D reconstruction.

## 3.2 Data Preprocessing

In Stage 2, data preprocessing is performed to prepare the required dataset for the semantic 3D reconstruction, containing RGB images, the corresponding semantic labels and the estimated camera poses. The collected footage is processed accordingly through 3 steps. At first, RGB frames from the captured footage are extracted. Sampling frequency for frame extraction is defined accordingly to ensure the required amount of overlap among the extracted consecutive frames. Next, the labeling of the RGB images is conducted in an automated manner, aiming to avoid the time-consuming and laborious manual labeling. To achieve that, a deep learning-based method that semantically annotates construction-related classes is deployed. In specific, the robust DeepLabv3+ [8] architecture pre-trained on ADE20K [39] dataset is employed, enabling the accurate pixel-wise annotation of the RGB images. Following this, camera poses calculation is conducted, determining the position and the orientation of the camera in 3D space for each image. To achieve that, Structure from Motion (SfM) [28] process is followed. The typical process initially requires extracting features per image to identify distinctive points in the 2D image plane. Then, feature matching is applied, to recognize the spatial correspondences between the images. In this context, two pretrained deep-learning models are utilized, namely SuperPoint [9] and SuperGlue [27] for feature extraction and feature matching, respectively. The employment of these models provides accurate results and a balanced trade-off with regard to computational time. Next, the position of the acquired features in 3D space is calculated, following the triangulation method. Last, camera poses iteratively estimated through an optimization process, guaranteeing the alignment between the observed 2D features and their corresponding 3D position. For the aforementioned process, the pinhole camera model is adapted.

## 3.3 Semantic NeRF Implementation

The final stage of the presented pipeline involves the utilization of the deployed dataset in order to train a Semantic-NeRF instance and acquire a 3D reconstruction of the scene. Contrary to NeRF-based models, Semantic-NeRF is capable of generating 3D representations enriched with semantic information from the input data.

In specific, the custom-built dataset contains the UAV captured RGB images, the corresponding semantic labels and the camera poses. This dataset is utilized to train the deployed Semantic-NeRF from scratch. Through the training process, the model learns to predict the radiance (color and density) and the semantic class for a given point and view direction in the 3D space. Upon completion of training, the 3D representation of the scene can be extracted. To this end, 3D grids of the scene are passed to the trained model to acquire the network predictions. Next, the marching cubes algorithm is applied to extract the geometric meshes. The final output is a detailed 3D model of the construction scene, where depicted instances are semantically annotated. This semantic 3D reconstruction provides a comprehensive overview of the examined area, enabling its exploitation for various construction-related operations and applications.

**Fig. 2.** General view of the examined highway overpass construction site [6].

## 4    Experimental Configurations and Results

In this section, the deployment of the proposed end-to-end pipeline to a specific construction site is presented. More specifically, the developed framework was tested over a highway overpass located in the Metropolitan Area of Barcelona (see Fig. 2), aiming to create the semantic 3D representation of one of the bridge pillars. The overall length of the bridge is estimated at approximately 846 meters, connecting two of the main highways of the city's network, crossing over rivers, roads and railway lines. Given this information, using hand-held LiDAR devices to obtain 3D representations would significantly escalate both the cost and the time required for scanning, presenting substantial challenge in terms of resources and efficiency. In our case, acquiring the 3D model of each pillar in an automated way is considered of crucial importance to enable efficient monitoring of the infrastructure.

Based on the developed framework, data collection was conducted as outlined in Stage 1, with the use of a DJI Mavic Pro UAV model. Its trajectory was designed accordingly to cover the target object from varying viewing points and distances. Toward this direction, the examined bridge pillar was captured during the UAV mission, leading to a high-resolution video of the scene. An illustration of the executed UAV path is presented in Fig. 5b. The acquired footage had a duration of 4 min and 50 s (4:50 min), with a resolution of $3840 \times 2160$ pixels and recorded at a frame rate of 30 frames per second (fps). In order to compose the required dataset to build the semantic 3D model, data preprocessing was conducted according to Stage 2 of the proposed framework. In this context, after experimentation, we determined that extracting 200 frames evenly distributed throughout the entire captured video, was the optimal choice to achieve a satisfactory balance among frames overlap and dataset size. Samples of the extracted frames can be seen in Fig. 3, which depict the bridge pillar from different viewing angles.

**Fig. 3.** UAV image samples from the cumstom-built dataset, illustrating the examined construction from different viewing angles.

**Fig. 4.** Results of the automatic semantic annotation of collected RGB images. With light red color is illustrated the detected pillar instance, while background is colored with black. (Color figure online)

Next, to feed the semantic-NeRF model with 2D semantic labels and end up with a 3D annotated model, pixel labeling of each extracted image was automatically performed using a segmentation model. More specifically, the 200 RGB images were fed to the deployed DeepLabv3+ model, classifying pixels belonging to pillar as "column" and all the other pixels as "background". Through this process, we acquired the semantic pixel-level labels of the collected data. Results of the labeling process are demonstrated in Fig. 4.

In the final step of Stage 2, camera poses were estimated for the extracted 200 RGB images following the aforementioned process. To achieve that, we utilized Hierarchical Localization (HLOC) [26] toolbox exploiting the SuperPoint and SuperGlue pretrained models for feature extraction and feature matching, respectively. For the triangulation process based on the acquired features and the iterative optimization, we employed COLMAP [28] tool, leading to the estimated intrinsic camera parameters and the camera poses for the collected images. In Fig. 5, the estimated camera poses in 3D space are demonstrated.

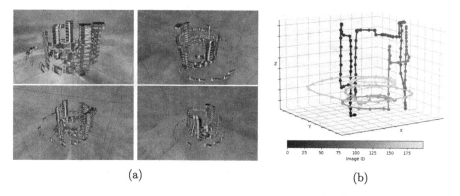

(a)                                                                          (b)

**Fig. 5.** Results of the camera poses estimation for the custom-built dataset. (a) different viewing angles of the estimated camera poses visualized via nerfstudio [29]. Each image's estimated capture location is represented in 3D space. (b) UAV trajectory based on estimated capture positions

Once the data preprocessing of Stage 2 was completed, the derived custom-built dataset was utilized to train the Semantic-NeRF model for the specific scene. To this end, the 200 aerial images among with the generated semantic labels and the camera poses were fed to the deployed Semantic-NeRF model. Following the configuration in [38], the training images were resized to $320 \times 240$. The model was trained for $200k$ steps, with Adam optimizer and learning rate equal to $5e^{-4}$. The overall process was conducted in a GeForce RTX 3090 GPU with 24GB memory.

The semantic 3D reconstruction of the pillar was acquired from the trained model, based on the methodology presented in Sect. 3.3. Figure 6 presents different viewing angles of the acquired 3D representation. One can notice that the examined pillar instance is efficiently reconstructed and clearly distinguishable from the background class. Details regarding the shape and the surface recesses of the captured pillar are highly visible in the extracted 3D reconstruction. Some noisy patterns are also depicted in the background area. This effect is generated due to the nature of the background, which contains several instances in various scales and sizes that are not captured as detailed as the pillar instance during the UAV mission. These elements are not considered as of equal importance, and thus, are not adequately reconstructed. Overall, the acquired 3D representation is a valuable asset, enclosing in detail the geometry of the scene and containing high-level semantic information that can be valuable for downstream construction monitoring tasks.

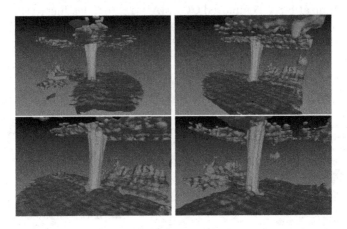

**Fig. 6.** Different viewing angles of the semantic 3D representation of the examined construction area. With orange color is depicted the pillar instance, while background information is colored with green. (Color figure online)

## 5    Conclusion

In this paper, we propose an innovative framework that combines the economy and speed of UAV imagery acquisition with the capabilities of semantic NeRF to facilitate the creation of detailed 3D models of construction sites.

Our experiments, which focused particularly on a highway bridge construction site, highlighted the possibility of autonomously creating 3D models with texture and semantic classifications directly from raw UAV visual data. This not only serves as an important step towards creating valuable digital twin representations of the as-built condition, enhancing various monitoring and construction management applications, but also marks a shift away from the limitations of traditional point clouds, adding semantic depth to reconstructed objects.

Despite the promising results, the approach has a few notable limitations. The need for data preprocessing introduces an additional step before the actual analysis can begin. Also, the data labeling process is not entirely foolproof, posing a risk of potential mislabels that might compromise the results. Another constraint, inherent to the case-specific training requirements of semantic NeRFs, is the necessity to deploy the entire training process for each individual scene. These challenges highlight critical areas requiring further improvement and development to enhance both the efficiency and reliability of the system in future iterations. Furthermore, the paper acknowledges the need for further validation to establish the quantitative accuracy of the generated models. In our work, we mainly focused on demonstrating the integration of various technologies into an end-to-end framework for real-case scenarios in the construction industry. Looking ahead, we aim to broaden the scope of our research by incorporating a more diverse range of semantic classes relevant to construction, further optimizing each stage of the pipeline, and extending our experiments to a wider variety of construction sites, including comprehensive evaluation metrics and comparative analyses. This, we believe, will make a significant contribution to the growing field of digitalization of construction applications, offering a low-cost and automated approach to transferring as-built conditions to digital twins or XR applications.

**Acknowledgments.** This work was supported by the EC-funded research and innovation programme H2020 ASHVIN: "Digitising and transforming the European construction industry" under the grant agreement No.958161.
**EU disclaimer:** This publication reflects only author's view and the European Commission is not responsible for any uses that may be made of the information it contains.

# References

1. Adams, S.M., Friedland, C.J.: A survey of unmanned aerial vehicle (UAV) usage for imagery collection in disaster research and management. In: 9th International Workshop on Remote Sensing for Disaster Response, vol. 8, pp. 1–8 (2011)
2. Anwar, N., Izhar, M.A., Najam, F.A.: Construction monitoring and reporting using drones and unmanned aerial vehicles (UAVs). In: The Tenth International Conference on Construction in the 21st Century (CITC-2010), pp. 2–4 (2018)
3. Ashour, R., et al.: Site inspection drone: a solution for inspecting and regulating construction sites. In: 2016 IEEE 59th International Midwest Symposium on Circuits and Systems (MWSCAS), pp. 1–4. IEEE (2016)
4. Barron, J.T., et al.: MIP-NERF: a multiscale representation for anti-aliasing neural radiance fields. In: Proceedings of the IEEE/CVF International Conference on Computer Vision, pp. 5855–5864 (2021)
5. Chacón, R., et al.: On the digital twinning of load tests in railway bridges. Case study: high speed railway network, extremadura, Spain. In: Bridge Safety, Maintenance, Management, Life-Cycle, Resilience and Sustainability, pp. 819–827. CRC Press (2022)
6. Chacón, R., Ramonell, C., Puig-Polo, C., Mirambell, E.: Geometrical digital twinning of a tapered, horizontally curved composite box girder bridge. ce/papers **5**(4), 52–58 (2022)

7. Chen, L.C., Papandreou, G., Kokkinos, I., Murphy, K., Yuille, A.L.: Deeplab: semantic image segmentation with deep convolutional nets, atrous convolution, and fully connected crfs. IEEE Trans. Pattern Anal. Mach. Intell. **40**(4), 834–848 (2017)

8. Chen, L.C., Zhu, Y., Papandreou, G., Schroff, F., Adam, H.: Encoder-decoder with atrous separable convolution for semantic image segmentation. In: Proceedings of the European Conference on Computer Vision (ECCV), pp. 801–818 (2018)

9. DeTone, D., Malisiewicz, T., Rabinovich, A.: Superpoint: self-supervised interest point detection and description. In: Proceedings of the IEEE Conference on Computer Vision and Pattern Recognition Workshops, pp. 224–236 (2018)

10. Ezequiel, C.A.F., et al.: UAV aerial imaging applications for post-disaster assessment, environmental management and infrastructure development. In: 2014 International Conference on Unmanned Aircraft Systems (ICUAS), pp. 274–283. IEEE (2014)

11. Fu, X., et al.: Panoptic nerf: 3d-to-2d label transfer for panoptic urban scene segmentation. In: 2022 International Conference on 3D Vision (3DV), pp. 1–11. IEEE (2022)

12. Han, K.K., Golparvar-Fard, M.: Potential of big visual data and building information modeling for construction performance analytics: an exploratory study. Autom. Constr. **73**, 184–198 (2017)

13. Hu, Q., et al.: Randla-net: efficient semantic segmentation of large-scale point clouds. In: Proceedings of the IEEE/CVF Conference on Computer Vision and Pattern Recognition, pp. 11108–11117 (2020)

14. Huang, H.-P., Tseng, H.-Y., Lee, H.-Y., Huang, J.-B.: Semantic view synthesis. In: Vedaldi, A., Bischof, H., Brox, T., Frahm, J.-M. (eds.) ECCV 2020. LNCS, vol. 12357, pp. 592–608. Springer, Cham (2020). https://doi.org/10.1007/978-3-030-58610-2_35

15. Koulalis, I., Dourvas, N., Triantafyllidis, T., Ioannidis, K., Vrochidis, S., Kompatsiaris, I.: A survey for image based methods in construction: from images to digital twins. In: Proceedings of the 19th International Conference on Content-based Multimedia Indexing, pp. 103–110 (2022)

16. Kundu, A., et al.: Panoptic neural fields: a semantic object-aware neural scene representation. In: Proceedings of the IEEE/CVF Conference on Computer Vision and Pattern Recognition, pp. 12871–12881 (2022)

17. Landrieu, L., Simonovsky, M.: Large-scale point cloud semantic segmentation with superpoint graphs. In: Proceedings of the IEEE Conference on Computer Vision and Pattern Recognition, pp. 4558–4567 (2018)

18. Li, Y., Liu, C.: Applications of multirotor drone technologies in construction management. Int. J. Constr. Manag. **19**(5), 401–412 (2019)

19. Liu, P., et al.: A review of rotorcraft unmanned aerial vehicle (UAV) developments and applications in civil engineering. Smart Struct. Syst. **13**(6), 1065–1094 (2014)

20. Martin-Brualla, R., Radwan, N., Sajjadi, M.S., Barron, J.T., Dosovitskiy, A., Duckworth, D.: Nerf in the wild: neural radiance fields for unconstrained photo collections. In: Proceedings of the IEEE/CVF Conference on Computer Vision and Pattern Recognition, pp. 7210–7219 (2021)

21. Mildenhall, B., Srinivasan, P.P., Tancik, M., Barron, J.T., Ramamoorthi, R., Ng, R.: Nerf: representing scenes as neural radiance fields for view synthesis. Commun. ACM **65**(1), 99–106 (2021)

22. Mo, Y., Wu, Y., Yang, X., Liu, F., Liao, Y.: Review the state-of-the-art technologies of semantic segmentation based on deep learning. Neurocomputing **493**, 626–646 (2022)

23. Outay, F., Mengash, H.A., Adnan, M.: Applications of unmanned aerial vehicle (UAV) in road safety, traffic and highway infrastructure management: recent advances and challenges. Transport. Res. Part A: Policy Practice **141**, 116–129 (2020)
24. Qi, C.R., Su, H., Mo, K., Guibas, L.J.: Pointnet: deep learning on point sets for 3D classification and segmentation. In: Proceedings of the IEEE Conference on Computer Vision and Pattern Recognition, pp. 652–660 (2017)
25. Qi, C.R., Yi, L., Su, H., Guibas, L.J.: Pointnet++: deep hierarchical feature learning on point sets in a metric space. Adv. Neural Inf. Process. Syst. **30** (2017)
26. Sarlin, P.E., Cadena, C., Siegwart, R., Dymczyk, M.: From coarse to fine: robust hierarchical localization at large scale. In: Proceedings of the IEEE/CVF Conference on Computer Vision and Pattern Recognition, pp. 12716–12725 (2019)
27. Sarlin, P.E., DeTone, D., Malisiewicz, T., Rabinovich, A.: Superglue: learning feature matching with graph neural networks. In: Proceedings of the IEEE/CVF Conference on Computer Vision and Pattern Recognition, pp. 4938–4947 (2020)
28. Schonberger, J.L., Frahm, J.M.: Structure-from-motion revisited. In: Proceedings of the IEEE Conference on Computer Vision and Pattern Recognition, pp. 4104–4113 (2016)
29. Tancik, M., et al.: Nerfstudio: a modular framework for neural radiance field development. arXiv preprint arXiv:2302.04264 (2023)
30. Tsouros, D.C., Triantafyllou, A., Bibi, S., Sarigannidis, P.G.: Data acquisition and analysis methods in UAV-based applications for precision agriculture. In: 2019 15th International Conference on Distributed Computing in Sensor Systems (DCOSS), pp. 377–384. IEEE (2019)
31. Ulku, I., Akagündüz, E.: A survey on deep learning-based architectures for semantic segmentation on 2d images. Appl. Artif. Intell. **36**(1), 2032924 (2022)
32. Vacca, A., Onishi, H.: Drones: military weapons, surveillance or mapping tools for environmental monitoring? the need for legal framework is required. Transport. Res. Procedia **25**, 51–62 (2017)
33. Vora, S., et al.: Nesf: neural semantic fields for generalizable semantic segmentation of 3d scenes. arXiv preprint arXiv:2111.13260 (2021)
34. Wang, Z., Wu, S., Xie, W., Chen, M., Prisacariu, V.A.: Nerf-: neural radiance fields without known camera parameters. arXiv preprint arXiv:2102.07064 (2021)
35. Zhang, K., Riegler, G., Snavely, N., Koltun, V.: Nerf++: analyzing and improving neural radiance fields. arXiv preprint arXiv:2010.07492 (2020)
36. Zhao, H., Shi, J., Qi, X., Wang, X., Jia, J.: Pyramid scene parsing network. In: Proceedings of the IEEE Conference on Computer Vision and Pattern Recognition, pp. 2881–2890 (2017)
37. Zheng, L., et al.: Active scene understanding via online semantic reconstruction. In: Computer Graphics Forum, vol. 38, pp. 103–114. Wiley Online Library (2019)
38. Zhi, S., Laidlow, T., Leutenegger, S., Davison, A.J.: In-place scene labelling and understanding with implicit scene representation. In: Proceedings of the IEEE/CVF International Conference on Computer Vision, pp. 15838–15847 (2021)
39. Zhou, B., Zhao, H., Puig, X., Fidler, S., Barriuso, A., Torralba, A.: Scene parsing through ade20k dataset. In: Proceedings of the IEEE Conference on Computer Vision and Pattern Recognition, pp. 633–641 (2017)

# Multimodal 3D Object Retrieval

Maria Pegia[1,2]([✉])(iD), Björn Þór Jónsson[2](iD), Anastasia Moumtzidou[1](iD),
Sotiris Diplaris[1](iD), Ilias Gialampoukidis[1](iD), Stefanos Vrochidis[1](iD),
and Ioannis Kompatsiaris[1](iD)

[1] Centre for Research and Technology Hellas, Information Technologies Institute,
Thessaloniki, Greece
{mpegia,moumtzid,diplaris,heliasgj,stefanos,ikom}@iti.gr
[2] Reykjavik University, Reykjavík, Iceland
{mariap22,bjorn}@ru.is

**Abstract.** Three-dimensional (3D) retrieval of objects and models plays
a crucial role in many application areas, such as industrial design, medical
imaging, gaming and virtual and augmented reality. Such 3D retrieval
involves storing and retrieving different representations of single objects,
such as images, meshes or point clouds. Early approaches considered only
one such representation modality, but recently the CMCL method has
been proposed, which considers multimodal representations. Multimodal
retrieval, meanwhile, has recently seen significant interest in the image
retrieval domain. In this paper, we explore the application of state-of-
the-art multimodal image representations to 3D retrieval, in comparison
to existing 3D approaches. In a detailed study over two benchmark 3D
datasets, we show that the MuseHash approach from the image domain
outperforms other approaches, improving recall over the CMCL approach
by about 11% for unimodal retrieval and 9% for multimodal retrieval.

**Keywords:** 3D retrieval · Supervised learning · 3D data

## 1 Introduction

In recent years, advances in three-dimensional (3D) modeling tools [15], 3D scanning technology [9], and consumer devices with 3D sensors [2] have made it easier
to create, share, and access 3D large collections of content, influencing various
domains, from entertainment and gaming to healthcare [27], archaeology [1],
computer-aided design (CAD) [7] and autonomous systems [6]. A large number
of 3D models have now become available on the Web [16]. Users can freely download, modify and build upon 3D models that suit their requirements. This not
only saves costs and time in product design but also enhances product reliability and quality. Sifting through the vast number of available models to find the
right one quickly and accurately is challenging. This is where 3D model retrieval
techniques come into play. These techniques enable users to retrieve models in
various ways, for example based on model class or model similarity.

S. Rudinac et al. (Eds.): MMM 2024, LNCS 14557, pp. 188–201, 2024.
https://doi.org/10.1007/978-3-031-53302-0_14

A fundamental issue for efficient 3D model retrieval at scale is the data representation of the 3D models [5]. Figure 1 outlines the approaches considered in the literature: voxels (data points on a grid in the model space); point clouds (data points of interest in the model space); meshes (networks of triangles that approximate the shape); and multi-view images that visually represent the model. Of these, 3D mesh models stand out due to their capacity to capture intricate details and structural aspects. Recently, however, the CMCL method [10] combines the latter three representation modalities into a unified representation, leveraging the center information from each modality. This integration has resulted in improved retrieval accuracy and robustness. Nevertheless, combining modalities in 3D data representation poses challenges due to their inherent complexities.

Multimodal representation has also recently received attention in the domain of image retrieval [17,19]. While all images have a visual component, that can for example be described using semantic feature vectors, some collections may also have textual, temporal or spatial information. This information could be combined in various ways for more accurate retrieval, depending on the intended application. Early approaches considered cross-modal retrieval, often aiming to learn a unified feature space for the visual and textual modalities. More recently, Label-Attended Hashing (LAH) [30] and MuseHash [17] have considered the fusion of multiple modalities. It is therefore of interest to consider whether the approaches developed in the domain of image retrieval could be applicable in the related, yet significantly different, domain of 3D model retrieval.

The main contributions of this paper can be summarized as follows:

- We adapt state-of-the-art methods from the image retrieval domain to the 3D model retrieval domain.
- In comparison with the state-of-the-art 3D retrieval methods on two class-based retrieval benchmarks, the MuseHash approach generally performs best. It enhances unimodal recall by up to 11% and multimodal recall by 9%, surpassing the CMCL approach.
- Furthermore, we explore various combinations of the 3D mesh, point cloud, and visual (multi-view) modalities, showing that using all three modalities achieves the highest accuracy.

The remainder of this paper is organized as follows: Sect. 2 provides an overview of the relevant state-of-the-art research in the domains of 3D model and image retrieval. Section 3 then details how image retrieval methods are adapted to 3D model retrieval. Section 4 presents and analyzes the experimental results for both unimodal and multimodal retrieval, before concluding in Sect. 5.

## 2 Related Work

### 2.1 Unimodal Retrieval

Volumetric data representations (Fig. 1) have been crucial in 3D data analysis and retrieval, prompting the development of diverse techniques:

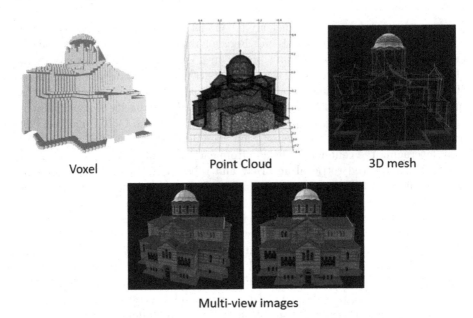

Voxel          Point Cloud          3D mesh

Multi-view images

**Fig. 1.** Examples of different 3D data representations.

**Voxels** Divide the 3D space into a grid, assigning values to voxels [11,23]. Common in medical design, but computationally demanding for large data. Maturana et al. [14] introduced a volumetric occupancy network called VoxNet to achieve robust 3D object recognition. Wang et al. [24] proposed an Octree-based CNN for 3D shape classification.

**Multi-view images** Capture 2D images for 3D reconstruction [13,21,22]. Useful for diverse viewpoints, but quality relies on captured views. Lin et al. [12] proposed two self-attention modules, View Attention Module and Instance Attention Module for building the representation of a 3D object as the aggregation of three features: original, view-attentive, and instance-attentive.

**Point clouds** Depict objects using individual points, which is particularly applicable in robotic [21,22,28]. Sparse and irregular points pose challenges. Qi et al. [18] introduced PointNet, a network architecture that effectively harnesses unordered point clouds and offers a comprehensive end-to-end solution for classification/retrieval tasks. DGCNN [26] employs dynamic graph convolution for point cloud processing, though challenges persist due to point cloud sparsity and irregularity.

**3D meshes** Describe surface geometry with vertices, edges, and faces [4,21, 22]. While they find applications in graphics and design, they also come with computational and storage complexities. MeshNet [3] transforms mesh data into a list of faces, calculating two types of information for each face: a spatial vector based on center data and a structural vector using center, normal, and neighbor information. These features are then merged using a multi-layer perceptron. In contrast, MeshCNN [8] applies convolution and pooling operations to mesh edges and the edges of connected triangles. When pooling is needed, it collapses edges while preserving the overall mesh structure.

Each representation method has its strengths and drawbacks. Voxels offer a structured approach for occupancy and property representation, but can be resource-intensive. Multi-view images leverage multiple perspectives for reconstruction, but accuracy depends on captured views. Point Clouds are storage-efficient and precise, but sparsity and irregularity pose challenges. 3D Meshes capture complex shapes and details, yet computational intensity and storage demands are notable. Deciding between the aforementioned representations methods, depends on the factors like accuracy, efficiency, and suitability.

In our research, we empasize mesh data, because it performs best due to its rich information representation [10]. Specifically, for our unimodal experimentation, we chose the most recent methods, MeshNet and MeshCNN as our preferred candidates based on [10].

## 2.2   Multimodal 3D Object Retrieval

The evolution of 3D data retrieval has spurred the exploration of multimodal approaches, which leverages diverse views to enhance accuracy and versatility. An important development is the fusion of multiple 3D views, capitalizing on their respective strengths in geometric precision and surface details. This integration answers the call for more potent retrieval systems capable of capturing intricate object traits while preserving precise shapes. By harmonizing these distinct viewpoints, retrieval techniques gain proficiency in object recognition, proving effective across an array of practical applications.

In the recent Cross-Modal Center Loss (CMCL) [10] approach, point clouds, meshes, and multi-view images are integrated into one unified framework. In this framework, the 3D modalities are combined to collectively train representations and identify optimal features. Various loss functions, including cross-entropy and mean-square-error, are employed to refine and enhance the performance of the framework. CMCL can be computationally intensive due to the integration of multiple 3D modalities into a single framework, potentially requiring significant computational resources. Performance of CMCL can also vary depending on the dataset, as it is sensitive to the characteristics of each modality.

## 2.3   Multimodal Image Retrieval

In multimodal image retrieval methods, various strategies combining different modalities are explored, with focus on hashing methods due to fast queries and less memory consumption. For instance, there are methods like Discrete Online Cross-modal Hashing (DOCH) [31], which creates high-quality hash codes for different data types by using both the likeness between data points and their detailed meanings. Fast Cross-Modal Hashing (FCMH) [25] adds an extra element to estimate the binary code, making it better by reducing mistakes. Label-Attended Hashing (LAH) [30] initially generates embeddings for images and

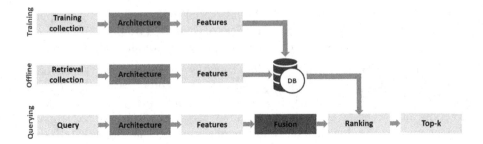

**Fig. 2.** Overview of abstract framework.

label co-occurrence separately. Following this, it employs a graph convolutional network to combine label features with image features, enhancing the model's capabilities. Nevertheless, it is worth noting that LAH learns hash functions from real data samples.

Based on recent research [17], we choose to adapt MuseHash as the state-of-the-art (SOTA) method and its competitor LAH [30] to 3D retrieval due to their effectiveness in other domain-specific datasets. MuseHash estimates semantic probabilities and statistical properties during the retrieval process, enhancing its performance in capturing meaningful relationships within the data. It not only demonstrates its prowess in multimodal retrieval but also aligns perfectly with the complexities of 3D data.

In our exploration, we compare MuseHash and LAH with CMCL to evaluate their performance in multimodal query scenarios, ultimately solidifying Muse-Hash as the prime candidate for adapting to 3D retrieval. Our study focuses on the potential of volumetric retrieval, leveraging the flexibility and detail offered by different representations of data. By exploring the challenges and opportunities of 3D retrieval, we aim to advance the field using multiple modalities.

## 3   Methodology

To formally address the problem, we define the following scenario: Given a query object denoted as $Q$ and a database $DB$ comprising a collection of 3D objects represented using varying views, such as images and meshes, the fundamental objective is to perform effective retrieval. This retrieval process aims to identify objects within $DB$ that share similarities with the query $Q$. The process involves a meticulous analysis of the distinctive features characterizing $Q$ and the subsequent comparison of these features with corresponding attributes of objects within $DB$ to ascertain pertinent matches.

Figure 2 provides a visual representation of the conceptual framework that underlies our research. This framework comprises three distinct phases: training, offline, and querying. During the training phase, data is input into a specific architecture, resulting in the generation of feature vectors. In the offline phase, features are extracted from a retrieval set and subsequently stored in a database for future reference. In the online phase, the architecture is applied to queries, and relevant results are retrieved from the database. To visually distinguish the areas where each studied model was applied, we use an ochre color. Additionally, the presence of a blue color indicates instances where the model has a multimodal capability; otherwise, it is absent.

When it comes to 3D retrieval methods, both MeshNet and MeshCNN fall under the category of unimodal mesh-based techniques. CMCL, on the other hand, stands out as a cross-modal 3D retrieval method. It adopts a different approach by concurrently learning a shared space for various features from different sources, utilizing MeshNet, DGCNN, and ResNet for mesh, point-cloud, and image modalities, respectively.

In our adaptation of image retrieval techniques for 3D retrieval, we apply supervised hashing methods within the architecture to generate hash code features. Specifically, we've chosen LAH and MuseHash due to their proven effectiveness across various data types, as emphasized in recent research by [17]. LAH, initially designed for unimodal image retrieval, learns hash codes by applying a non-linear hash function to these mesh features using features from MeshNet. MuseHash, on the other hand, leverages the same models as CMCL to extract features from all modalities. Subsequently, MuseHash employs Bayesian ridge regression to learn hash functions, mapping feature vectors to the Hamming space, thereby enabling both unimodal and multimodal queries.

In our experiments, we transformed the different modalities into feature vectors to prepare them for the hashing methods. For the visual modality, we conducted an averaging process involving 180 multi-view image feature vectors extracted from ResNet50's fc-7 layer, generating a 2048-D vector. As for the point cloud and mesh modalities, we obtained 256-D vectors directly from the final layers of DGCNN and MeshNet, respectively.

## 4    Experiments

In this section, we begin by describing the datasets used for evaluation and providing an overview of the experimental setup. We then present detailed experimental results for a variety of modalities and hash code lengths.

### 4.1    Datasets

The evaluation of our method and the comparison with existing SOTA methods is done on the two publicly available datasets. Table 1 summarizes the properties of the datasets (including the abbreviations used in tables to save space).

**BuildingNet_v0** The BuildingNet_v0 [20] provides high-quality annotations and diverse building types (like church, palace etc.)

**ModelNet40** The ModelNet40 [29] is a large-scale 3D CAD model dataset, offering a wide range of object categories (like car, bottle, etc.).

**Table 1.** Two benchmark datasets used in experiments.

| Dataset | Abbr. | Ground Truth Labels | Collection Sizes | | | |
|---|---|---|---|---|---|---|
| | | | Whole | Retrieval | Training | Test |
| BuildingNet_v0 | BNv0 | 60 | 2000 | 1900 | 500 | 100 |
| ModelNet40 | MN40 | 40 | 12311 | 11696 | 4843 | 615 |

## 4.2  Experimental Settings

In our experiments, we evaluate the performance of two different types of methods using various metrics. Specifically, we consider the hashing methods, MuseHash [17] and the LAH[1] [30], where we examine the impact of different hash code lengths ($d_c$ = 16, 32, 64, 128). For each volumetric method, we vary the number of epochs (epochs = 10, 50, 100, 150) used for computing each metric. We use the proposed training and testing size by the authors [20,29] for each dataset as suggested by the authors.

We compare our approach with two state-of-the-art 3D mesh methods MeshNet[2] [3], MeshCNN[3] [8], and one cross-modal 3D retrieval method CMCL[4] [10] in terms of mean Average Precision (mAP), precision at k (prec@k), recall at k (recall@k), f-score at k (fscore@k), accuracy and training time.

In the paper, a 5-fold cross-validation methodology was employed in all of the experiments for a more robust evaluation. We measured runtime of the experiments per epoch or per hash code length. Additionally, all 3D retrieval implementations used big amounts of memory, while MuseHash uses a small amount of memory (only some bits). In the following tables, the symbol '*' indicates that MuseHash, has demonstrated statistical significance compared to the other methods, as determined by a t-test.

## 4.3  Unimodal Retrieval Results

To study the performance of the multimodal approaches in unimodal situations, we compare all the aforementioned methods using only the mesh queries over the mesh modality. The results of those methods over the two datasets are given in Table 2 and Table 3 for mAP and accuracy, respectively. In this scenario, MuseHash outperforms all state-of-the-art methods.

The MuseHash algorithm performs better than other methods for different hash lengths and epochs on both datasets. It shows the highest mAP and accuracy scores, highlighting its effectiveness for 3D retrieval tasks using only the mesh view. Particularly on the BuildingNet_v0 dataset, MuseHash achieves the best accuracy, surpassing other methods. Additionally, the CMCL approach

---

[1] https://github.com/IDSM-AI/LAH.
[2] https://github.com/iMoonLab/MeshNet.
[3] https://github.com/ranahanocka/MeshCNN.
[4] https://github.com/LongLong-Jing/Cross-Modal-Center-Loss.

**Table 2.** MAP results for ModelNet40 and BuildingNet_v0 with different code lengths or number of epochs and mesh modality.

| Dataset | No. Epochs | MeshNet [3] | MeshCNN [8] | CMCL [10] | Hash Length | LAH [30] | MuseHash [17] |
|---|---|---|---|---|---|---|---|
| MN40 | 10 | 0.6801* | 0.6726* | 0.7097* | 16 | 0.7811* | **0.8010** |
| | 50 | 0.6954* | 0.6900* | 0.7099* | 32 | 0.7889* | **0.8056** |
| | 100 | 0.7091* | 0.6711* | 0.7103* | 64 | 0.8001* | **0.8101** |
| | 150 | 0.6654* | 0.6502* | 0.6695* | 128 | 0.8058* | **0.8122** |
| BNv0 | 10 | 0.6201* | 0.6007* | 0.6511* | 16 | 0.7629* | **0.7723** |
| | 50 | 0.6350* | 0.6226* | 0.6520* | 32 | 0.7701 | **0.7791** |
| | 100 | 0.6552* | 0.6449* | 0.6670* | 64 | 0.7754* | **0.7834** |
| | 150 | 0.6550* | 0.6501* | 0.6623* | 128 | 0.7821* | **0.7883** |

**Table 3.** Accuracy results for ModelNet40 and BuildingNet_v0 with different code lengths or number of epochs and mesh modality.

| Dataset | No. Epochs | MeshNet [3] | MeshCNN [8] | CMCL [10] | Hash Length | LAH [30] | MuseHash [17] |
|---|---|---|---|---|---|---|---|
| MN40 | 10 | 0.8091* | 0.7511* | 0.7916* | 16 | 0.9221* | **0.9431** |
| | 50 | 0.8363* | 0.8002* | 0.8001* | 32 | 0.9278* | **0.9488** |
| | 100 | 0.8422* | 0.8101* | **0.9791** | 64 | 0.9312* | 0.9500 |
| | 150 | 0.8490* | 0.8091* | **0.9895** | 128 | 0.9401 | 0.9510 |
| BNv0 | 10 | 0.7882* | 0.7716* | 0.7910* | 16 | 0.9189* | **0.9323** |
| | 50 | 0.8025* | 0.7922* | 0.8001* | 32 | 0.9207* | **0.9344** |
| | 100 | 0.8337* | 0.8267* | 0.8510* | 64 | 0.9255* | **0.9390** |
| | 150 | 0.8405* | 0.8373* | 0.8601* | 128 | 0.9345* | **0.9401** |

achieves top accuracy on the ModelNet40 dataset with more epochs, yet its mAP performance lags behind. This implies CMCL's proficiency in classification but potential challenges in organizing relevant retrieval results. Apart from that, image retrieval methods can perform better in 3D retrieval task from current SOTA 3D retrieval methods.

### 4.4 Multimodal Retrieval Results

For the multimodal case, we consider the combination of point clouds, meshes, and multi-view images. The results of these techniques for both mAP and accuracy are detailed in Table 4 for ModelNet40 and BuildingNet_v0 dataset.

Specifically, Table 4 highlights the mAP and accuracy results for ModelNet40 dataset across different hash lengths, epochs, and query modalities. MuseHash demonstrates competitive performance in the majority of scenarios. MuseHash exhibits a distinct advantage in accuracy when queries involve both visual and point cloud modalities. While CMCL also exhibits competitive results, Muse-Hash's efficacy in handling diverse query modalities showcases its adaptability and robustness across different data representations.

In addition, the multimodal variant of MuseHash, which incorporates both mesh and image modalities, demonstrates substantial performance improvements with longer code lengths (from 16 to 32), particularly for larger code lengths (64

**Table 4.** MAP and accuracy results for ModelNet40 and BuildingNet_v0 with different code lengths or number of epochs and query modalities.

| Dataset | Query | No. Epochs | CMCL [10] | | Hash Length | MuseHash [17] | |
|---|---|---|---|---|---|---|---|
| | | | mAP | Accuracy | | mAP | Accuracy |
| MN40 | Visual Mesh | 10 | 0.6911* | 0.9012* | 16 | **0.8184** | **0.9501** |
| | | 50 | 0.7010* | 0.9045* | 32 | **0.8201** | **0.9578** |
| | | 100 | 0.7122* | 0.9091* | 64 | **0.8234** | **0.9601** |
| | | 150 | 0.7415* | 0.9129* | 128 | **0.8212** | **0.9525** |
| | Visual Point Cloud | 10 | 0.6710* | 0.7661* | 16 | **0.7712** | **0.9423** |
| | | 50 | 0.6912* | 0.7712* | 32 | **0.7821** | **0.9489** |
| | | 100 | 0.7010* | 0.7891* | 64 | **0.7823** | **0.9510** |
| | | 150 | 0.7122* | 0.7922* | 128 | **0.7840** | **0.9517** |
| | Mesh Point Cloud | 10 | 0.6910* | 0.8992* | 16 | **0.7882** | **0.9345** |
| | | 50 | 0.7039* | 0.9042* | 32 | **0.7910** | **0.9422** |
| | | 100 | 0.7128* | 0.9188* | 64 | **0.7900** | **0.9577** |
| | | 150 | 0.7231* | 0.9201* | 128 | **0.7854** | **0.9611** |
| | Visual Mesh Point Cloud | 10 | 0.7097* | 0.7916* | 16 | **0.8051** | **0.9611** |
| | | 50 | 0.7099* | 0.8001* | 32 | **0.7976** | **0.9601** |
| | | 100 | 0.7103* | **0.9791** | 64 | **0.7923** | 0.9583 |
| | | 150 | 0.6695* | **0.9895** | 128 | **0.7911** | 0.9550 |
| BNv0 | Visual Mesh | 10 | 0.6911* | 0.8011* | 16 | **0.7810** | **0.9423** |
| | | 50 | 0.7010* | 0.8091* | 32 | **0.7912** | **0.9455** |
| | | 100 | 0.7122* | 0.8123* | 64 | **0.8010** | **0.9589** |
| | | 150 | 0.7415* | 0.8231* | 128 | **0.8091** | **0.9610** |
| | Visual Point Cloud | 10 | 0.6801* | 0.7938* | 16 | **0.7701** | **0.9244** |
| | | 50 | 0.6881* | 0.7957* | 32 | **0.7734** | **0.9301** |
| | | 100 | 0.6910* | 0.8010* | 64 | **0.7791** | **0.9451** |
| | | 150 | 0.7001* | 0.8139* | 128 | **0.7801** | **0.9510** |
| | Mesh Point Cloud | 10 | 0.6761* | 0.7810* | 16 | **0.7610** | **0.9181** |
| | | 50 | 0.6810* | 0.7910* | 32 | **0.7691** | **0.9201** |
| | | 100 | 0.6902* | 0.8031* | 64 | **0.7701** | **0.9221** |
| | | 150 | 0.6971* | 0.8091* | 128 | **0.7688** | **0.9200** |
| | Visual Mesh Point Cloud | 10 | 0.6511* | 0.7910* | 16 | **0.7790** | **0.9021** |
| | | 50 | 0.6520* | 0.8001* | 32 | **0.7800** | **0.8900** |
| | | 100 | 0.6670* | 0.8510* | 64 | **0.7881** | **0.8991** |
| | | 150 | 0.6623* | 0.8601* | 128 | **0.7912** | **0.8920** |

and 128). However, further increasing the code length does not lead to significant performance gains. This observation highlights an optimal code length range where MuseHash excels in capturing intricate multimodal relationships.

Turning to the BuildingNet_v0 dataset results in Table 4, MuseHash outperforms in all multimodal cases the CMCL approach. In general, MuseHash has higher value on mAP as the code length increases and reaches the highest value when it uses visual and mesh view as a query and for code length 128.

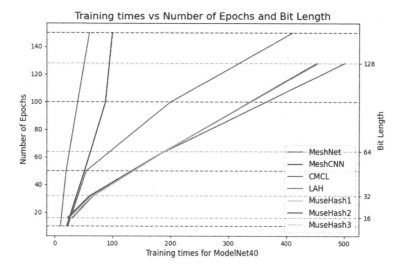

**Fig. 3.** Comparison of training times for all methods in minutes.

## 4.5 Analysis of Runtime Requirements

Figure 3 represents the training time (in minutes) for various methods, including MeshNet, MeshCNN, CMCL, LAH and three variants of MuseHash, across different training epochs or code lengths. Each line in the plot corresponds to a specific method, and the $x$-axis represents the training time in minutes, while the left $y$-axis the number of training epochs and the right $y$-axis the code length used in the training process. The black and grey dotted lines correspond to the values of each method for a specific epoch or code length, respectively. In the figure, three MuseHash variants (MuseHash1, MuseHash2, and MuseHash3) are evaluated for different code lengths, corresponding to mesh, mesh and visual, and mesh, visual and point clouds, respectively.

MeshNet and MeshCNN exhibit relatively shorter training times compared to CMCL and MuseHash variants, with CMCL requiring significantly more time for the same number of epochs. Among the MuseHash variants, MuseHash1 consistently shows the shortest training time across different code lengths, making it the most computationally efficient option. As the code length increases, all variants of MuseHash experience longer training times due to more complex computations and increased memory demands. In summary, the Fig. 3 compares training times among different methods. MuseHash1 exhibits the shortest training times initially, but as the code length increases, the training times become longer. Regarding the inference times all methods have the roughly the same performance.

**Table 5.** Comparison of all methods based on Precision at k (k = 10, 25, 50) for different number of epochs or code lengths on ModelNet40 dataset

| Method | Variable | Precision@k | | | Recall@k | | | Fscore@k | | |
|---|---|---|---|---|---|---|---|---|---|---|
| | Epochs | 10 | 25 | 50 | 10 | 25 | 50 | 10 | 25 | 50 |
| MeshNet [3] | 10 | 0.6510 | 0.6560 | 0.6410 | 0.6802 | 0.6533 | 0.6602 | 0.6653 | 0.6546 | 0.650 |
| | 50 | 0.6810 | 0.6712 | 0.6678 | 0.7011 | 0.7051 | 0.7187 | 0.6909 | 0.6877 | 0.6923 |
| | 100 | 0.6901 | 0.6854 | 0.6802 | 0.7029 | 0.7011 | 0.7089 | 0.6964 | 0.6932 | 0.6943 |
| | 150 | 0.7010 | 0.6910 | 0.6824 | 0.7091 | 0.7123 | 0.7189 | 0.7055 | 0.7015 | 0.7002 |
| MeshCNN [8] | 10 | 0.5822 | 0.5701 | 0.5623 | 0.5791 | 0.5607 | 0.5689 | 0.5806 | 0.6011 | 0.6178 |
| | 50 | 0.6001 | 0.5803 | 0.5734 | 0.5998 | 0.6011 | 0.6183 | 0.5999 | 0.5905 | 0.5950 |
| | 100 | 0.6245 | 0.6183 | 0.6002 | 0.6011 | 0.6190 | 0.6189 | 0.6126 | 0.6186 | 0.6094 |
| | 150 | 0.6221 | 0.6112 | 0.6009 | 0.6005 | 0.6123 | 0.6230 | 0.6111 | 0.6117 | 0.6118 |
| CMCL [10] | 10 | 0.8290 | 0.7679 | 0.7142 | **0.9985** | 0.9943 | 0.9968 | 0.9011 | 0.8666 | 0.8321 |
| | 50 | 0.8291 | 0.7687 | 0.7147 | 0.9883 | 0.9943 | **0.9968** | 0.9018 | 0.8671 | 0.8325 |
| | 100 | **0.8298** | 0.7687 | 0.7149 | 0.9884 | **0.9944** | **0.9968** | **0.9019** | 0.8671 | 0.8326 |
| | 150 | 0.8283 | 0.7677 | 0.7142 | 0.9865 | **0.9944** | **0.9968** | 0.9013 | 0.8665 | 0.8322 |
| | Code Length | 10 | 25 | 50 | 10 | 25 | 50 | 10 | 25 | 50 |
| LAH [30] | 16 | 0.6190 | 0.6179 | 0.6242 | 0.9215 | 0.9243 | 0.9268 | 0.7405 | 0.7407 | 0.7460 |
| | 32 | 0.6202 | 0.6287 | 0.6347 | 0.9383 | 0.9343 | 0.9461 | 0.7468 | 0.7516 | 0.7597 |
| | 64 | 0.6298 | 0.6287 | 0.6349 | 0.9584 | 0.9444 | 0.9468 | 0.7601 | 0.7549 | 0.7601 |
| | 128 | 0.6281 | 0.6271 | 0.6242 | 0.9265 | 0.9344 | 0.9468 | 0.7487 | 0.7505 | 0.7524 |
| MuseHash1 [17] | 16 | 0.6412 | 0.6501 | 0.6623 | 0.9567 | 0.9689 | 0.9781 | 0.7454 | 0,7781 | 0.7898 |
| | 32 | 0.6589 | 0.6620 | 0.6778 | 0.9612 | 0.9723 | 0.9612 | 0.7818 | 0.7877 | 0.8018 |
| | 64 | 0.6601 | 0.6789 | 0.6801 | 0.9667 | 0.9712 | 0.9789 | 0.7845 | 0,7992 | 0.8026 |
| | 128 | 0.6791 | 0.7123 | 0.7256 | 0.9701 | 0.9734 | 0.9601 | 0.7989 | 0.8229 | 0.8265 |
| MuseHash2 [17] | 16 | 0.6571 | 0.6810 | 0.7020 | 0.9612 | 0.9723 | 0.9865 | 0.7806 | 0.8010 | 0.8203 |
| | 32 | 0.6910 | 0.7001 | 0.7112 | 0.9546 | 0.9612 | 0.9667 | 0.8017 | 0.8101 | 0.8195 |
| | 64 | 0.7662 | 0.7405 | 0.7156 | 0.9712 | 0.9781 | 0.9801 | 0.8566 | 0.8429 | 0.8272 |
| | 128 | 0.8010 | **0.8588** | **0.8423** | 0.9865 | 0.9902 | 0.9923 | 0.8841 | **0.9198** | **0.9112** |
| MuseHash3 [17] | 16 | 0.6480 | 0.6501 | 0.6589 | 0.9523 | 0.9678 | 0.9621 | 0.7712 | 0.7778 | 0.7821 |
| | 32 | 0.6510 | 0.6678 | 0.6781 | 0.9678 | 0.9698 | 0.9512 | 0.7784 | 0.7910 | 0.7918 |
| | 64 | 0.6782 | 0.6789 | 0.6834 | 0.9701 | 0.9700 | 0.9634 | 0.7983 | 0.7988 | 0.7996 |
| | 128 | 0.7012 | 0.6910 | 0.6901 | 0.9701 | 0.9623 | 0.9603 | 0.8140 | 0.8044 | 0.8038 |

## 4.6    Discussion

Table 5 presents a comprehensive comparison of various methods based on Precision@k, Recall@k, and Fscore@k for k = 10, 25, 50 and for different code lengths or number of epochs on the ModelNet40 dataset. Thus, the table provides a detailed understanding of how well the retrieval methods rank and retrieve relevant items. Moreover, the selected metrics reveal how well the methods rank and capture relevant data in the top-k results.

The methods are categorized into two groups: MeshNet, MeshCNN, and CMCL, each evaluated for different numbers of epochs. Additionally, there are the three predefined variants of the MuseHash (MuseHash1, MuseHash2, MuseHash3) method and the LAH method evaluated for different code lengths.

Upon careful observation, it is evident that multimodal approaches excel in comparison to MeshCNN and MeshNet, revealing limitations in the architecture or feature representation of the latter two methods when operating exclusively within the mesh view. While CMCL occasionally achieves superior outcomes, considering the trade-off between performance gains and training time, Muse-Hash emerges as the more efficient choice. Additionally, MuseHash's capability to incorporate multiple modalities (e.g., mesh and image) into a unified hash code enhances its retrieval accuracy and diversity. The efficiency of MuseHash becomes particularly valuable in scenarios with extensive datasets and resource constraints, where fast and accurate similarity searches are paramount.

## 5  Conclusion

In this paper, we have leveraged the state-of-the-art methods from image retrieval to the domain of 3D object retrieval. In particular, we have adapted the recently proposed multimodal MuseHash method to support queries within volumetric data. The MuseHash method exploits the inner relations between different modalities. Our experiments show that MuseHash outperforms in most cases three state-of-the-art methods in both unimodal and multimodal queries across two different domain-specific benchmark image collections.

**Acknowledgment.** This work was supported by the EU's Horizon 2020 research and innovation programme under grant agreement H2020-101004152 XRECO.

## References

1. Brutto, M.L., Meli, P.: Computer vision tools for 3D modelling in archaeology. HDE **1**, 1–6 (2012)
2. Dummer, M.M., Johnson, K.L., Rothwell, S., Tatah, K., Hibbs-Brenner, M.K.: The role of VCSELs in 3D sensing and LiDAR. In: OPTO (2021)
3. Feng, Y., Feng, Y., You, H., Zhao, X., Gao, Y.: MeshNet: mesh neural network for 3D shape representation. In: AAAI, Honolulu, Hawaii (2019)
4. Garland, M., Heckbert, P.S.: Simplifying surfaces with color and texture using quadric error metrics. In: VC (1998)
5. Gezawa, A.S., Zhang, Y., Wang, Q., Yunqi, L.: A review on deep learning approaches for 3D data representations in retrieval and classifications. IEEE Access **8**, 57566–57593 (2020)
6. Ha, Q., Yen, L., Balaguer, C.: Robotic autonomous systems for earthmoving in military applications. Autom. Constr. **107**, 102934 (2019)
7. Han, Y.S., Lee, J., Lee, J., Lee, W., Lee, K.: 3D CAD data extraction and conversion for application of augmented/virtual reality to the construction of ships and offshore structures. Int. J. Comput. Integr. Manuf. **32**(7), 658–668 (2019)
8. Hanocka, R., Hertz, A., Fish, N., Giryes, R., Fleishman, S., Cohen-Or, D.: MeshCNN: a network with an edge. ACM Trans. Graph. (ToG) **38**(4), 1–12 (2019)
9. Javaid, M., Haleem, A., Singh, R.P., Suman, R.: Industrial perspectives of 3D scanning: features, roles and it's analytical applications. Sensors Int. **2**, 100114 (2021)
10. Jing, L., Vahdani, E., Tan, J., Tian, Y.: Cross-modal center loss for 3D cross-modal retrieval. In: Proceedings of the IEEE/CVF Conference on Computer Vision and Pattern Recognition, pp. 3142–3151 (2021)

11. Klokov, R., Lempitsky, V.: Escape from cells: deep Kd-networks for the recognition of 3D point cloud models. In: Proceedings of the IEEE International Conference on Computer Vision, pp. 863–872 (2017)
12. Lin, D., et al.: Multi-view 3D object retrieval leveraging the aggregation of view and instance attentive features. Knowl. Based Syst. **247**, 108754 (2022)
13. Maglo, A., Lavoué, G., Dupont, F., Hudelot, C.: 3D mesh compression: survey, comparisons, and emerging trends. ACM Comput. Surv. (CSUR) **47**(3), 1–41 (2015)
14. Maturana, D., Scherer, S.: VoxNet: a 3D convolutional neural network for realtime object recognition. In: 2015 IEEE/RSJ International Conference on Intelligent Robots and Systems (IROS), pp. 922–928. IEEE (2015)
15. Mohr, E., Thum, T., Bär, C.: Accelerating cardiovascular research: recent advances in translational 2D and 3D heart models. Eur. J. Heart Fail. **24**(10), 1778–1791 (2022)
16. Pal, P., Ghosh, K.K.: Estimating digitization efforts of complex product realization processes. Int. J. Adv. Manuf. Technol. **95**, 3717–3730 (2018)
17. Pegia, M., et al.: MuseHash: supervised Bayesian hashing for multimodal image representation. In: Proceedings of the 2023 ACM International Conference on Multimedia Retrieval, pp. 434–442 (2023)
18. Qi, C.R., Su, H., Mo, K., Guibas, L.J.: PointNet: deep learning on point sets for 3D classification and segmentation. In: CVPR, Honolulu, HI, USA (2017)
19. Rahate, A., Walambe, R., Ramanna, S., Kotecha, K.: Multimodal co-learning: challenges, applications with datasets, recent advances and future directions. Inf. Fus. **81**, 203–239 (2022)
20. Selvaraju, P., et al.: BuildingNet: learning to label 3D buildings. In: Proceedings of the IEEE/CVF International Conference on Computer Vision, pp. 10397–10407 (2021)
21. Su, H., Maji, S., Kalogerakis, E., Learned-Miller, E.: Multi-view convolutional neural networks for 3D shape recognition. In: Proceedings of the IEEE International Conference on Computer Vision, pp. 945–953 (2015)
22. Su, J.-C., Gadelha, M., Wang, R., Maji, S.: A deeper look at 3D shape classifiers. In: Leal-Taixé, L., Roth, S. (eds.) ECCV 2018. LNCS, vol. 11131, pp. 645–661. Springer, Cham (2019). https://doi.org/10.1007/978-3-030-11015-4_49
23. Tatarchenko, M., Dosovitskiy, A., Brox, T.: Octree generating networks: efficient convolutional architectures for high-resolution 3D outputs. In: Proceedings of the IEEE International Conference on Computer Vision, pp. 2088–2096 (2017)
24. Wang, P.S., Liu, Y., Guo, Y.X., Sun, C.Y., Tong, X.: O-CNN: octree-based convolutional neural networks for 3D shape analysis. ACM Trans. Grap. (TOG) **36**(4), 1–11 (2017)
25. Wang, Y., Chen, Z.D., Luo, X., Li, R., Xu, X.S.: Fast cross-modal hashing with global and local similarity embedding. IEEE Trans. Cybern. **52**(10), 10064–10077 (2021)
26. Wang, Y., Sun, Y., Liu, Z., Sarma, S.E., Bronstein, M.M., Solomon, J.M.: Dynamic graph CNN for learning on point clouds. ACM Trans. Graph. (ToG) **38**(5), 1–12 (2019)
27. Willemink, M.J., et al.: Preparing medical imaging data for machine learning. Radiology **295**(1), 4–15 (2020)
28. Wu, W., Qi, Z., Fuxin, L.: PointConv: deep convolutional networks on 3D point clouds. In: Proceedings of the IEEE/CVF Conference on Computer Vision and Pattern Recognition, pp. 9621–9630 (2019)
29. Wu, Z., et al.: 3D ShapeNets: a deep representation for volumetric shapes. In: CVPR, pp. 1912–1920 (2015)

30. Xie, Y., Liu, Y., Wang, Y., Gao, L., Wang, P., Zhou, K.: Label-attended hashing for multi-label image retrieval. In: IJCAI, pp. 955–962 (2020)
31. Zhan, Y.W., Wang, Y., Sun, Y., Wu, X.M., Luo, X., Xu, X.S.: Discrete online cross-modal hashing. Pattern Recogn. **122**, 108262 (2022)

# An Integrated System for Spatio-temporal Summarization of 360-Degrees Videos

Ioannis Kontostathis⬥, Evlampios Apostolidis(✉)⬥, and Vasileios Mezaris⬥

Information Technologies Institute - Centre for Research and Technology, Hellas,
6th km Charilaou - Thermi Road, 57001 Thessaloniki, Greece
{ioankont,apostolid,bmezaris}@iti.gr

**Abstract.** In this work, we present an integrated system for spatio-temporal summarization of 360-degrees videos. The video summary production involves the detection of salient events in the 360-degrees videos, and their synopsis in a concise summary. The analysis relies on state-of-the-art methods for saliency detection in 360-degrees video (ATSal and SST-Sal) and video summarization (CA-SUM). It also contains a camera motion detection mechanism (CMDM) that decides which saliency detection method will be used, as well as a 2D video production component that is responsible for creating a conventional 2D video containing the detected salient events in the 360-degrees video. Quantitative evaluations using two datasets for 360-degrees video saliency detection (VR-EyeTracking, Sports-360) show the accuracy and positive impact of the developed CMDM, and justify our choice to use two different methods for detecting saliency. A qualitative analysis using content from these datasets, gives further insights about the functionality of the CMDM, shows the pros and cons of each used saliency detection method and demonstrates the advanced performance of the trained summarization method against a more conventional approach.

**Keywords:** 360-degrees video · saliency detection · video summarization · equirectangular projection · cubemap projection

## 1 Introduction

Over the last years, we are experiencing a rapid growth of 360° videos. This growth is mainly fueled by the increasing engagement of users with advanced 360° video recording devices, such as GoPro and GearVR, the ability to share such content via video sharing platforms (such as YouTube and Vimeo) and social networks (such as Facebook), as well as the emergence of extended reality technologies. The rapid growth of 360° video content comes with an increasing need for technologies which facilitate the creation of a short summary that

This work was supported by the EU Horizon Europe and Horizon 2020 programmes under grant agreements 101070109 TransMIXR and 951911 AI4Media, respectively.

S. Rudinac et al. (Eds.): MMM 2024, LNCS 14557, pp. 202–215, 2024.
https://doi.org/10.1007/978-3-031-53302-0_15

conveys information about the most important activities in the video. Such technologies would allow viewers to quickly get a grasp about the content of the 360° video, thus significantly facilitating their browsing in large collections.

Despite the increasing popularity of 360° video content, the research on the summarization of this content is still limited. A few methods focus on piloting the viewer through the unlimited field of view of the 360° video, by controlling the camera's position and field of view, and generating an optimal camera trajectory [9,10,16,22,23,25]. These methods perform a spatial summarization of the 360° video by focusing on the most salient regions and ignoring areas of the 360° video that are less interesting. Nevertheless, the produced normal Field-Of-View (NFOV) video might contain redundant information as it presents the detected activities in their full duration; thus it cannot be seen as a condensed summary of the video content. A couple of recent methods target both spatial and temporal summarization of 360° videos [12,27]. However, both methods assume the existence of a single important activity [27] or narrative [12] in order to create a video highlight and a story-based video summary, respectively.

In this paper, we propose an approach that considers both the spatial and the temporal dimension of the 360° video (contrary to [9,16,22,23]) and takes into account several different activities that might take place in parallel (differently to [12,27]), in order to form a summary of the video content. The developed system is compatible with 360° videos captured using either a static or a moving camera. Moreover, the duration of the output video summary can be adjusted according to the users' needs, thus facilitating the production of different summaries for a given 360° video. Our main contributions are as follows:

- We propose a new approach for spatio-temporal summarization of 360° videos, that relies on a combination of state-of-the-art deep learning methods for saliency detection and video summarization.
- We develop a mechanism that classifies a 360° video according to the use of a static or a moving camera, and a method that forms a conventional 2D video showing the detected salient events in the 360° video.
- We build an integrated system that performs spatio-temporal summarization of 360° video in an end-to-end manner.

## 2   Related Work

One of the early attempts to offer a more natural-looking NFOV video that focuses on the interesting events of a panoramic 360° video, was made by Su et al. (2016) [23]. Their method learns a discriminative model of human-captured NFOV web videos and utilizes this model to identify candidate view-points and events of interest in a 360° video. Then, it stitches them together through optimal human-like camera motions and creates a new NFOV presentation of the 360° video content. In their following work, Su et al. (2017) [22], allowed to control the camera's field of view dynamically, applied an optimization approach that iteratively refines and makes it tractable, and added a mechanism to encourage visual diversity. Hu et al. (2017) [9], formulated the selection of the most interesting parts of a 360° video as a viewer piloting task. They described a deep-learning-based agent which extracts a set of candidate objects from each frame using an object detector [17], and selects the main object using a trainable RNN. Given this object and the previously selected viewing angles, the agent shifts the current viewing angle to the next preferred one. In the same direction, Qiao et al. (2017) [16], presented the multi-task DNN (MT-DNN) method that predicts viewport-dependent saliency over 360° videos based on both video content and viewport location. Each different task relates to a different viewport and for each task MT-DNN uses a combination of CNN and ConvLSTM to model both spatial and temporal features at specific viewport locations. The output of all tasks is taken into account by MT-DNN to estimate saliency at any viewport location. Yu et al. (2018) [27], addressed the problem of 360° video highlight detection using a trainable deep ranking model that produces a spherical composition score map per video segment and determines which view can be considered as a highlight via a sliding window kernel. Their method performs spatial summarization by finding out the best NFOV subshot per 360° video segment, and temporal summarization by selecting the N top-ranked NFOV subshots as highlights. Lee et al. (2018) [12] proposed an approach for story-based summarization of 360° videos. They used a trainable deep ranking network to score NFOV region proposals cropped from the input 360° video, and performed temporal summarization using a memory network to correlate past and future information, based on the assumption that the parts of a story-based summary share a common story-line. Kang et al. (2019) [10] presented an interactive 360° video navigation system which applies optical flow and saliency detection to find a virtual camera path with the most salient events in the video and generate a NFOV video. In a similar direction, Wang et al. (2020) [25] developed a tool for 360° video navigation and playback on 2D displays, which applies a spatial-aware transitioning between multiple NFOV views that track potentially interesting targets or events. Finally, Li et al. (2023) [13] and Setayesh et al. (2023) [18] deal with the task of viewport prediction in the context of 360° video multicast and streaming over wireless networks, respectively. The above literature review shows that there is an ongoing interest towards the automated selection of the most important parts of a 360° video; however, there is a lot of room for further advancements in order to facilitate the creation of descriptive and engaging summaries for such videos.

## 3    Proposed Approach

An overview of the proposed approach is given in Fig. 1. The input 360° video is initially subjected to equirectangular projection (ERP) to form a set of omnidirectional planar frames (called ERP frames in the following). This set of frames is then analysed by a camera motion detection mechanism (CMDM) which decides on whether the 360° video was captured by a static or a moving camera. Based on this mechanism's output, the ERP frames are then forwarded to one of the integrated methods for saliency detection, which produce a set of frame-level

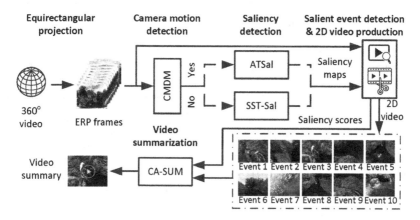

**Fig. 1.** An overview of the proposed approach. Bold-font text indicates the different processing steps. Dashed lines denote alternative paths of the processing pipeline.

saliency maps. The use of such a mechanism in combination with two different methods for saliency detection was based on our study of the relevant literature and the observation that most methods either deal with or are more effective on 360° videos that have been captured by one of the aforementioned video recording conditions. The ERP frames and the extracted saliency maps are given as input to a component that detects salient events (related to different parts of one or more activities shown in the 360° video), computes a saliency score per frame, and produces a 2D video presenting the detected events. The 2D video along with its frames' saliency scores are fed to a video summarization method which estimates the importance of each event and forms the video summary. The following subsections describe in more details each processing component.

### 3.1   Camera Motion Detection Mechanism (CMDM)

As noted in [11], the main action zones on ERP frames are most commonly near the equator. Taking this into account, the decision on whether the 360° video was captured by a static or a moving camera is based on the analysis of the north and south regions of the ERP frames, as depicted in Fig. 2. Such regions should exhibit limited variation across a sequence of frames when the 360° video is captured by a static camera. So, given the entire sequence of ERP frames of an input video, the camera motion detection mechanism focuses on the aforementioned frame regions and computes the phase correlation between each pair of consecutive frames in order to estimate their relative translative offset (using frequency-domain data and the fast Fourier transform). If the computed scores frequently exceed a threshold $t_0$, then the mechanism declares the use of a moving camera; otherwise, it indicates the use of a static one.

## 3.2 Saliency Detection

In the case of images, the visual saliency of an object refers to its state or quality of standing out in relation to its neighboring items. However, when analyzing video, visual saliency has to deal also with moving objects, as motion is a very salient cue for bottom-up attention [8]. Based on a literature review and experimentation with a few approaches, we use two state-of-the-art methods for saliency detection in 360° videos, called ATSal [5] and SST-Sal [3], that take into account both spatial and temporal (motion) features of the visual content. ATSal [5] employs two parallel models: i) an attention model that encodes the global visual features of each ERP frame, and ii) the SalEMA expert model that extracts the temporal characteristics of the 360° video using cubemap projection (CMP) frames. The attention model consists of a fine-tuned VGG16 [19] encoder-decoder with an intermediate attention mechanism. The SalEMA expert is composed of the SalEMA-Poles model, which is responsible for the north and south regions of the 360° video, and the SalEMA-Equator model which focuses on the front, back, left and right regions of the 360° video. The attention model processes the ERP frames sequentially, while the SalEMA expert model analyzes the corresponding CMP frames in mini-batches of 10 frames. The final saliency map of each frame of the 360° video is formulated after performing a pixel-wise multiplication between the outputs of the attention and the SalEMA expert models. SST-Sal [3] processes the input sequence of ERP frames in mini-batches of 20 frames and adopts an encoder-decoder approach that takes into account both temporal and spatial information at feature encoding and decoding time. The encoder is formed by a Spherical Convolutional LSTM (ConvLSTM) and a spherical max pooling layer, and extracts the spatio-temporal features from an input sequence of ERP frames. The decoder is composed of a Spherical ConvLSTM and an up-sampling layer, and leverages the latent information to predict a sequence of saliency maps. The utilized ConvLSTMs extend the functionality of LSTMs by applying spherical convolutions, thus allowing to account for the introduced distortion when projecting (ERP) the 360° frames onto a 2D plane, while extracting spatial features from the videos.

## 3.3 Salient Event Detection and 2D Video Production

This component takes as input the ERP frames and their associated saliency maps (extracted by the utilized saliency detection method), and produces a conventional 2D video that contains the detected salient events in the 360° video. As presented in Algorithm 1, this procedure starts by identifying the salient regions of each ERP frame. To form such a region in a given ERP frame, our method focuses on points of the associated saliency map that surpass an intensity value $t_1$, converts their coordinates to radians and clusters them using the DBSCAN algorithm [6] and a predefined distance $t_2$. Following, the salient regions that are spatially related across a sequence of frames are grouped together, thus establishing a spatial-temporally-correlated sub-volume of the 360° video. For this, taking into account the entire frame sequence, our method examines whether the

---

**Algorithm 1.** Salient event detection and 2D video production

---

**Require:** $N$ is the number of frames/saliency maps, $R$ is the number of salient regions, $L$ is the number of salient regions per frame, $S$ is the number of defined sub-volumes, $LS$ is the length (in frames) of each sub-volume, $FS$ is the number of finally formed sub-volumes, $FL$ is the length (in frames) of the finally formed sub-volumes.

**Output:** 2D Video with the detected salient events of the 360° video

"*Define the salient regions in each frame by clustering salient points with intensity higher than $t_1$, using DBSCAN clustering with distance (in radians) equal to $t_2$:*"

1: **for** $i = 0$ to $N$ **do**
2:     $Salient\_regions_i = F_1(Saliency\_Map_i, t_1, t_2)$

"*Define spatial-temporally-correlated 2D sub-volumes by grouping together spatially related regions (distance less that $t_3$) across a sequence of frames:*"

3: **for** $i = 1$ to $N$ **do**
4:     **for** $j = 0$ to $L_i$ **do**
5:         $SubVolumes_{i,j} = F_2(Salient\_regions_{i,j}, t_3)$

"*Mitigate abrupt changes in the visual content of sub-volumes by adding possibly missing frames (up to $t_4$; otherwise define a new sub-volume):*"

6: **for** $m = 0$ to $S$ **do**
7:     **for** $n = 0$ to $LS_m$ **do**
8:         $Final\_SubVolumes_{m,n} = F_3(SubVolumes_{m,n}, t_4)$

"*Produce the 2D video by extracting the FOV for the salient regions of each ERP frame of the finally-formed sub-volume:*"

9: **for** $k = 0$ to $FS$ **do**
10:     **for** $l = 0$ to $FL_k$ **do**
11:         $F_4(Final\_SubVolumes_{k,l})$

---

salient regions of the $f_i$ frame are close enough to the salient regions of one of the previous frames ($f_{i-1}...f_1$). If this distance is less than $t_3$, then the spatially-correlated regions over the examined sequence of frames ($f_1...f_i$) are grouped and form a sub-volume; otherwise, it creates a new sub-volume for each of the salient regions in the $f_i$ ERP frame. Given the fact that a spatially-correlated salient region can be found in non-consecutive frames (e.g., appearing in frames $f_t$ and $f_{t+2}$), the applied grouping might result in sub-volumes that are missing one or more frames. To mitigate abrupt changes in the visual content of the formulated sub-volumes, our method adds the missing frames withing each sub-volume. Moreover, in the case that a sub-volume contains a large sequence of missing frames (higher in length than $t_4$) our method splits this sub-volume and considers that the aforementioned sequence of frames does not contain a salient event. For each ERP frame of the defined sub-volume, the developed method extracts the FOV of its salient regions and creates a frame with aspect ratio 4:3 that is centered to the extracted FOV, thus producing a short spatio-temporally-coherent 2D video fragment with the same aspect ratio. Finally, the output 2D video is formed by stitching the created 2D video fragments in chronological order, while a saliency score is computed for each frame as the average saliency score of the identified salient regions in it.

## 3.4   Video Summarization

The temporal summarization of the generated 2D is performed using a variant of the CA-SUM method [2]. This method integrates an attention mechanism that is able to concentrate on specific parts of the attention matrix (that correspond to different non-overlapping video fragments of fixed length) and make better estimates about their significance by incorporating knowledge about the uniqueness and diversity of the relevant frames of the video. The utilized variant of CA-SUM has been trained by taking into account also the saliency of the video frames. In particular, the computed saliency scores for the frames of the 2D video are used to weight the extracted representations of the visual content of these frames. Hence, the utilized video summarization model is trained based on a set of representations that incorporate information about both the visual content and the saliency of each video frame. At the output, this model produces a set of frame-level importance scores. After considering the different sub-volumes of the 2D video as different video fragments, fragment-level importance scores are computed by averaging the importance scores of the frames that lie within each fragment. These fragment-level scores are then used to select the key-fragments given a target summary length $L$. Following the typical paradigm in the relevant literature [1], we perform this selection by solving the Knapsack problem.

## 4   Experiments

### 4.1   Datasets and Implementation Details

**Datasets:** To train and evaluate ATSal and SST-Sal, we utilized 206 videos of the VR-EyeTracking dataset [5] that were used in [26] (two videos were excluded due to the limited clarity in their ground-truth saliency maps). VR-EyeTracking is composed of 147 and 59 short (up to 60 s) videos that have been captured by static and moving camera, respectively. Their visual content is diverse, covering indoor scenes, outdoor activities and music shows. To train the attention model of ATSal, we used 85 high definition ERP images of the Salient360! dataset [7] and 22 ERP images of the Sitzman dataset [20]. These images include diverse visual content captured in both indoor (e.g., inside a building) and outdoor (e.g., a city square) scenes. For further evaluation, we used the 104 videos of the Sports-360 dataset [28]. These videos last up to 60 seconds and show activities from five different sport events. 84 of them have been captured by a static camera, while the remaining ones (18 videos) were recorded using a moving camera. Finally, for training the employed video summarization model we used 100 2D videos that were produced and scored in terms of frame-level saliency, based on the method described in Sect. 3.3. These videos relate to 57 videos of the VR-EyeTracking, 37 videos of the Sports-360, and 6 videos of the Salient360! dataset.

**Implementation Details:** The parameter $t_0$ of the CMDM, that relates to the computed translative offset between a pair of consecutive ERP frames, was set equal to 0.5. The attention model of ATSal was trained for 90 epochs using a set of 2140 ERP images, that was produced after applying the common data augmentation techniques (e.g., rotation, flipping, mirroring) to the 85 and 22 ERP images of the Salient360! and Sitzman datasets, respectively. 1840 images of the produced set were used for training and the remaining ones (300 images) for validation. Training was performed on mini-batches of 80 images, using the Adam optimizer with learning rate equal to $10^{-5}$ and weight decay equal to $10^{-5}$. The overall ATSal method was trained using 140 videos of the VR-EyeTracking dataset, while 66 videos were used as validation set. The utilized videos were decomposed into frames and training was performed on mini-batches of 10 consecutive frames, using the Adam optimizer with learning rate equal to $10^{-5}$ and weight decay equal to $10^{-6}$. To fine-tune SalEMA-Poles and SalEMA-Equators, we ran 20 training epochs using mini-batches of 80 and 10 frames, respectively, using the Binary Cross Entropy optimizer from [14] with learning rate equal to $10^{-6}$. Concerning SST-Sal, the number of hidden-layers was set equal to 9 and the number of input-channels was set equal to 3. Training was performed on mini-batches of 20 consecutive frames for 100 epochs using the Adam optimizer, with a starting learning rate equal to $10^{-3}$ and an adjusting factor equal to 0.1. 92 videos of the VR-EyeTracking dataset were used as a training set and 55 videos were used as a validation set. Concerning the 2D video production step: i) $t_1$ can range in $[0, 255]$ and pixels with saliency score lower than 150 were not considered as important; ii) $t_2$ represents the Haversine distance [15] that affects the number of possibly identified salient regions within a frame by the applied the DBSCAN algorithm, and was set equal to 1.2; $t_3$ is associated to the Euclidean distance between salient regions in consecutive ERP frames, and given the input frames' resolution it was set equal to 100; and $t_4$ affects the duration of the created sub-volumes, and was set equal to 100. Finally, to train the utilized variant of the CA-SUM video summarization model, deep representations were obtained from sampled frames of the videos (2 frames per second) using GoogleNet [24]. The block size of the concentrated attention mechanism was set equal to 20. The learning rate and the L2 regularization factor were equal to $5 \cdot 10^{-4}$, and $10^{-5}$, respectively. For network initialization we used the Xavier uniform initialization approach with gain $= \sqrt{2}$ and biases $= 0.1$. Training was performed for 400 epochs in a full-batch mode using the Adam optimizer and 80 videos of the formed dataset; the remaining ones were used for model selection and testing. Finally, the created summary does not exceed 15% of the 2D video's duration. The code is publicly-available at: https://github.com/IDT-ITI/CA-SUM-360.

## 4.2   Quantitative Results

We evaluated the performance of the developed CMDM and the used methods for saliency detection, both solely and in combination. The comparative assessment of the employed variant of the CA-SUM method against other methods for 360° video story-based summarization [12] and highlight detection [27] was not possible, as the code and datasets of the latter are not publicly-available.

To assess the accuracy of the developed CMDM, we used 329 videos from the VR-EyeTracking, Sports-360 and Salient360! datasets. 232 of these videos

**Table 1.** Performance (Accuracy in percentage) of the CMDM.

|  | Static camera | Moving camera | Total |
|---|---|---|---|
| Number of videos | 232 | 97 | 329 |
| Correctly classified | 200 | 92 | 292 |
| Accuracy | 86.21% | 94.85% | 88.75% |

**Table 2.** Ablation study about the use of the CMDM.

|  | CC | SIM |
|---|---|---|
| ATSal Only | 0.290 | 0.241 |
| SST-Sal Only | 0.377 | 0.279 |
| CMDM & ATSal or SST-Sal | **0.379** | **0.280** |

**Table 3.** Performance of the trained ATSal and SST-Sal models on videos of the VR-EyeTracking (upper part) and Sports-360 (lower part) datasets.

| VR-EyeTracking | | | | | | |
|---|---|---|---|---|---|---|
|  | Static View Videos (55) | | Moving View Videos (11) | | Total Videos (66) | |
|  | CC↑ | SIM↑ | CC↑ | SIM↑ | CC↑ | SIM↑ |
| ATSal | **0.336** | **0.240** | **0.230** | **0.172** | **0.322** | **0.229** |
| SST-Sal | 0.309 | 0.167 | 0.168 | 0.106 | 0.285 | 0.157 |
| Sports-360 | | | | | | |
|  | Static View Videos (86) | | Moving View Videos (18) | | Total Videos (104) | |
|  | CC↑ | SIM↑ | CC↑ | SIM↑ | CC↑ | SIM↑ |
| ATSal | 0.270 | 0.251 | 0.270 | 0.243 | 0.270 | 0.249 |
| SST-Sal | **0.464** | **0.372** | **0.273** | **0.283** | **0.436** | **0.358** |

were captured by a static camera and 97 videos were recorded using a moving camera. The results in Table 1 show that our mechanism correctly classifies a video in more than 88% of the cases, while its accuracy is even higher (close to 95%) in the case of 360° videos recorded using a moving camera. The slightly lower performance in the case of static camera (approx. 86%) relates to mis-classifications due to changes in the visual content of the north and south regions of the 360° video, caused by the appearance of a visual object right above or bellow the camera (see example (d) in Fig. 2); however, the latter is not a desired, and thus common, case when recording a 360° video using a static camera.

Following, to quantify the performance of ATSal and SST-Sal we used the Pearson Correlation Coefficient (CC) and the Similarity (SIM) measures, as proposed in [4]. Moreover, we trained models of ATSal and SST-Sal from scratch, aiming to fine-tune their hyper-parameters and optimize their performance; these models are not the same with the ones released by the authors of the relevant works. The results in Table 3 show that ATSal performs better on videos of the VR-EyeTracking dataset and SST-Sal is more effective on videos of the Sports-360 dataset; so, practically we have a tie. However, the performance of ATSal on videos captured by a moving camera is significantly higher than the

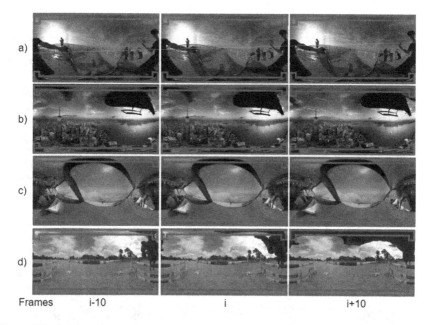

Frames        i-10                    i                    i+10

**Fig. 2.** Examples of ERP frames from videos of the Sports-360 and VR-EyeTracking datasets. Highlighted in red, their north and south regions that are used by the CMDM. (Color figure online)

performance of SST-Sal on VR-EyeTracking, and slightly lower but comparable with the performance of SST-Sal on Sports-360. The opposite can be observed in the case of videos recorded using a static camera, where SST-Sal performs clearly better than ATSal on Sports-360 and comparatively good on VR-EyeTracking. Based on these findings, we integrate both ATSal and SST-Sal in the saliency detection component of the system, and utilize them to analyze 360° videos captured using a moving and a static camera, respectively.

Finally, to evaluate the impact of the utilized CMDM, we formed a large set of test videos (by merging the 104 test videos of the Sports-360 dataset with the 66 test videos from the VR-EyeTracking dataset) and considered three different processing options: a) the use of ATSal only, b) the use of SST-Sal only, and c) the use of both methods in combination with the developed CMDM. The results in Table 2 document the positive impact of the CMDM, as its use in combination with the integrated saliency detection methods results in higher performance.

### 4.3 Qualitative Results

Figure 2 shows indicative examples of ERP frames and their north and south regions (highlighted in red) that are processed by the CMDM. In the first two examples the mechanism detects changes in the aforementioned regions and correctly identifies the use of a moving camera for video recording. However, in the remaining two examples the CMDM fails to make a correct decision. In the third example, despite the fact that the video was recorded from the cockpit

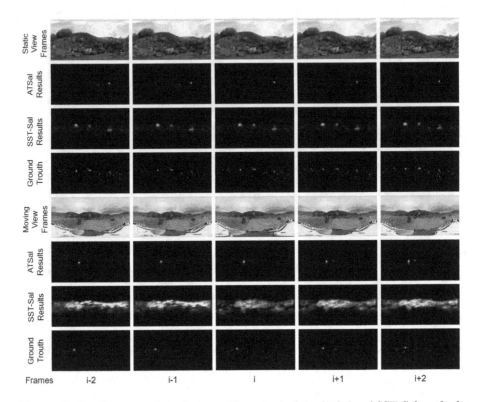

**Fig. 3.** Qualitative comparisons between the output of the ATSal and SST-Sal methods on frame sequences of videos from the VR-EyeTracking dataset.

of a moving helicopter, the CMDM was not able to detect sufficient motion in the regions that focuses on. In the last example, the appearance of a horse right above the static camera led to noticeable changes in the observed regions, thus resulting in the erroneous detection of a moving camera.

Figure 3 presents two sequences of ERP frames, the produced saliency maps by ATSal and SST-Sal, and the ground-truth salience maps. Starting from the top, the first frame sequence was extracted from a 360° video captured using a static camera. From the associated saliency maps we can observe that SST-Sal performs clearly better compared to ATSal and creates saliency maps that are very close to the ground-truth; on the contrary, ATSal fails to detect several salient points. For the second frame sequence, which was obtained from a 360° video recorded using a moving camera, we see the exact opposite behavior. ATSal defines saliency maps that are very similar with the ground-truth, while the saliency maps of SST-Sal method contain too much noise. The findings of our qualitative analysis are aligned with our observations in the quantitative analysis, justifying once again the use of ATSal (SST-Sal) as the best option for analysing 360° videos captured using a moving (static) camera.

The top part of Fig. 4 gives a frame-based overview of the produced 2D video for a 360° video after selecting one frame per detected salient event. As

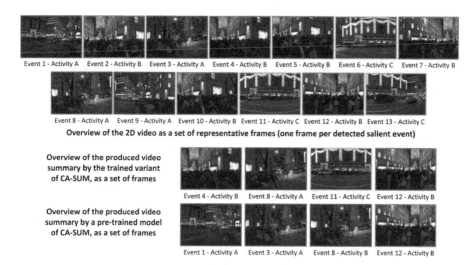

Fig. 4. A frame-based overview of the produced 2D video for a 360° video of the VR-EyeTracking dataset, and the produced summaries by two methods.

shown, the detected events relate to different activities that can happen either sequentially (activities A and C) or in parallel (activity B with A and C). The lower part of Fig. 4 presents the produced summaries by: a) the trained saliency-aware variant of the CA-SUM method using the produced 2D videos for 360° videos of the VR-EyeTracking and Sports-360 and Salient360! datasets, and b) a pre-trained model of CA-SUM using conventional videos from the TVSum dataset [21] for video summarization. As can been seen, the trained variant produces a more complete video summary after including parts of the video showing the gathered people in a square with a Christmas tree (activity B), the persons right in front of the avenue that take some photos (activity A), and the illuminated shopping mall behind the avenue (activity C). On the contrary, the pre-trained CA-SUM model focuses more on video fragments showing the avenue (activity A) and completely ignores parts presenting the shopping mall. This example shows that taking into account the saliency of the visual content is important when summarizing 360° videos, as it allows the production of more comprehensive and thus useful video summaries.

## 5    Conclusions

In this paper, we described an integrated solution for summarizing 360° videos. To create a video summary, our system initially processes the 360° video to indicate the use of a static or moving camera for its recording. Based on the output, the 360° video is then forwarded to one of the integrated state-of-the-art methods for saliency detection (ATSal and SST-Sal), which produce a set of frame-level saliency maps. The latter are utilized to form spatio-temporally-correlated sub-volumes of the 360° video which relate to different salient events in the 360° video, and produce a 2D video that shows these events. Finally, a saliency-aware

variant of a state-of-the-art video summarization method (CA-SUM) analyses
the produced 2D video and formulates the video summary by selecting the most
important and informative events. Quantitative and qualitative evaluations using
two datasets for saliency detection in 360° videos (VR-EyeTracking, Sports-360),
demonstrated the performance of different components of the system and documented their relative contribution in the 360° video summarization process.

# References

1. Apostolidis, E., Adamantidou, E., Metsai, A.I., Mezaris, V., Patras, I.: Video summarization using deep neural networks: a survey. Proc. IEEE **109**(11), 1838–1863
   (2021). https://doi.org/10.1109/JPROC.2021.3117472
2. Apostolidis, E., Balaouras, G., Mezaris, V., Patras, I.: Summarizing videos using
   concentrated attention and considering the uniqueness and diversity of the video
   frames. In: Proceedings of the 2022 International Conference on Multimedia
   Retrieval, ICMR 2022, pp. 407–415. Association for Computing Machinery, New
   York, NY, USA (2022). https://doi.org/10.1145/3512527.3531404
3. Bernal-Berdun, E., Martin, D., Gutierrez, D., Masia, B.: SST-Sal: a spherical
   spatio-temporal approach for saliency prediction in 360 videos. Comput. Graph.
   **106**, 200–209 (2022). https://doi.org/10.1016/j.cag.2022.06.002
4. Bylinskii, Z., Judd, T., Oliva, A., Torralba, A., Durand, F.: What do different
   evaluation metrics tell us about saliency models? IEEE Trans. Pattern Anal. Mach.
   Intell. **41**(3), 740–757 (2019). https://doi.org/10.1109/TPAMI.2018.2815601
5. Dahou, Y., Tliba, M., McGuinness, K., O'Connor, N.: ATSal: an attention based
   architecture for saliency prediction in 360° videos. In: Del Bimbo, A., et al. (eds.)
   ICPR 2021. LNCS, vol. 12663, pp. 305–320. Springer, Cham (2021). https://doi.
   org/10.1007/978-3-030-68796-0_22
6. Ester, M., Kriegel, H.P., Sander, J., Xu, X.: A density-based algorithm for discovering clusters in large spatial databases with noise. In: Proceedings of the Second
   International Conference on Knowledge Discovery and Data Mining, pp. 226–231.
   AAAI Press (1996)
7. Gutiérrez, J., David, E.J., Coutrot, A., Da Silva, M.P., Callet, P.L.: Introducing un
   salient360! Benchmark: a platform for evaluating visual attention models for 360°
   contents. In: 2018 10th International Conference on Quality of Multimedia Experience (QoMEX), pp. 1–3 (2018). https://doi.org/10.1109/QoMEX.2018.8463369
8. Haidar Sharif, M., Martinet, J., Djeraba, C.: Motion saliency. Encycl. Multimedia,
   442–444 (2008)
9. Hu, H.N., Lin, Y.C., Liu, M.Y., Cheng, H.T., Chang, Y.J., Sun, M.: Deep 360 pilot:
   learning a deep agent for piloting through 360° sports videos. In: Proceedings of the
   IEEE Conference on Computer Vision and Pattern Recognition (CVPR) (2017)
10. Kang, K., Cho, S.: Interactive and automatic navigation for 360° video playback.
    ACM Trans. Graph. **38**(4) (2019). https://doi.org/10.1145/3306346.3323046
11. Lebreton, P., Raake, A.: GBVS360, BMS360, ProSal: extending existing saliency
    prediction models from 2D to omnidirectional images. Sig. Process. Image Com-
    mun. **69**, 69–78 (2018). https://doi.org/10.1016/j.image.2018.03.006
12. Lee, S., Sung, J., Yu, Y., Kim, G.: A memory network approach for story-based
    temporal summarization of 360° videos. In: Proceedings of the IEEE Conference
    on Computer Vision and Pattern Recognition (CVPR) (2018)

13. Li, J., Han, L., Zhang, C., Li, Q., Liu, Z.: Spherical convolution empowered viewport prediction in 360 video multicast with limited FoV feedback. ACM Trans. Multimedia Comput. Commun. Appl. **19**(1) (2023). https://doi.org/10.1145/3511603
14. Linardos, P., Mohedano, E., Nieto, J.J., O'Connor, N.E., Giró-i-Nieto, X., McGuinness, K.: Simple vs complex temporal recurrences for video saliency prediction. CoRR abs/1907.01869 (2019). https://arxiv.org/abs/1907.01869
15. Nichat, M.: Landmark based shortest path detection by using a* algorithm and haversine formula (2013)
16. Qiao, M., Xu, M., Wang, Z., Borji, A.: Viewport-dependent saliency prediction in 360° video. IEEE Trans. Multimedia **23**, 748–760 (2021). https://doi.org/10.1109/TMM.2020.2987682
17. Ren, S., He, K., Girshick, R., Sun, J.: Faster R-CNN: towards real-time object detection with region proposal networks. In: Cortes, C., Lawrence, N., Lee, D., Sugiyama, M., Garnett, R. (eds.) Advances in Neural Information Processing Systems, vol. 28. Curran Associates, Inc. (2015)
18. Setayesh, M., Wong, V.W.: A content-based viewport prediction framework for 360° video using personalized federated learning and fusion techniques. In: 2023 IEEE International Conference on Multimedia and Expo (ICME), pp. 654–659 (2023). https://doi.org/10.1109/ICME55011.2023.00118
19. Simonyan, K., Zisserman, A.: Very deep convolutional networks for large-scale image recognition. In: International Conference on Learning Representations (2015)
20. Sitzmann, V., et al.: Saliency in VR: how do people explore virtual environments? IEEE Trans. Visual Comput. Graphics **24**(4), 1633–1642 (2018). https://doi.org/10.1109/TVCG.2018.2793599
21. Song, Y., Vallmitjana, J., Stent, A., Jaimes, A.: TVSum: summarizing web videos using titles. In: 2015 IEEE/CVF Conference on Computer Vision and Pattern Recognition (CVPR), pp. 5179–5187 (2015). https://doi.org/10.1109/CVPR.2015.7299154
22. Su, Y.C., Grauman, K.: Making 360° video watchable in 2D: learning videography for click free viewing. In: Proceedings of the IEEE Conference on Computer Vision and Pattern Recognition (CVPR) (2017)
23. Su, Y.C., Jayaraman, D., Grauman, K.: Pano2Vid: automatic cinematography for watching 360 videos. In: Proceedings of the Asian Conference on Computer Vision (ACCV) (2016)
24. Szegedy, C., et al.: Going deeper with convolutions. In: 2015 IEEE/CVF Conference on Computer Vision and Pattern Recognition (CVPR), pp. 1–9 (2015). https://doi.org/10.1109/CVPR.2015.7298594
25. Wang, M., Li, Y.J., Zhang, W.X., Richardt, C., Hu, S.M.: Transitioning360: content-aware NFoV virtual camera paths for 360° video playback. In: 2020 IEEE International Symposium on Mixed and Augmented Reality (ISMAR), pp. 185–194 (2020). https://doi.org/10.1109/ISMAR50242.2020.00040
26. Xu, Y., et al.: Gaze prediction in dynamic 360° immersive videos. In: 2018 IEEE/CVF Conference on Computer Vision and Pattern Recognition, pp. 5333–5342 (2018). https://doi.org/10.1109/CVPR.2018.00559
27. Yu, Y., Lee, S., Na, J., Kang, J., Kim, G.: A deep ranking model for spatio-temporal highlight detection from a 360 video. In: Proceedings of the 2018 AAAI Conference on Artificial Intelligence (2018)
28. Zhang, Z., Xu, Y., Yu, J., Gao, S.: Saliency detection in 360° videos. In: Ferrari, V., Hebert, M., Sminchisescu, C., Weiss, Y. (eds.) ECCV 2018. LNCS, vol. 11211, pp. 504–520. Springer, Cham (2018). https://doi.org/10.1007/978-3-030-01234-2_30

# Brave New Ideas

# Mutant Texts: A Technique for Uncovering Unexpected Inconsistencies in Large-Scale Vision-Language Models

Mingliang Liang$^{(\boxtimes)}$, Zhouran Liu, and Martha Larson$^{(\boxtimes)}$

Radboud University, Nijmegen, Netherlands
{m.liang,z.liu,m.larson}@cs.ru.nl

**Abstract.** Recently, Vision-Language Models (VLMs) trained on large-scale noisy data have shown strong generalization abilities on many downstream tasks. In this paper, we introduce a new technique for uncovering unexpected inconsistencies in VLMs, which lead to the formulation of new research questions on how to improve VLMs. Specifically, we propose that performance on original texts should be compared with that of 'mutant texts', carefully-designed variants of the original texts. In contrast to text perturbations used to study robustness, 'mutant texts' represent large changes in the original texts that impact semantics. We present two types of example mutant texts: one-word-only (OWO) mutants, which replace the original text with one of the words it contains and plus-one-word (POW) mutants, which add a word to the original text. The mutant texts allow us to discover the existence of dominating words in texts that correspond to images. The embedding of a dominating words is closer to the image embedding than the embedding of the entire original text. The existence of dominating words reflects underlying inconsistency in a VLM's embedding space, a possible source of risk for bias undetected without the mutant text technique.

**Keywords:** Vision-language models · Mutant texts · Dominating words · Cross-modal representation · Joint Representation space

## 1 Introduction

Vision-Language Models (VLMs) provide a strong basis for downstream tasks [8, 10,13,24–26,34,36]. Researchers consider a VLM to be of high quality if it supports good performance in downstream tasks. However, currently, downstream tasks are treated as static and isolated. In this paper, we propose that researchers can gain deeper understanding of VLMs by studying tasks in pairs and comparing their performance. We introduce the idea of *mutant texts*, which are versions of the original texts that include a carefully-designed mutation. By studying the difference in performance between the original and various types of mutant

S. Rudinac et al. (Eds.): MMM 2024, LNCS 14557, pp. 219–233, 2024.
https://doi.org/10.1007/978-3-031-53302-0_16

texts, researchers can test expectations of the VLM. When expectations fail to be upheld, a type of mutant texts has been successful in generating a new research question about VLMs that will point out directions in which they can be improved.

| Image | Text | Mutant Text |
|-------|------|-------------|
|   | a computer, telephone, music player and lamp on a wooden desk. **(0.1663)** | desk **(0.2092)** |
|   | White bathroom with a tile wall an opened window a toilet and sink and a bathtub with a curtain. **(0.2224)** | bathroom **(0.2514)** |
|   | A number of train tracks with a man standing near a tower. **(0.0556)** | train **(0.1496)** |

**Fig. 1.** The full text (middle) of an image-text pair includes *more information* about the corresponding image (left) than the mutant text consisting of one-word-only (OWO) does (right). However, we see for the examples here that the embedding of the full text is *less similar* to the embedding of the image than mutant text is (see similarities in parentheses). We call such examples, *dominating words*. (VLM is OpenCLIP pre-trained on LAION-400M with ViT-B/32 image encoder.)

Figure 1 illustrates a key example of a type of mutant text that we will discuss in this paper. The columns labeled 'Image' and 'Text' provide three image-text pairs, which are drawn from our data set. The original task is cross-modal retrieval, which involves calculating similarity score of the embedding of the text to the image embedding in the vision-language embedding (VLE) space of the VLM. The similarity score is given in parentheses in the "Text" column. The mutant-text version of this task is called "one word only" (OWO). We create a mutant version of the original text by retaining only a single word that occurred in the original text. We then calculate similarity score of the embedding of this single word to the image embedding in VLE space. The similarity score is given in parentheses in the final column.

The mutant text is interesting because it is designed on the basis of a specific expectation about how the VLE space should behave if it is a good, consistent space. The expectation is that the full text reflects a more specific semantic description of the image than a single word does. For this reason, we anticipate that the embedding of the text is closer to the image than the embedding of any single word it contains. The examples in Fig. 1 demonstrate that, unexpectedly, this is not the always case. Instead, there exist words, such as the three shown in the figure, whose embeddings are *closer* to the image embedding, than the embedding of the entire sentence, as illustrated in this examples. We call such words *dominating words*. The existence of dominating words is unexpected in

VLE space. It is important to note the problem that dominating words pose for downstream tasks. Specifically, for cross-modal retrieval, an image will retrieve one-word texts at top ranks, since those texts are close to the image in the embedding space than more descriptive texts.

The idea of mutant texts is brave because their goal is to create research questions, and not to answer them. Only by asking new research questions can we discover the fundamental strengths and weaknesses of VLMs. On the surface, our work seems similar to two different lines of research. We discuss each in turn, pointing out how our idea is different. First, recent research has introduce perturbations into texts before carrying out cross-modal tasks [23]. This research is aimed at measuring robustness, and the authors are careful that any changes that they make to the texts do not affect the semantics of the text. In our mutant text approach, we are interested in uncovering buried flaws in the VLM by comparing the performance of original and mutant texts. Mutant texts are specifically designed to be unnatural and to change the semantics of the original text. They do not measure robustness, but rather their goal is help researchers find hidden holes and biases. Second, recent research has studied the negative social bias that is inherent in VLMs [33]. In this work, the type of bias is known in advance, and the VLM is tested for this bias. In our work, we do not focus on negative social biases that are known in advance, but rather seek to uncover other potential biases, e.g., the dominating behavior of the words in the right-hand column of Fig. 1.

In this paper, we carry out cross-modal retrieval experiments on pre-trained VLMs that utilize a dual-stream encoder to align visual and language semantic representations in a shared embedding space. The focus of our work is CLIP [24] and LiT [36], and not ALIGN [10], since it is not open source. Our work makes the following contributions:

- We propose that researchers should not be afraid to test the impact of radical, semantic-changing modifications of texts. Comparing the performance of VLMs on original texts and on the changed, *mutant texts* leads to new research questions.
- We present two examples of mutant texts that uncover unexpected inconsistencies in large-scale vision-language embedding models. First, "one-word-only" (OWO), already illustrated in Fig. 1, and, second, "plus-one-word" (POW).
- We demonstrate the existence of *dominating words* and show that this unexpected phenomenon centers on nouns. We carry out experiments that provide insight into the impact of dominating words.
- We discuss other possible types of mutant text that could be studied in future work.

The paper is organized as follows. In Sect. 2, we cover the necessary background and related work and in Sect. 3 we introduce our experimental framework. Section 4 covers the "one-word-only" (OWO) mutant texts, showing that they reveal the existence of dominating words, which are predominantly nouns.

Section 5 explores dominating words with respect to "plus-one-word" (POW) mutant texts, in which we experiment with the impact of adding dominating words to texts. Finally, Sect. 6 presents an outlook to how researchers can build on the idea of mutant texts to formulate and investigate new research questions related to VLMs in the future.

## 2   Related Work

In this section, we provide the necessary background on Vision-Language Models (VLMs). Learning visual-semantic embeddings using natural language supervision, which maps images and their corresponding textual descriptions into a shared embedding space, gives models the ability to carry out zero-shot prediction, i.e., generate a prediction for a new class that they have not been explicitly trained on. DeViSE is the first work to learn the visual-semantic embeddings from language supervision using a deep learning network [7]. DeViSE used a skip-gram language model to output the word embeddings, which are used as supervised labels to train a visual model by hinge rank loss [7,32]. The skip-gram text model is pre-trained on a corpus of 5.7 million documents extracted from wikipedia.org. This allows the model to predict unseen categories by using the semantic knowledge learned from the language. This seminal work aimed to explore the capabilities of VLMs within the context of limited computing power. Another model that has been successful in this respect is VSE++, which learns visual-semantic embedding for cross-modal retrieval with a hard negative mining rank loss function [6,27]. VSE++ was trained on the MS COCO and Flickr 30K datasets [6,17,35], which were annotated by humans to describe the details of the images. Obtaining manually annotated data has proven to be not only costly, but also difficult on a large scale.

In recent work, more and more researchers focus on CLIP-like models, which pre-train VLMs with data consisting of large-scale noisy image-text pairs from the Internet [10,13,24,36]. These VLMs are pre-trained to align visual and language representations in a shared embedding space using a dual stream encoder. The presentation of images and text is learned by contrastive learning loss, which pushes positive image-text pairs closer to each other and separates negative image-text pairs [21]. This approach has been shown to be effective for representation learning of both images and text in previous work [2,6,7,9]. These dual stream VLMs have also been used to improve the performance of cross-modal tasks, such as using the frozen text encoder of CLIP to generate high-resolution images [25,26,34], and using frozen image encoder of CLIP to retrieve knowledge from Wikidata in order to improve the performance of Visual Question Answering (VQA) task [8]. The baseline model is CLIP [24], which learns to align the distributions of images and text in the embedding space so that semantically similar images and text are close together. This consistency is achieved through a training process, where CLIP is trained on a large-scale data of image-text pairs to learn the relationships between their semantics. Although CLIP was originally pre-trained using private data, OpenCLIP has made the model accessible to the public by pre-training it with the public LAION dataset and open-source Open-CLIP repository [3,28]. LiT allows text encoders to learn better representations from a frozen ViT model pre-trained on JFT-3B dataset [4,36]. The model shows better performance because the image encoder is pre-trained on more than 300 million labeled images.

Unlike CLIP, a two-stream model incorporating image and text encoders, ViLBERT, OSCAR, METER, and ViLT are single-stream models. In these models, images and text are combined and fed into the cross-attention layer for richer multimodal interaction, resulting in superior performance at the expense of higher computational cost [5,12,14–16]. Since these models cannot generate semantic embeddings of images and text, they are not in the scope of this paper.

Despite the remarkable performance of VLMs in various visual-semantic embedding tasks, in the VLM domain, there has been limited exploration of the shared embedding space of these models to gain a deeper understanding of their functioning. Therefore, in this work, we focus on the shared embedding space of VLMs to reveal the workings of VLMs and we report some semantic inconsistencies in VLMs.

## 3   Experimental Setup

We carry out a series of cross-modal retrieval experiments on CLIP-type VLMs. The main model we use is OpenCLIP that is pre-trained on the LAION-400M dataset. The image encoder is ViT-B/32 [3,28]. We select this model as our main model because it provides excellent performance and is pre-trained on the same scale of data as the original CLIP.

The other CLIP-type models that we test are the original CLIP [24] and LiT [36]. CLIP is pre-trained on private-400M and LiT is pre-trained on YFCC-100M [31], while the image encoders are ViT-B/32 and ViT-B/16 [4] respectively.

In order to create mutant texts and to test the models we use MS COCO (Microsoft Common Objects in Context) [17], a widely-used image recognition and captioning data set containing 113K images with 5 captions per image that have been provided by human annotators. Note that the ratio of detected duplicates between LAION dataset and MS COCO is 1.12% [3]. Cross-modal retrieval with MS COCO involves retrieving relevant images given a text that serves as a textual query or retrieving relevant texts given an image query.

Following [11], the MS COCO data set is split into 113,287 training images, 1K validation images, and 5K test images. For our initial experiments with mutant texts, we randomly extract 10K samples from the MS COCO training data in order to be experimenting with a large and varied set. Then, we carry out retrieval experiments on the standard 5K test set. In all experiments, we used cosine distances to calculate similarity scores between image embeddings and the text embeddings produced by the VLM. Retrieval is carried out by ranking the results on the basis of this score.

For visualization of the embeddings a lower-dimensional, we use t-SNE [18] to visualise the embeddings space of VLM. With the visualization we seek to understand the nature of dominating words.

## 4    One-Word-Only (OWO) Mutant Texts

If the VLM embedding space is consistent, we expect that the original full text will always be closer to the corresponding image than a one-word-only (OWO) mutant text consisting of only a single word from the original text. We investigate cases where this expectation fails, as in the examples in Fig. 1.

**Table 1.** Percentage reports the percentage of texts that contain a dominating noun (verb, or adjective), given that the text contains at least one noun (verb, or adjective). A higher percentage of texts involve dominating nouns than dominating verbs or adjectives. (VLM is OpenCLIP pre-trained on LAION-400M with ViT-B/32 image encoder.)

| Types | Percentage | Top Words |
|---|---|---|
| Noun | 17.5% | man, train, giraffe, tennis, sign, bus |
| Verb | 5.2% | is, are, sitting, holding, surfing |
| Adjective | 3.1% | young, large, close, small, white |

**Table 2.** Percentage drop in similarity scores of images when different types of text are retained compared to the original text. 'Retention' is the percentages of words retained from the original text. (VLM is OpenCLIP pre-trained on LAION-400M with ViT-B/32 image encoder.)

| Noun | Verb | Adjective | Others | Retention | Score Change |
|---|---|---|---|---|---|
| ✗ | ✗ | ✗ | ✓ | 44.4% | 32.0% |
| ✓ | ✗ | ✗ | ✗ | 35.6% | 13.2% |
| ✗ | ✓ | ✗ | ✗ | 14.9% | 34.9% |
| ✗ | ✗ | ✓ | ✗ | 14.2% | 34.1% |
| ✓ | ✓ | ✗ | ✗ | 47.3% | 10.5% |
| ✓ | ✗ | ✓ | ✗ | 43.9% | 10.5% |
| ✓ | ✓ | ✓ | ✗ | 55.6% | 7.4% |

### 4.1    Dominating Words

We investigate OWO mutant texts for the case the single remaining word is an adjective, verb, or noun. Other parts of speech, such as determiners or conjunctions, we leave for future work. This experiment is carried out on 10K text-image pairs randomly selected from the MS COCO training set. First, we select the texts that contain at least one noun. We test each OWO mutant text to see if its embedding is closer to the embedding of the corresponding image than the embedding of the full text. If it is, we have identified a dominating noun. We calculate the percentage of texts that contain at least one dominating noun. We do the same for verbs and adjectives. Note that about 99.99% of the texts contain at least one noun and about 79.6% and 59.3% of the texts contain at least one verb and adjective.

The results are reported in Table 1. We define the most dominating words as words that are dominating in a large proportion of the sentences in which they occur. For each case, we have listed the top most dominating words. We see that nouns are the most pervasive dominating words. It is tempting to link this fact to the fact that nouns correspond to objects depicted in the image, and as such may be easier for the VLM to learn. However, we see that "is", a verb without semantic content, is one of the top dominating verbs. As such, OWO mutant words have given rise to a unexpected and curious inconsistency in VLMs that deserves further study.

**Table 3.** The percentage reports the proportion of texts that contain a dominating noun. The final column provides examples the nouns that are most frequently dominating in the sentences in which they occur.

| Model | Data | Image Encoder | Percent | Top Number Dominating Nouns |
|---|---|---|---|---|
| CLIP | Private-400M | ViT-B/32 | 8.0% | train, kitchen, giraffe, pizza, skateboard |
| OpenCLIP | LAION-400M | ViT-B/32 | 17.5% | man, train, giraffe, tennis, sign, bus |
| | LAION-2B | ViT-B/32 | 16.4% | train, giraffe, tennis, bathroom, food |
| | LAION-400M | ViT-B/16 | 14.0% | train, tennis, pizza, baseball, giraffe |
| | LAION-2B | ViT-B/16 | 18.0% | train, tennis, giraffe, pizza, room |
| | LAION-2B | ViT-L/14 | 17.7% | train, giraffe, skateboard, street, baseball |
| | LAION-2B | ViT-H/14 | 19.8% | train, tennis, giraffe, baseball, food |
| LiT | YFCC-100M | ViT-B/16 | 20.5% | bathroom, kitchen, tennis, desk, giraffe |

We carry out another experiment that reveals the importance of nouns. In Table 2, we analyze different combination of different types of words. When we retain all the nouns, verbs, and adjectives from the text, approximately 55.6% of the original content remains. In comparison to the initial text, there is a 7.4% drop in the similarity scores with the image. The first row of the Table 2 also shows the effect of removing nouns, verbs, and adjectives, and the similarity scores of the images and text change significantly when all nouns, verbs, and adjectives are removed. The other rows of Table 2 reveal that, while the nouns of a text dominate the semantics of the text, verbs and adjectives also play a role in learning the semantics of the text. In the remainder of the paper, we zero in to focus on nouns of the text.

## 4.2 Dominating Nouns

The first step in exploring the unexpected phenomenon of dominating nouns is to check whether they exist across multiple CLIP-like models. In Table 3, we show the percent of texts that contain a dominating noun for a variety of different models. We note that when the VLM is pre-trained on different datasets [24,28, 29,31] a somewhat different percentage of texts contain at least one dominating noun (8.0%, 17.5%, 16.4% and 20.5%). Notably, despite these differences, the top dominating nouns across these settings remain similar, such as "giraffe," "bathroom," and "kitchen."

It is also interesting to ask how strong the domination effect is. To explore the strength we calculated the percent by which the similarity score between the dominating noun and the image exceeds the similarity score between the original text and the image for our main mode (OpenCLIP trained on LAION-400M with ViT-B/32). The increase in similarity score was calculated to be 10.1%.

**Table 4.** The cases of training data from LAION [28] (100K samples). As we can see in these cases, the image pairs of dominating nouns are more similar and occur alone than the non-dominating nouns. For giraffes in particular, we extracted only 6 out of 100K samples from LAION-400M, and most of the images were of single giraffes.

| Concepts | Images |
|----------|--------|
| cat | |
| giraffe | |

**Table 5.** The most dominating nouns extract from MS COCO training set (random select 10K samples). The second column displays the raw number of dominating words, while the third column represents how often a word is dominating when it occurs. (VLM is OpenCLIP pre-trained on LAION-400M with ViT-B/32 image encoder.)

| Concept | Number | Ratio (%) |
|---------|--------|-----------|
| skier | 10 | 27.78 |
| bedroom | 11 | 26.83 |
| refrigerator | 13 | 25.49 |
| giraffes | 17 | 21.25 |
| boat | 20 | 20.62 |
| boats | 11 | 18.97 |
| kitchen | 36 | 17.91 |
| elephants | 11 | 17.19 |
| plane | 16 | 17.02 |
| zebras | 13 | 16.88 |

In Table 4 we depict some samples from the LAION training data 'cat', which is not frequently dominating and 'giraffe' a top-dominating noun. We notice that giraffe images often depict a single giraffes, where as cats co-occur with other objects. However, this trend is far from systematic leaving the question opens of why similarities occur between cases involving dominating nouns across CLIP-type models and data sets. The training data of OpenCLIP (LAION-400M) [28] shows a higher percentage of texts that have dominating nouns (27.9%).

In Table 5, we report the top most dominating nouns, together with the raw count of the number of times the word occurs in the data set and the ratio of the number of texts in which a word is dominating to the number of texts in which a word appears. Interestingly, there is no apparent trend between the number of times a word occurs and the fact that it is a highly dominating word.

To further figure out why some specific nouns dominate the semantic of text, we visualize image and text embedding spaces of OpenCLIP separately and we label the embeddings with the nouns: If a noun appears in the text, we label the image and text with the noun. As is shown in Fig. 2, the embeddings of text and image that involve the most dominating nouns (bathroom, giraffe, . . . ) are more highly clustered than the less-dominating nouns (man, girl, . . . ). This suggests that the model has learned to associate these dominating nouns with certain visual or semantic features, and that these associations are consistent across the image and text modalities.

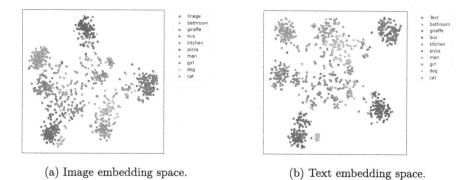

(a) Image embedding space.                    (b) Text embedding space.

**Fig. 2.** We visualise the image and text embedding spaces of CLIP [18, 24] separately to check for dominating and non-dominating nouns. The embedding of text and image include the dominating nouns more possible in the same cluster than non-dominating nouns. (VLM is OpenCLIP pre-trained on LAION-400M with ViT-B/32 image encoder.)

Finally, we run cross-modal retrieval tests to observe the impact of dominating nouns. In Table 6, the results calculated with respect to the original texts are in the first row. The other rows report the case that texts containing a specific most-dominating noun (bathroom, giraffe, bus) or a specific less-dominating noun (dog, cat, man) are replaced by an OWO mutant text. We see from Table 6 that replacing texts containing dominating nouns (upper part of the table) with the words themselves did not substantially affect the model's performance, indicating that the semantic of the text remained unchanged. When we replace the less-dominating nouns (lower part of the table), the performance of the model decreases more than for the dominating nouns. This results points to the most-dominating nouns being inconsistencies in the VLM that need to be addressed in order to improve the quality of the VLE space.

**Table 6.** We replace the whole text with a single word if that word appears in the text. ✗ is no replacement and word-R stands for the words for which we tested replacement. The top three words are strongly dominating, the bottom three not. (VLM is OpenCLIP pre-trained on LAION-400M with ViT-B/32 image encoder.)

| Word | Text retrieval | | | Image retrieval | | | mR |
|---|---|---|---|---|---|---|---|
| | R@1 | R@5 | R@10 | R@1 | R@5 | R@10 | |
| ✗ | 52.4 | 76.4 | 84.6 | 34.3 | 60.0 | 70.6 | 63.1 |
| bathroom-R | 51.7 | 75.5 | 83.6 | 33.7 | 58.9 | 69.3 | 62.3 |
| giraffe-R | 51.8 | 75.1 | 83.0 | 33.8 | 58.9 | 69.3 | 62.2 |
| bus-R | 51.5 | 75.2 | 83.1 | 33.6 | 58.7 | 69.0 | 61.9 |
| dog-R | 50.3 | 73.5 | 81.6 | 32.5 | 57.2 | 67.5 | 60.4 |
| cat-R | 50.5 | 73.8 | 81.8 | 32.6 | 57.1 | 67.3 | 60.4 |
| man-R | 47.4 | 71.0 | 78.6 | 28.3 | 49.6 | 58.4 | 55.6 |

## 5   Plus-One-Word (POW) Mutant Texts

If the VLM embedding space is consistent, we expect that adding an arbitrary word to a text will move the text embedding, but not to an unduly large degree. The arbitrary word will not fit the semantics of the text, and the rest of the words should compensate by helping to balance its effects. We create "plus-one-word" (POW) mutant texts in order to determine if the most-dominating words have an overwhelming effect on a text. POW mutant texts will also show if the most dominating words are dominating because of the contexts in which they occur, or if there is some inconsistency in the VLM specific to dominating words.

**Table 7.** Plus one word (POW). "Word" is the word that was added to the text. "mR" is the average of the image-text retrieval results. (VLM is OpenCLIP pre-trained on LAION-400M with ViT-B/32 image encoder.)

| Word | Text retrieval | | | Image retrieval | | | mR |
|---|---|---|---|---|---|---|---|
| | R@1 | R@5 | R@10 | R@1 | R@5 | R@10 | |
| bathroom | 46.9 | 71.9 | 80.4 | 23.3 | 46.1 | 57.1 | 54.3 |
| giraffe | 48.6 | 73.5 | 82.0 | 24.9 | 46.7 | 57.4 | 55.5 |
| bus | 49.6 | 73.4 | 82.0 | 25.4 | 49.3 | 60.5 | 56.7 |
| kitchen | 50.0 | 73.6 | 81.7 | 25.8 | 49.1 | 60.0 | 56.7 |
| pizza | 50.0 | 73.6 | 81.7 | 25.8 | 49.1 | 60.0 | 56.7 |
| girl | 49.0 | 73.8 | 81.7 | 27.7 | 52.2 | 63.6 | 58.0 |
| dog | 49.0 | 73.6 | 81.9 | 24.6 | 48.6 | 60.1 | 56.3 |
| cat | 50.0 | 74.1 | 82.1 | 26.8 | 50.9 | 61.9 | 57.6 |
| man | 49.9 | 74.0 | 82.0 | 28.8 | 53.8 | 65.2 | 58.9 |

To create POW mutant texts, nouns are added to the end of the original text. In the Table 7, the results of CLIP indicate that the most dominating nouns (top half) can degrade the model's performance more than randomly chosen nouns (bottom half). Apparently, the model is unduly impacted by dominating nouns, even when they appear in irrelevant texts. In comparison to text retrieval, image retrieval tends to exhibit a greater decrease in performance.

**Table 8.** Randomly swapping the image and text embedding of image-text pairs in the test set, leading to a substantial performance drop. (VLM is OpenCLIP pre-trained on LAION-400M with ViT-B/32 image encoder.).

| Model | Replace | Text retrieval | | | Image retrieval | | |
|---|---|---|---|---|---|---|---|
| | | R@1 | R@5 | R@10 | R@1 | R@5 | R@10 |
| OpenCLIP | ✗ | 52.4 | 76.4 | 84.6 | 34.3 | 60.0 | 70.6 |
| | ✓ | 10.5 | 12.3 | 23.4 | 4.7 | 23.7 | 39.7 |

# 6    Conclusion and Outlook

We propose that researchers investigating the quality of VLMs should make use of mutant texts, which introduce large differences, including semantic changes, into the original texts. By comparing VLM performance with mutant texts and original text, researchers can uncover inconsistencies of VLMs and formulate new research questions that will help to improve them. We introduce two types of mutant texts "one-word-only" (OWO) and "plus-one-word" (POW) for cross-modal retrieval of CLIP-type VLMs and discuss the phenomenon of dominating words that these mutant texts allow us to uncover and explore.

## 6.1    More Types of Mutant Text

Our goal has been to introduce mutant texts and to argue that they provide an useful technique in uncovering inconsistencies in VLMs that can lead to new research questions. Future work will invent more mutant texts. In order to provide an impression of the future potential, we discuss a few other possibilities that we have explored briefly here.

**Shuffling Words.** We conducted experiments on another type of mutant text created by shuffling words in the text. Surprisingly, this mutant text yielded similar results to the case in which we removed all words except for nouns. Specifically, 84.3% of the texts showed a decrease in their similarity scores, while the remaining 15.7% saw an increase. On average, the drop in similarity score was 0.024, which represents an 8.0% difference from the original score.

This experiment suggests that VLMs are learning something about word order, because destroying the word order leads to a drop in performance. However, in the 15.7% cases in which an increase occurs points to the VLM not being completely consistent in this respect. Shuffled mutant texts open up important questions about the extent to which VLM actually moves beyond bag-of-words type representations to capture the syntax of language.

**Swapping Embeddings.** We randomly swap the text and the image embeddings of one text-image pair in the data set, and measure the impact on cross-modal retrieval. Results are reported in Table 8. The swapping leads to considerable performance deterioration, which is surprising given that only one pair has been swapped. What is happening is that in the VLM space the image embeddings are clustered together and the text embeddings are clustered together. If one of the text embeddings is replaced by an image embedding, query image embeddings will match that image embedding because they are more similar to it than any other text embedding. Similarly, if one of the image embeddings is replaced by a text embeddings, query text embeddings will match that text embedding. These results suggest that CLIP-like VLMs avoid learning the semantics of image-text pairs and focus on pushing mismatched pairs further apart. We note that the same drop in performance is not observed for VSE++, offering an interesting path for future work.

**From Words to Phrases.** In this work, we have created mutant texts by changing single words. However, nouns are often more than single words. For instance, "motor bike" is a multi-word noun. Also, noun phrases are interesting to study. Our initial experiments with one-noun-phrase-only mutant texts revealed that simple noun phrases with no adjectives, e.g., "a giraffe" are dominating. The phrases contained roughly the same nouns that were found to be dominating in the OWO experiments. Other phrases are interesting to look at in the future.

## 6.2   The Big Picture

Our aim is to encourage researchers to move away from comparing original texts with perturbed texts with no semantic change towards comparing original texts with mutant texts, that are designed to reveal inconsistencies in the space. We can continue as a research community to develop downstream tasks that test VLM embeddings. However, it would be more effective if we could test the quality of the space directly. The idea of mutant texts is a move in that direction. A large number of mutant texts can be created on the basis of a single downstream task. Also future work can test mutant versions of text from the training data.

We note that testing the quality of the VLM embedding space directly would move the multimedia research field into alignment with Natural Language Processing (NLP). NLP researchers create Distributed Semantic Models (DSMs) which learn semantic information from text corpus, such as Word2Vec with skip-gram or continuous bag of words (CBOW) and GloVe [19,20,22]. After training the corpus of text, DSMs can generate embeddings of words that represent their semantics, which conventionally has tried to test text embedding spaces independently of downstream tasks. Understanding semantic consistency directly in the embedding space has been the subject of research. Brown et al. [1] used tested different semantic relationships between different types of words on eight popular DSMs. The mutant texts we use here encompass both semantic relationships (i.e., the whole text should be more specific than a single words) and also information about which relationships should not occur (i.e., one word should not unduly determine similarity).

We point out that the idea of mutant texts can also be extended to images. Here, the idea would be to make a large change to an image that changes how the image looks and use this to study the way the VLM represents images. We might consider a recent study focused on changes to images [30] to be a step in that direction. Specifically, this study discovered that when a red circle is drawn on the image, the model's attention is naturally directed towards the region where the circle is located. Remarkably, the VLMs still understand the whole image, retaining the global information. Their findings indicate that VLMs possess the ability to identify and understand rare events, like a red circle, within large-scale datasets. In the future, the technique of mutant texts can also be used to zero in on rare occurrences and prevent models from models learning undesirable behaviours or biases.

# References

1. Brown, K.S., et al.: Investigating the extent to which distributional semantic models capture a broad range of semantic relations. Cogn. Sci. **47**(5), e13291 (2023)
2. Chen, T., Kornblith, S., Norouzi, M., Hinton, G.E.: A simple framework for contrastive learning of visual representations. In: ICML, vol. 119, pp. 1597–1607 (2020)
3. Cherti, M., et al.: Reproducible scaling laws for contrastive language-image learning. In: CVPR, pp. 2818–2829 (2023)
4. Dosovitskiy, A., et al.: An image is worth $16 \times 16$ words: transformers for image recognition at scale. In: ICLR (2021)
5. Dou, Z., et al.: An empirical study of training end-to-end vision-and-language transformers. In: CVPR, pp. 18145–18155 (2022)
6. Faghri, F., Fleet, D.J., Kiros, J.R., Fidler, S.: VSE++: improving visual-semantic embeddings with hard negatives. In: BMVC, p. 12 (2018)
7. Frome, A., et al.: DeViSE: a deep visual-semantic embedding model. In: Burges, C.J.C., Bottou, L., Ghahramani, Z., Weinberger, K.Q. (eds.) NeurIPS, pp. 2121–2129 (2013)
8. Gui, L., Wang, B., Huang, Q., Hauptmann, A., Bisk, Y., Gao, J.: KAT: a knowledge augmented transformer for vision-and-language. In: The Conference of the North American Chapter of the Association for Computational Linguistics: Human Language Technologies, pp. 956–968 (2022)
9. He, K., Fan, H., Wu, Y., Xie, S., Girshick, R.B.: Momentum contrast for unsupervised visual representation learning. In: CVPR, pp. 9726–9735 (2020)

10. Jia, C., et al.: Scaling up visual and vision-language representation learning with noisy text supervision. In: Meila, M., Zhang, T. (eds.) ICML, vol. 139, pp. 4904–4916 (2021)
11. Karpathy, A., Li, F.: Deep visual-semantic alignments for generating image descriptions. In: CVPR, pp. 3128–3137 (2015)
12. Kim, W., Son, B., Kim, I.: ViLT: vision-and-language transformer without convolution or region supervision. In: Meila, M., Zhang, T. (eds.) ICML, vol. 139, pp. 5583–5594 (2021)
13. Li, J., Li, D., Xiong, C., Hoi, S.C.H.: BLIP: bootstrapping language-image pretraining for unified vision-language understanding and generation. In: Chaudhuri, K., Jegelka, S., Song, L., Szepesvári, C., Niu, G., Sabato, S. (eds.) ICML, vol. 162, pp. 12888–12900 (2022)
14. Li, L.H., Yatskar, M., Yin, D., Hsieh, C.J., Chang, K.W.: VisualBERT: a simple and performant baseline for vision and language. arXiv preprint (2019)
15. Li, L.H., Yatskar, M., Yin, D., Hsieh, C.J., Chang, K.W.: What does BERT with vision look at? In: ACL, pp. 5265–5275 (2020)
16. Li, X., et al.: OSCAR: object-semantics aligned pre-training for vision-language tasks. In: Vedaldi, A., Bischof, H., Brox, T., Frahm, J.-M. (eds.) ECCV 2020. LNCS, vol. 12375, pp. 121–137. Springer, Cham (2020). https://doi.org/10.1007/978-3-030-58577-8_8
17. Lin, T.-Y., et al.: Microsoft COCO: common objects in context. In: Fleet, D., Pajdla, T., Schiele, B., Tuytelaars, T. (eds.) ECCV 2014. LNCS, vol. 8693, pp. 740–755. Springer, Cham (2014). https://doi.org/10.1007/978-3-319-10602-1_48
18. Van der Maaten, L., Hinton, G.: Visualizing data using t-SNE. J. Mach. Learn. Res. 9(11), 2579–2605 (2008)
19. Mikolov, T., Chen, K., Corrado, G., Dean, J.: Efficient estimation of word representations in vector space. In: ICLR (2013)
20. Mikolov, T., Sutskever, I., Chen, K., Corrado, G.S., Dean, J.: Distributed representations of words and phrases and their compositionality. In: NeurIPS, pp. 3111–3119 (2013)
21. van den Oord, A., Li, Y., Vinyals, O.: Representation learning with contrastive predictive coding. arXiv preprint (2018)
22. Pennington, J., Socher, R., Manning, C.: GloVe: global vectors for word representation. In: EMNLP, pp. 1532–1543 (2014)
23. Qiu, J., et al.: Are multimodal models robust to image and text perturbations? arXiv preprint (2022)
24. Radford, A., et al.: Learning transferable visual models from natural language supervision. In: Meila, M., Zhang, T. (eds.) ICML, vol. 139, pp. 8748–8763 (2021)
25. Ramesh, A., Dhariwal, P., Nichol, A., Chu, C., Chen, M.: Hierarchical text-conditional image generation with CLIP latents. arXiv preprint (2022)
26. Rombach, R., Blattmann, A., Lorenz, D., Esser, P., Ommer, B.: High-resolution image synthesis with latent diffusion models. In: CVPR (2022)
27. Schroff, F., Kalenichenko, D., Philbin, J.: FaceNet: a unified embedding for face recognition and clustering. In: CVPR, pp. 815–823 (2015)
28. Schuhmann, C., et al.: LAION-5B: an open large-scale dataset for training next generation image-text models. In: NeurIPS, vol. 35, pp. 25278–25294 (2022)
29. Schuhmann, C., et al.: LAION-400M: open dataset of CLIP-filtered 400 million image-text pairs. arXiv preprint: abs/2111.02114 (2021)
30. Shtedritski, A., Rupprecht, C., Vedaldi, A.: What does CLIP know about a red circle? Visual prompt engineering for VLMs. In: ICCV, pp. 11987–11997 (2023)

31. Thomee, B., et al.: YFCC100M: the new data in multimedia research. Commun. ACM **59**, 64–73 (2016)
32. Weston, J., Bengio, S., Usunier, N.: Large scale image annotation: learning to rank with joint word-image embeddings. Mach. Learn. **81**, 21–35 (2010)
33. Wolfe, R., Banaji, M.R., Caliskan, A.: Evidence for hypodescent in visual semantic AI. In: ACM Conference on Fairness, Accountability, and Transparency (2022)
34. Yasunaga, M., et al.: Retrieval-augmented multimodal language modeling. arXiv preprint (2022)
35. Young, P., Lai, A., Hodosh, M., Hockenmaier, J.: From image descriptions to visual denotations: new similarity metrics for semantic inference over event descriptions. Trans. Assoc. Comput. Linguist. **2**, 67–78 (2014)
36. Zhai, X., et al.: LiT: zero-shot transfer with locked-image text tuning. In: CVPR (2022)

# Exploring Artificial Intelligence for Advancing Performance Processes and Events in Io3MT

Romulo Vieira[1]([⊠]), Debora Muchaluat-Saade[1], and Pablo Cesar[2,3]

[1] MídiaCom Lab, Fluminense Federal University, Niterói, RJ, Brazil
{romulo_vieira,debora}@midiacom.uff.br
[2] Centrum Wiskunde & Informatica, Amsterdam, The Netherlands
garcia@cwi.nl
[3] TU Delft, Amsterdam, The Netherlands

**Abstract.** The Internet of Multisensory, Multimedia and Musical Things (Io3MT) is a new concept that arises from the confluence of several areas of computer science, arts, and humanities, with the objective of grouping in a single place devices and data that explore the five human senses, besides multimedia aspects and music content. In the context of this brave new idea paper, we advance the proposition of a theoretical alignment between the emerging domain in question and the field of Artificial Intelligence (AI). The main goal of this endeavor is to tentatively delineate the inceptive trends and conceivable consequences stemming from the fusion of these domains within the sphere of artistic presentations. Our comprehensive analysis spans a spectrum of dimensions, encompassing the automated generation of multimedia content, the real-time extraction of sensory effects, and post-performance analytical strategies. In this manner, artists are equipped with quantitative metrics that can be employed to enhance future artistic performances. We assert that this cooperative amalgamation has the potential to serve as a conduit for optimizing the creative capabilities of stakeholders.

**Keywords:** Internet of Multisensory · Multimedia and Musical Things · Artificial Intelligence · Networked Performances

## 1 Introduction

The Internet of Things (IoT) [5] denotes a collection of embedded systems imbued with sensing and/or actuation capabilities, which collaborate with neighboring devices to execute specific tasks. The recent instantiation of these concepts has engendered noteworthy progress in the development and enhancement of applications spanning diverse domains such as home automation, smart cities, industry 4.0, e-health, security, and geospatial analysis, among others. Nevertheless, artistic and musical applications have hitherto received relatively scant attention in comparison to the aforementioned domains [4,28,29]. The existing instances within this realm predominantly reflect the technological and artistic

S. Rudinac et al. (Eds.): MMM 2024, LNCS 14557, pp. 234–248, 2024.
https://doi.org/10.1007/978-3-031-53302-0_17

preferences of their creators, thereby impeding the seamless integration of diverse multimedia content and the utilization of the inherent advantages offered by their counterparts.

To alleviate these limitations and move computing as much as possible to combine these domains, we have articulated an area called the Internet of Multisensory, Multimedia and Musical Things (Io3MT) [32]. This field is envisioned to serve as a unified platform where these three distinct content types can be seamlessly integrated, rendering them interchangeable and devoid of hierarchical constraints. The theoretical underpinnings of this area draw upon a confluence of various interdisciplinary subjects, including but not limited to Networked Music Performance (NMP) [27], Multiple Sensorial Media (Mulsemedia) [16], Interactive Art [13], among others.

In light of these considerations, Io3MT holds the promise of exerting a substantial influence and offering valuable support in the development of a diverse array of entertainment applications, such as multisensory cinema, digital television, games, and interactive museums and galleries. It is noteworthy that the current research endeavors within this domain are primarily concentrated on networked artistic performances.

Artistic performances can be understood through various lenses. For instance, Ervin Goffman [18] categorizes them within the context of theatrical paradigms, viewing every action performed by an individual during an interaction as a form of performance. Alternatively, Peter Berger and Thomas Luckmann [7] propose an interactive construction perspective, suggesting that these actions are not bound by predefined social, cultural, or natural prescriptions. From a practical and technical standpoint, a performance is considered the culmination of extensive theoretical and practical preparation. In this context, elements such as repertoire, arrangement, and gestures come together to materialize as auditory, aesthetic expressions, dances, movements, orality, vocality, and narratives, whether they are presented in recorded or live formats. These performances enable the audience to experience visual and interactive forms of expression.

Traditionally, artistic performance has always incorporated intertextual elements, featuring an inherent heterogeneity that often remains implicit in discourse. In the context of contemporary society's pervasive technological influence, new dimensions of the performative act have emerged, including the utilization of Artificial Intelligence (AI) as a tool, medium, mediator, or partner [9]. In response to this evolving landscape, this paper endeavors to provide an exploratory examination of the techniques, potentialities, and open challenges associated with the incorporation of AI within performances guided by the principles of Io3MT. This discussion can be used as a start point to build Io3MT environments using real-time AI techniques to enhance multisensory, multimedia and musical performances.

The remainder of the paper is divided as follows. Section 2 examines prior literature addressing the usage of AI in the generation of components essential to artistic performances. This encompasses areas such as AI-driven musical composition for accompaniment, the generation of lyrics, and choreographic movements aimed at enhancing the overall presentation. In Sect. 3, we expound upon the fundamental architectural and functional tenets underpinning Io3MT. Following this, Sect. 4 engages in a discourse regarding the ramifications arising from the intersection of these two domains of knowledge. Finally, Sect. 5 offers a succinct summation of the key findings and conclusions derived from this research endeavor.

## 2   Related Work

Artificial intelligence tools, particularly those of a generative nature, have witnessed extensive utilization for expediting and enhancing the production of diverse forms of creative content, encompassing domains such as music, animations, storytelling, and visual arts. Within this context, the ensuing section is dedicated to examining scholarly contributions that leverage these AI techniques to generate elements that substantially contribute to the realization and enrichment of artistic performances.

A current area of prominence within the domain of music composition is Generative Music, which is also referred to as Algorithmic Composition or Musical Metacreation [10]. Generative Music entails the creation of musical pieces facilitated by computational algorithms or autonomous systems guided by predetermined rules, parameters, and algorithms. It affords the capacity to generate compositions that dynamically evolve over time, adapting to different variables including environmental factors, audience interaction, or stochastic occurrences. Consequently, this approach has the potential to yield music compositions characterized by uniqueness in each performance iteration.

An example in this context is the Mind Band [26] initiative, a cross-media platform harnessing artificial intelligence to craft musical compositions by interpreting emojis, images, and vocalizations. This system operates on the foundation of the valence-arousal model, which translates emotional cues identified within user-contributed symbols or vocal inputs into musical compositions. The generation of these musical pieces is facilitated by a Variational Autoencoder-Generative Adversarial Network (VAE-GAN) architecture [11].

Their model consists of three main parts: input, music generation, and output. The first part receives multimedia information for transformation into music. The second part, at the core of the project, generates the music content using a combination of Variational Autoencoder (VAE) and Generative Adversarial Network (GAN) techniques, where the VAE decoder also serves as the GAN generator. Those techniques are trained together with shared parameters, and element-wise losses apply to both the encoder and decoder, while the discriminator loss also involves the encoder. After that iterative process, the third phase produces various musical compositions as outputs.

An additional instance of a tool devised for algorithmic composition is exemplified by XiaoIce Band [34], designed to address concerns related to the assurance of harmonic coherence within multichannel musical compositions. In pursuit of this objective, the authors have introduced a model capable of producing melodies and orchestrations for various musical instruments while upholding the integrity of chord progressions and rhythmic structures. In this endeavor, they have formulated a Cross Melody and Rhythm Generation Model (CMRGM) and a Multi-Instrument Co-Arrangement Model (MICA) through the application of multi-task learning principles to both components.

Similarly, the MuseGAN project [12] aligns itself with the objective of generating multichannel music compositions. However, it diverges from the previous approach by replacing multi-task learning with a methodology grounded in GANs. Notably, MuseGAN introduces an objective evaluation framework that scrutinizes both the intra-track and inter-track characteristics of the generated musical compositions. A distinctive feature of this proposal is its capacity to extrapolate from a human-composed musical track provided as input. MuseGAN can autonomously generate four additional tracks to complement the original audio, thereby expanding its potential for active participation in a performance context.

Exploring further into works related to musical improvisation and accompaniment, it is worth mentioning the GEN JAM [11], which employs genetic algorithms to construct a computational model capable of improvising jazz-style compositions. Similarly, Papadopoulos and Wiggins' proposal [25] also utilizes genetic algorithms for the same purpose, albeit the difference of generating improvised melodies within a predefined chord progression. This approach includes the use of a fitness function to automatically assess the quality of improvisations, considering factors like melodic shape, note duration, and intervals between notes. It is worth noting that GEN JAM [11] relies on human evaluations to assess improvisation quality, unlike Papadopoulos and Wiggins' approach [25].

Two systems, DIRECTOR MUSICES [15] and Widmer's project [11], employ rules for tempo, dynamics, and articulation transformations in musical performances. DIRECTOR MUSICES follows a set of prescribed rules, including differentiation, grouping, and set rules, while Widmer's project achieves similar transformations through machine learning. In a similar way, the research conducted by Bresin [8] utilizes artificial neural networks, which go beyond conventional feed-forward networks by incorporating feedback connections for improved modeling of temporal and amplitude variations in musical output.

Artistic performances involve various aesthetic elements, including choreographed dance movements. AI techniques are utilized to create such choreography, exemplified by the Full-Attention Cross-modal Transformer (FACT) network [20]. FACT generates three-dimensional (3D) dance movements based on auditory input and initial motion encoding. It combines these inputs within a cross-modal network to capture the relationship between sound and motion. Given a musical piece and a brief initial motion sequence of about 2 s, FACT can generate lifelike dance sequences comprising joint rotations and translations. The model is trained to predict N upcoming motion sequences and, during testing, autonomously generates continuous motion, effectively learning the intricate connection between music and motion. This capability allows for diverse dance sequences tailored to different musical compositions.

AI-Lyricist [22] is a system capable of generating meaningful lyrics for melodies, receiving a set of words and a MIDI file as input. That task involves several challenges, including automatically identifying the melody and extracting a syllable template from multichannel music, generating creative lyrics that match the input music style and syllable alignment, and satisfying vocabulary constraints. To address those challenges, the system is divided into four modules: i) Music structure analyzer: to derive the music structure and syllable model from a given MIDI file; ii) Letter generator: based on Sequence Generative Adversarial Networks (SeqGAN) optimized by multi-adversarial training through policy gradients with twin discriminators for text quality and syllable alignment; iii) Deeply coupled lyric embedding model: to project music and lyrics into a joint space, enabling a fair comparison of melody and lyric constraints; and iv) Polisher module: to satisfy vocabulary restrictions, applying a mask to the generator and replacing the words to be learned. The model was trained on a dataset of over 7,000 song-lyric pairs, enhanced with manually annotated labels in terms of theme, sentiment, and genre.

In light of the aforementioned context, the primary goal of this study is to provide a theoretical and exploratory exposition of the key Artificial Intelligence (AI) methodologies applicable to artistic processes and events grounded in the principles of Io3MT. This investigation contemplates the specific facets that may be encompassed, including the automatic generation of musical and visual content, machine-facilitated musical accompaniment, the automated recommendation and extraction of sensory effects, the composition of lyrical content and choreographic movements. Moreover, this work also seeks to highlight the inherent challenges in these scenarios, and it delineates the forthcoming areas of research that we intend to pursue.

## 3    The Internet of Multisensory, Multimedia and Musical Things (Io3MT) Design

The Internet of Multisensory, Multimedia, and Musical Things (Io3MT) [32] represents an emergent domain of research characterized by the inclusion of heterogeneous objects capable of establishing connections to the Internet, thereby participating in the continual exchange of multisensory, multimedia, and musical data. The fundamental aim underlying this interconnectivity is to promote seamless interaction and the accrual of valuable information, consequently fostering the realization and automation of an array of services. These services span from those oriented towards artistic endeavors and entertainment applications to educational and therapeutic contexts. From a functional perspective, Io3MT seeks to streamline the integration and real-time communication among these interconnected devices. Furthermore, it endeavors to enable a specific class of data to either act upon or exert influence over another. For instance, this could entail scenarios where a sequence of musical chords induces alterations in the color palette of a visual display, or the exhibition of an image triggers the emission of scents reminiscent of the corresponding depicted environment.

Within this context, it becomes evident that Io3MT adheres to a horizontal framework with distinctive characteristics. These attributes include low latency, signifying the system's ability to swiftly report events within a predefined timeframe. Additionally, low jitter is crucial, indicating the acceptable variability in data packet arrival times. Reliability plays a pivotal role, underscoring the system's effectiveness in delivering data with minimal packet loss. Furthermore, Io3MT environments necessitate an adequate bandwidth to accommodate the diverse multimedia content they handle. The adoption of multi-path networking is vital, allowing the system to utilize multiple communication routes to mitigate overloading of a single path and reduce the risk of network failure. Alongside these aspects, fault tolerance, lightweight implementation, and efficient service coordination are essential components. These environments must possess the capability to handle and synchronize various signal types, spanning audio, video, images, sensory stimuli, and more. Moreover, they should facilitate the active participation of individuals with differing levels of musical expertise, thereby enriching and streamlining creative processes within such contexts.

As is customary in well-established IoT models and within various tools in the field of computer science, the adoption of a layered architecture is prevalent. This approach offers a significant advantage in ensuring that each layer's services are executed in accordance with the prerequisites established by the preceding layers. Typically, the lower the hierarchical layer, the more straightforward and efficient the services it provides, with increasingly intricate functionalities emerging as one ascends through the layers [14]. In line with this convention, the proposed architecture for Io3MT is structured into five distinct levels. The lowest of these layers is designated as the Things Layer. Functioning as the virtualization layer, it bears the responsibility of converting analog data originating from physical devices into digital formats. This transformation initiates the generation of substantial volumes of data that subsequently traverse throughout the Io3MT environment.

Subsequently, the Link Layer emerges with its primary objective being the facilitation of device connectivity to the overarching network. Devices integrated within this layer exhibit sensorial capabilities, processing data derived from the physical realm before transmitting it to centralized servers for further analysis and dissemination.

The third tier, referred to as the Network Layer, shoulders the responsibility for the foundational infrastructure upon which the Io3MT ecosystem is constructed. Within this layer, critical Quality of Service (QoS) parameters pertinent to the measurement of data transmission performance are taken into consideration. These parameters encompass attributes such as latency, bandwidth, packet loss, jitter, among others.

Next in line, the Middleware Layer assumes a pivotal role within the architectural framework. This level shoulders the critical responsibility of processing incoming data, making informed decisions, and providing essential services. Its multifaceted functions encompass tasks such as data translation to ensure compatibility across different formats, resource discovery to identify available assets, synchronization for temporal alignment, identity management for secure user and device identification, concurrency and transition control for managing simultaneous operations and state transitions, persistence for reliable data storage and retrieval, and stream processing to enable real-time analysis and manipulation of data streams to meet dynamic requirements and applications.

The fifth and ultimate stratum within the architectural framework is the Application Layer, charged with the responsibility of furnishing the services as solicited by customers or users. This layer can be further subcategorized into two distinct levels. The first of these is the service unit, tasked with offering functionalities pertaining to information management, data analysis, and decision-making tools. The second sub-division is the applications unit, which scrutinizes the services requested by end-users, subsequently processing data originating from the physical world into cyber manifestations to facilitate the accomplishment of specific tasks. Figure 1 presents a synthesis of this discussion, with the main technologies of each of these layers, illustrated for educational purposes.

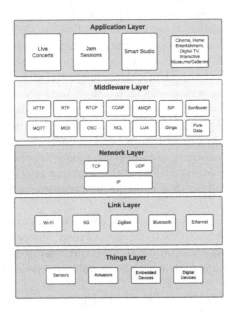

**Fig. 1.** Layered architecture of Io3MT.

Regarding artistic creation, the environment should be designed to deliver an immersive and captivating experience for the audience. It should also facilitate seamless integration and user participation, with resources accessible to those lacking artistic or computational expertise. To fulfill these objectives, several prerequisites must be met. Firstly, the presentation should be made more accessible to a wide range of users. Additionally, it should capture information from the audience and amalgamate it into a coherent artistic narrative, ensuring artistic security. Furthermore, the environment should encourage active partic- ipation, captivating the audience and fostering an environment that promotes initiation. Moreover, it should establish a clear and transparent relationship between participants' gestures and the outcomes achieved, providing a compre- hensible and engaging experience. This approach allows the system to remain open and editable, enabling the combination of various elements and offering users opportunities to acquire and refine skills over time. Any imperfections within the system should not be attributed to functional errors but rather to potentially unsatisfactory results from a musical and aesthetic standpoint.

The constituent entities of these immersive environments are commonly referred to as multisensory, multimedia, and musical things (3MT). These objects are distinguished by their capability to engage in at least one of the specified actions, which may involve either generating or responding to content within these designated categories. They fulfill roles as providers of services and informa- tion, encompassing a diverse range of components, including hardware, software, sensors, actuators, cloud-based services, as well as electroacoustic and electronic musical instruments, along with smart devices. The structures of these devices may manifest either in physical or digital forms.

Moreover, these entities include embedded electronics, wireless communica- tion, sensing, and actuation capabilities. They collaborate with other devices to perform tasks within project constraints. They require unique identification, scalability, persistence, reliability, and a loosely coupled architecture. Due to their artistic nature, these devices must prioritize aesthetics, expressiveness, and ergonomics. They should be user-friendly, flexible, efficient, and produc- tive. Accessibility is vital for users with various abilities, and communicability is essential for conveying their purpose and design rationale while responding to user input [31].

The stakeholder can be categorized into six distinct groups. The initial cate- gory comprises musicians, performers, and artists who bear the responsibility of generating a diverse array of services. These services encompass a wide spectrum of creative expressions, spanning from musical compositions, live programming, dance performances, digital painting, graphic projections, 3D printing endeavors, and collaborative projects with robotic systems, among others. This category is inclusive of participants with varying levels of proficiency, encompassing begin- ners, amateurs, and experts, all actively engaged in the creation of multimedia art.

Subsequently, there are designers, entrusted with the task of conceiving inter- active components encompassing electronic devices, software, and novel inter- faces tailored for artistical expression. These professionals contribute to the design and development of innovative tools and interfaces that enhance the artistic experience. Following this group, there are the audience members, rep- resenting the end-users who consume the artistic content generated during pre- sentations. These individuals actively engage with the system, either locally or remotely, facilitating the transmission and reception of multisensory and multimedia information. In doing so, they demonstrate the capacity to influence and modify the presented artistic outcomes through their interactions with the system.

Another significant category consists of members of the entertainment industry, comprising record labels, publishers, distributors, and manufacturers. These stakeholders are primarily concerned with the commercial aspects of applications within this domain, encompassing activities such as publishing, marketing, digital distribution, copyright management, product creation, and promotion, among others.

Lastly, the stakeholders encompass teachers and students who leverage the principles and theories derived from this field for educational purposes. They incorporate various disciplines related to science, technology, engineering, art, and mathematics (STEAM) into their educational curricula, utilizing insights from this domain to enhance learning and foster interdisciplinary engagement among students.

## 4    Discussion and Open Challenges

The preceding discussion highlights key AI techniques that enhance artistic performances. When considering the Io3MT layers, the Things Layer allows for real-time data analysis, aiding in equipment maintenance and gathering insights about the environment and audience reactions. This leads to adaptable environments and the potential for new artistic applications using computer vision and speech recognition, increasing device versatility [21]. The techniques that can be used in these tasks are as diverse as possible. The methodologies applicable to these endeavors encompass a wide spectrum, ranging from machine learning-based classification techniques that facilitate data categorization and label generation, to Recurrent Neural Networks (RNNs) adept at prognosticating failures. Additionally, signal processing algorithms are employed for noise filtration, feature extraction, and frequency analysis. Furthermore, Convolutional Neural Networks (CNNs) are leveraged for object recognition, motion detection, and the subsequent analysis and adaptation of content [17].

In the Link Layer, critical for establishing device connections and pre-processing physical data, integrating machine learning signal processing is crucial. These techniques assist in filtering, normalizing, pattern recognition, and anomaly detection, improving system efficiency. Deep learning methods like CNNs and RNNs are prominent for multimedia and time series data, but Isolation Forest, SVM for anomaly detection, and Reinforcement Learning for autonomous decisions are also valuable options [3, 24].

In the Network Layer, responsible for data transmission and QoS management, the integration of AI offers several advantages. These include intelligent routing through Artificial Neural Networks (ANNs) and genetic algorithms, latency prediction via recurrent neural networks, bandwidth control with reinforcement learning and deep learning, jitter and packet loss detection using decision trees and clustering methods, optimization facilitated by genetic algorithms, and enhanced security, with the application of techniques like SVM, reinforce-ment learning, and genetic algorithms [19, 33].

The Middleware Layer may be implemented by harnessing the capabilities of Natural Language Processing (NLP) for the translation or conversion of data characterized by distinct linguistic or semantic formats. In addition, machine learning techniques can be applied for the purpose of service recommendation and discovery, drawing upon historical data pertaining to individual partici-pants within the environment. Furthermore, the synchronization of temporal data originating from various sources can be accomplished through the utiliza-tion of RNNs [2].

Ultimately, the Application Layer, supporting a wide range of services, can effectively utilize various AI resources, including the valence-arousal model, VAEs, GANs, and cross-generation models for algorithmic composition. Genetic algorithms aid in real-time tracking and improvisation, while machine learn-ing and ANNs enhance performance. Cross-modal transformers are valuable for generating dance movements, and combining SeqGAN with multi-adversarial training enables automatic lyrics generation that matches musical tempo and more. A concise summary of this discussion is presented in Table 1.

**Table 1.** Synthesis of AI techniques that can be used in the Io3MT.

| Layer | Tasks | AI Techniques |
|---|---|---|
| Things Layer | Specifies physical devices | Machine Learning, RNNs, CNNs |
| Link Layer | Connect devices to the network and pre-process data | RNNS, CNNs, SVM, Isolation Forest, Reinforcement Learning |
| Network Layer | Data transmission | ANNs, RNNs, SVM, Genetic Algorithms, Reinforcement Learning, Decision Trees, Clustering Methods |
| Middleware Layer | System control and management | NLP, RNNs, Machine Learning |
| Application Layer | Support services | VAEs, GANs, ANNs, SeqGAN, Cross-generation Models, Cross-model Transformers, Genetic Algorithms, Machine Learning |

Henceforth, the workflow of Io3MT performances are poised to experience substantial enhancements through the integration of these procedures. Within the domain of audio processing, the real-time application of effects and cor-rections serves to elevate the auditory experience and the narrative dimension of artistic endeavors. This involves the capacity to dynamically adjust musi-cal elements such as notes, rhythm, dynamics, tonal characteristics, and visual aesthetics, with the aim of achieving a more harmonious resonance with the emotional responses of the audience.

Encompassing the entirety of artistic creation, the benefits engendered by the incorporation of artificial intelligence are manifold. These encompass the real-time generation of content, whether it pertains to music, visual effects, or dance movements. Additionally, AI facilitates the production of diverse variations of content within predefined thematic or stylistic parameters, as well as the capacity to deviate from established norms. Moreover, artificial intelligence can be effectively harnessed to recommend or extract sensory effects [1]. Such advancements significantly expand the potential applications within this artistic context. Consequently, these advancements yield highly adaptable and interactive experiences that respond to audience actions, thereby sustaining engagement and sustained interest throughout the artistic endeavor.

The utilization of these techniques within the creative process is conducive to the attainment of precise aesthetic outcomes and the elicitation of intended emotional responses. Simultaneously, it aids in the judicious selection of color palettes, shapes, and musical compositions in a holistic manner. This iterative approach serves to perpetually align the ultimate artistic output with the initially envisioned aesthetic objectives.

Regarding infrastructure considerations, artificial intelligence exhibits the potential to enhance QoS by endowing devices with cognitive capabilities and contextual awareness. This capability facilitates the optimization of network traffic, amelioration of packet routing, mitigation of packet losses, efficient management of power consumption, and augmentation of the adaptability of network devices. Likewise, the integration, synchronization, and translation of diverse forms of information within an Io3MT environment play a pivotal role in bolstering QoE. In this context, the ecosystem is amenable to tailoring according to audience preferences, thereby rendering it more resonant and memorable performance.

Also, it contributes to the realm of business intelligence by enabling the examination of demographic data, both prior to and following a performance, with the objective of deriving critical insights. These insights play a fundamental role in repertoire selection, the formulation of efficacious marketing strategies, and the generation of comprehensive performance reports.

Despite numerous advantages, there are significant challenges in the realm of technology and AI. From a technological perspective, issues like network latency in real-time content creation and the diverse nature of data pose obstacles. Socially, disparities in technology access, standards set by the dominant class, and environmental concerns are pressing issues [30]. In AI domain, security, biases, and misinformation are overarching concerns, including equitable access, privacy, and addressing AI-generated misinformation.

Conversely, AI in artistic contexts faces unique challenges, such as the complexity of selecting attributes during system training, hindering adaptability to diverse artistic narratives and increasing development time and costs. Challenges also arise in hybridization and rendering in Io3MT, where various signal types and multisensory integration complicate content creation. Structural challenges involve limitations in cohesive organization and mapping issues in handling diverse content representations. Reproducing automatically generated content is difficult due to technical mastery and human irreproducibility, with potential solutions including increased training competence and defining difficulty levels.

In light of these considerations, AI holds significant disruptive potential in the realm of art, particularly when integrated with Io3MT. Unlike traditional methods, it enables autonomous learning through pattern recognition, shifting the artist's role to that of a curator and AI system refiner. This change prompts a reevaluation of originality and authenticity in art, allowing for more fluid and less constrained artistic performance while challenging existing paradigms [6,23].

In summary, AI techniques and systems are progressively integrating into artistic endeavors. While challenges such as copyright issues, work misappropriation, and plagiarism persist, this domain, particularly when intertwined with Io3MT, holds the promise of advancing digital literacy, fostering critical thinking, and democratizing the means of art production and reproduction.

## 5  Conclusion

In the modern era of widespread technology, especially in entertainment, the Internet of Multisensory, Multimedia, and Musical Things (Io3MT) has emerged at the intersection of various technological domains. This burgeoning field is characterized by a quest for new experiences and knowledge accumulation. This paper aims to draw parallels between Io3MT and Artificial Intelligence (AI), exploring techniques and potential impacts.

Incorporating AI entails exploring optimal tools for visual displays, experimenting with agency in digital performances, and utilizing programming languages for combining multimedia and multisensory elements. Despite presenting challenges and risks, the potential gains in artistic exploration and innovation are significant.

In summary, this paper contributes to the understanding of Io3MT and AI, providing a comprehensive introduction for artists and scientists. It encourages further investigation and outlines a future focus on creating practical, collaborative environments that integrate these areas, emphasizing equity, accessibility, and validation through fluency metrics and security.

**Acknowledgment.** The authors would like to thank the support of the following funding agencies CAPES (Grant Number 88887.668167/2022-00), CAPES Print, CNPq, and FAPERJ.

# References

1. de Abreu, R.S., Mattos, D., Santos, J.D., Ghinea, G., Muchaluat-Saade, D.C.: Toward content-driven intelligent authoring of mulsemedia applications. IEEE Multimed. **28**(1), 7–16 (2021). https://doi.org/10.1109/MMUL.2020.3011383
2. Agarwal, P., Alam, M.: Investigating IoT middleware platforms for smart application development. In: Ahmed, S., Abbas, S.M., Zia, H. (eds.) Smart Cities—Opportunities and Challenges. LNCE, vol. 58, pp. 231–244. Springer, Singapore (2020). https://doi.org/10.1007/978-981-15-2545-2_21
3. Ahanger, T.A., Aljumah, A., Atiquzzaman, M.: State-of-the-art survey of artificial intelligent techniques for IoT security. Comput. Netw. **206**, 108771 (2022)
4. Alvi, S.A., Afzal, B., Shah, G.A., Atzori, L., Mahmood, W.: Internet of multimedia things: vision and challenges. Ad Hoc Netw. **33**, 87–111 (2015). https://doi.org/10.1016/j.adhoc.2015.04.006, https://www.sciencedirect.com/science/article/pii/S1570870515000876
5. Atzori, L., Iera, A., Morabito, G.: The internet of things: a survey. Comput. Netw. **54**, 2787–2805 (2010). https://doi.org/10.1016/j.comnet.2010.05.010
6. Benjamin, W.: The Work of Art in the Age of Mechanical Reproduction. Penguin UK (2008)
7. Berger, P.L., Luckmann, T.: The Social Construction of Reality: A Treatise in the Sociology of Knowledge. Anchor (1967)
8. Bresin, R.: Artificial neural networks based models for automatic performance of musical scores. J. New Music Res. **27**(3), 239–270 (1998)
9. Brooks, A.L., Brooks, E.: Interactivity, Game Creation, Design, Learning, and Innovation: 5th International Conference, ArtsIT 2016, and First International Conference, DLI 2016, Esbjerg, Denmark, 2–3 May 2016, Proceedings, vol. 196. Springer (2017)
10. Brown, A.: Generative music in live performance. In: Generate and Test: Proceedings of the Australasian Computer Music Conference 2005, pp. 23–26. Australasian Computer Music Association (2005)
11. De Mantaras, R.L., Arcos, J.L.: Ai and music: from composition to expressive performance. AI Mag. **23**(3), 43–43 (2002)
12. Dong, H.W., Hsiao, W.Y., Yang, L.C., Yang, Y.H.: MuseGAN: multi-track sequential generative adversarial networks for symbolic music generation and accompaniment. In: Proceedings of the AAAI Conference on Artificial Intelligence, vol. 32 (2018)
13. Edmonds, E., Turner, G., Candy, L.: Approaches to interactive art systems. In: Proceedings of the 2nd International Conference on Computer Graphics and Interactive Techniques in Australasia and South East Asia, pp. 113–117 (2004)
14. Floris, A., Atzori, L.: Managing the quality of experience in the multimedia internet of things: a layered-based approach. Sensors **16**, 2057 (2016). https://doi.org/10.3390/s16122057
15. Friberg, A., Colombo, V., Frydén, L., Sundberg, J.: Generating musical performances with director musices. Comput. Music. J. **24**(3), 23–29 (2000)
16. Ghinea, G., Timmerer, C., Lin, W., Gulliver, S.R.: Mulsemedia: state of the art, perspectives, and challenges. ACM Trans. Multimed. Comput. Commun. Appl. (TOMM) **11**(1s), 1–23 (2014)
17. Ghosh, A., Chakraborty, D., Law, A.: Artificial intelligence in internet of things. CAAI Trans. Intell. Technol. **3**(4), 208–218 (2018)

18. Goffman, E.: The arts of impression management. Organ. Identity Reader, 11–12 (2004)
19. Hodo, E., et al.: Threat analysis of IoT networks using artificial neural network intrusion detection system. In: 2016 International Symposium on Networks, Computers and Communications (ISNCC), pp. 1–6. IEEE (2016)
20. Li, R., Yang, S., Ross, D.A., Kanazawa, A.: AI choreographer: music conditioned 3D dance generation with AIST++. In: Proceedings of the IEEE/CVF International Conference on Computer Vision, pp. 13401–13412 (2021)
21. Lopez-Rincon, O., Starostenko, O., Ayala-San Martín, G.: Algorithmic music composition based on artificial intelligence: a survey. In: 2018 International Conference on Electronics, Communications and Computers (CONIELECOMP), pp. 187–193. IEEE (2018)
22. Ma, X., Wang, Y., Kan, M.Y., Lee, W.S.: Ai-lyricist: generating music and vocabulary constrained lyrics. In: Proceedings of the 29th ACM International Conference on Multimedia, pp. 1002–1011 (2021)
23. Mazzone, M., Elgammal, A.: Art, Creativity, and the Potential of Artificial Intelligence. In: Arts, vol. 8, p. 26. MDPI (2019)
24. Osuwa, A.A., Ekhoragbon, E.B., Fat, L.T.: Application of artificial intelligence in internet of things. In: 2017 9th International Conference on Computational Intelligence and Communication Networks (CICN), pp. 169–173. IEEE (2017)
25. Papadopoulos, G., Wiggins, G.: AI methods for algorithmic composition: a survey, a critical view and future prospects. In: AISB Symposium on Musical Creativity, vol. 124, pp. 110–117. Edinburgh, UK (1999)
26. Qiu, Z., et al.: Mind band: a crossmedia AI music composing platform. In: Proceedings of the 27th ACM International Conference on Multimedia, pp. 2231–2233 (2019)
27. Rottondi, C., Chafe, C., Allocchio, C., Sarti, A.: An overview on networked music performance technologies. IEEE Access **4**, 8823–8843 (2016). https://doi.org/10.1109/ACCESS.2016.2628440
28. Turchet, L., Fischione, C., Essl, G., Keller, D., Barthet, M.: Internet of musical things: vision and challenges. IEEE Access **6**, 61994–62017 (2018). https://doi.org/10.1109/ACCESS.2018.2872625
29. Turchet, L., et al.: The internet of sounds: convergent trends, insights, and future directions. IEEE Internet Things J. **10**(13), 11264–11292 (2023). https://doi.org/10.1109/JIOT.2023.3253602
30. Vieira, R., Barthet, M., Schiavoni, F.L.: Everyday use of the internet of musical things: intersections with ubiquitous music. In: Proceedings of the Workshop on Ubiquitous Music 2020, pp. 60–71. Zenodo, Porto Seguro, BA, Brasil (2020). https://doi.org/10.5281/zenodo.4247759
31. Vieira, R., Gonçalves, L., Schiavoni, F.: The things of the internet of musical things: defining the difficulties to standardize the behavior of these devices. In: 2020 X Brazilian Symposium on Computing Systems Engineering (SBESC), pp. 1–7 (2020). https://doi.org/10.1109/SBESC51047.2020.9277862
32. Vieira, R., Muchuluat-Saade, D., César, P.: Towards an internet of multisensory, multimedia and musical things (Io3MT) environment. In: Proceedings of the 4th International Symposium on the Internet of Sounds, pp. 231–238. IS2 2023, IEEE, Pisa, Italy (2023). https://doi.org/10.1145/XXX, in Press

33. Wang, Y., et al.: Network management and orchestration using artificial intelligence: overview of ETSI ENI. IEEE Commun. Stand. Magaz. **2**(4), 58–65 (2018)
34. Zhu, H., et al.: Xiaoice band: a melody and arrangement generation framework for pop music. In: Proceedings of the 24th ACM SIGKDD International Conference on Knowledge Discovery & Data Mining, pp. 2837–2846 (2018)

# Demonstrations

# Implementation of Melody Slot Machines

Masatoshi Hamanaka$^{(\boxtimes)}$ (iD)

RIKEN, Tokyo, Japan
masatoshi.hamanaka@riken.jp

**Abstract.** We previously developed the Melody Slot Machine with which many melody combinations generated with our melodic-morphing method have a similar musical structure, and the global structure of the melodies does not change when a portion of one of the melody time axes is replaced with another. The Melody Slot Machine requires the use of two iPads, a Mac mini, and a microcontroller. We therefore present several versions of the Melody Slot Machine, i.e., portable, on iPhone, HD, and Melody Slot Machine II.

**Keywords:** Melody Slot Machine · Melodic morphing · Generative theory of tonal music (GTTM)

## 1 Introduction

We previously developed the Melody Slot Machine, which uses our melodic-morphing method, that is based on the Generative Theory of Tonal Music (GTTM) [7,10,11]. In the GTTM, a time-span tree is a binary tree in which each branch is connected to each note (Fig. 1). The branches of the tree are connected closer to the root than those connected to structurally important notes. The main advantage of time-span trees is that they can be used to reduce notes. Specifically, reduced melodies can be extracted by cutting a time-span tree with a horizontal line and omitting the notes connected below the line. In melody reduction with the GTTM, these notes are essentially absorbed by structurally more important ones.

We previously proposed a melodic-morphing method that applies this reduction (Fig. 2) to generate a melody that is structurally intermediate between two input melodies [8]. This is done by combining two melodies after reducing their respective time-span trees.

Since multiple melodies generated with this method have similar musical structures, the global structure of the melodies does not change when a portion of one of the melody time axes is replaced with another.

## 2 Melody Slot Machine

The goal with the Melody Slot Machine is composing using GTTM analysis, recording, arranging, shooting, and implementing to provide a special experience by using commercially available devices rather than specialized ones.

S. Rudinac et al. (Eds.): MMM 2024, LNCS 14557, pp. 251–257, 2024.
https://doi.org/10.1007/978-3-031-53302-0_18

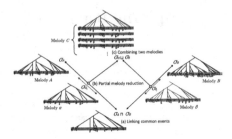

**Fig. 1.** Time-span tree     **Fig. 2.** Melodic-morphing method

In the Melody Slot Machine, melody segments in musical notation are displayed on a dial, and the melody to be played can be switched by rotating the dial (Fig. 3) [2][1].

Melody Slot Machine can generate a huge number of melody combinations. For example, in Fig. 3, 11 variation melodies are divided into 7 melody segments in the time direction, so there are $11^7 = 19,487,171$ melody combinations. Therefore, the user can feel as if they are composing a melody because a combination of melodies that they have not yet heard is generated through their own operation.

**Fig. 3.** Slot dial

The music performer's movements can be viewed on Pepper's ghost display (Fig. 4(a)). Recording is done in a reverb-suppressed room so that the reverb of the previous segmentation will not enter after the segmentation boundary. There are stereo speakers each for the front, left, and right, suspended from the truss, and the six channels are mixed down so that the performance sound can be heard from the instrument positions on the hologram (Fig. 4(b)). At the melody-segment boundaries, the previous segment fades out for a short time while the next segment fades in for a natural connection. Reverb is added when the sound is output.

---

[1] https://gttm.jp/hamanaka/en/melody-slot-machine/.

(a) Pepper's ghost display     (b) Speakers          (c) Portable version

**Fig. 4.** Melody Slot Machine

The video is shot in front of a black wall. Short interpolation video clips of the performer generated by AI are sandwiched into recorded videos segments of an actual performer, so the AI-generated performer moves seamlessly on the display [12,13].

The Melody Slot Machine requires the use of two iPads, one Mac mini, and a microcontroller (Arduino) [1]. One iPad displays a dial with a melody segment and outputs MIDI signals for changes in the dial (Fig. 5). When a switch built into the slot lever is pressed, the Arduino detects it and outputs a MIDI signal. The MIDI signals from the iPad and Arduino are received by the MIDI interface and sent to the Mac mini via a USB cable. When the slot lever is pressed and the MIDI signal reaches the Mac mini from the Arduino, the MIDI signal is transmitted to the iPad and the dial rotates and changes randomly. The dial and synchronization information is transmitted from the Mac mini to the other iPad, and the screen for the holographic display is shown.

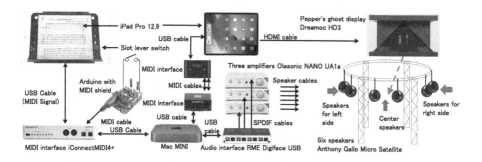

**Fig. 5.** Implementation of Melody Slot Machine

Since the Melody Slot Machine requires a large exhibition space and is difficult to display at many international conferences, we developed a portable version (Fig. 4(c)) requiring only two iPads. One iPad displays the slot dial, which, when operated, sends MIDI signals to the other iPad via a peer-to-peer network. The slot lever is connected via Bluetooth to the iPad displaying the slot dial, and when the slot is pressed, the dial rotates and changes randomly. The other iPad outputs music and video.

## 2.1   Melody Slot Machine on iPhone and HD

After we developed the Melody Slot Machine, we exhibited it at an international conference, but the Covid-19 Pandemic made it difficult to demonstrate it in person [9]. Therefore, we developed the Melody Slot Machine on iPhone and HD to enable many people to experience it by downloading an application.

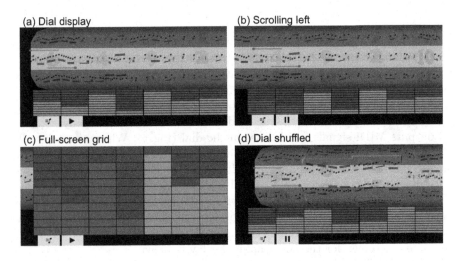

**Fig. 6.** Screenshot of Melody Slot Machine on iPhone

Figure 6 is a screenshot of the Melody Slot Machine on iPhone [6]. The horizontal axis is time, and each dial displays a melody segment in musical notation (Fig. 6(a)). By swiping up and down on each dial, the user can switch the segments. Due to the limited screen size of the iPhone, only four melody segments can be viewed simultaneously, and the currently playing segment can be viewed by automatically scrolling left as the musical piece progresses (Fig. 6(b)).

The dial changes for the entire musical piece are displayed in a grid at the bottom of the screen, corresponding to the numbers written on the dials. Swiping up from the bottom of the grid display brings up the full-screen grid (Fig. 6(c)). Touching the full-screen grid changes the selected grid. The change in the grid is linked to the dial, and the melody is played reflecting the change. Swiping down terminates the full-screen grid, and the dial appears again. When the iPhone is shaken up and down, each dial is shuffled to generate a new combination of melodies (Fig. 6(d)).

Figure 7a is a screenshot of Melody Slot Machine HD in which the symbols represent changes in melody variations [4]. For example, the musical-note symbol means that the same variation will continue. The cherry symbol indicates that the variations will change one after another.

Pressing the mode-switch buttons on the left and right of the screen displays the grid-tile screen, and the user can check and change the variations in the entire song (Fig. 7b). If the user uses the grid to change the melody combinations, the symbols on the slot screen will change accordingly.

The user presses the mode switch buttons or holds the iPad vertically to display the performer screen (Figs. 7c, d). This shows a performer playing new melody combinations determined from the slots or grids. Short video clips of the AI-generated performer are sandwiched into the recorded videos of an actual performer, so the AI-generated performer moves seamlessly.

## 3   Melody Slot Machine II

The ability to hear a melody varies from person to person, and musical novices may not notice that melody segments have been replaced. When multiple parts of a melody change, a high level of musical listening ability it required to notice all changes. Therefore, we are developing Melody Slot Machine II, which will use information from the multimodal interface to limit the number of musical parts that can be operated on a touchscreen simultaneously to one and emphasize the sound of the part the user is operating to make the changes in melodies easier to hear [5].

Melody Slot Machine II can vary the melody segments of two musical parts, e.g., vocal and saxophone (Fig. 8(a, b)). Since the objective is to notice melody changes through only listening to music, a grid-panel interface with time on the horizontal axis and melody variations (morphing levels) on the vertical axis is used instead of a dial with music notations displayed. The user concentrates on listening to each melody while operating the grid panel. However, adding new operations to emphasize the melody being manipulated makes it difficult to concentrate on listening to the melody.

We therefore developed Sound Scope Phone, which enables the user to enhance the sound of a specific part by detecting the direction of the user's

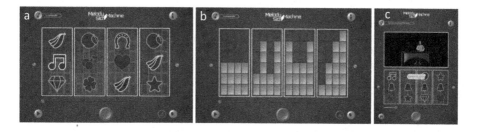

**Fig. 7.** Screenshots of Melody Slot Machine HD

head with the smartphone's front camera (Fig. 8(c)) [3][2] When the user pauses to listen carefully by cupping their hands around their ears, the sound of that part is further emphasized (Fig. 8(d)).

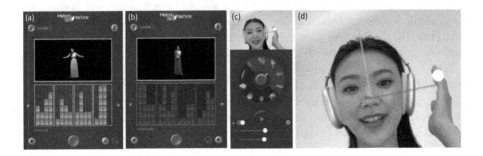

**Fig. 8.** Melody Slot Machine II and Sound Scope Phone

Melody Slot Machine II is equipped with Sound Scope Phone. Thus, the grid panel of the part the user is facing is active, and the panel of the part the user is not facing is inactive.

## 4   Experiment and Results

The versions of the Melody Slot Machine described above were used for 15 min each, in no particular order, by two musical novices and two performers with more than 10 years of experience. When the four players were asked which of the four versions they enjoyed, the two performers chose the original Melody Slot Machine and the two musical novices chose Melody Slot Machine II. This indicates that it is important for musical novices to be able to hear the sound of the parts. The two performers were also interested in the Melody Slot Machine on iPhone, but they pointed out that the screen was too small to operate the way they wanted. However, they were not interested in Melody Slot Machine HD or Melody Slot Machine II, which have improved operability, indicating that the user interface needs to be changed between musical novices and performers.

## 5   Conclusion

We presented several versions of the Melody Slot Machine. Experimental results indicate that musical novices and performers prefer different user interfaces. We will continue to improve the user interface for Melody Slot Machine.

**Acknowledgements.** This work was supported by JSPS KAKENHI Grant number 21H03572.

---

[2] https://gttm.jp/hamanaka/en/soundscopephone/.

# References

1. Arduino Reference - Arduino Reference – arduino.cc. https://www.arduino.cc/reference/en/. Accessed 25 Nov 2023
2. Hamanaka, M.: Melody slot machine: a controllable holographic virtual performer. In: Proceedings of the 27th ACM International Conference on Multimedia, MM 2019, New York, NY, USA, pp. 2468–2477. Association for Computing Machinery (2019). https://doi.org/10.1145/3343031.3350863
3. Hamanaka, M.: Sound scope phone: focusing parts by natural movement. In: ACM SIGGRAPH 2022 Appy Hour, SIGGRAPH 2022, New York, NY, USA. Association for Computing Machinery (2022). https://doi.org/10.1145/3532723.3535465
4. Hamanaka, M.: Melody slot machine HD. In: ACM SIGGRAPH 2023 Appy Hour, SIGGRAPH 2023, New York, NY, USA. Association for Computing Machinery (2023). https://doi.org/10.1145/3588427.3595356
5. Hamanaka, M.: Melody slot machine II: sound enhancement with multimodal interface. In: Companion Publication of the 25th International Conference on Multimodal Interaction, ICMI 2023 Companion, pp. 119–120, New York, NY, USA. Association for Computing Machinery (2023). https://doi.org/10.1145/3610661.3620663
6. Hamanaka, M.: Melody slot machine on iphone: dial-type interface for morphed melody. In: Adjunct Proceedings of the 36th Annual ACM Symposium on User Interface Software and Technology, UIST 2023 Adjunct, New York, NY, USA. Association for Computing Machinery (2023). https://doi.org/10.1145/3586182.3615764
7. Hamanaka, M., Hirata, K., Tojo, S.: Implementing "a generative theory of tonal music". J. New Music Res. 35(4), 249–277 (2006)
8. Hamanaka, M., Hirata, K., Tojo, S.: Implementation of melodic morphing based on generative theory of tonal music. J. New Music Res. 51(1), 86–102 (2023). https://doi.org/10.1080/09298215.2023.2169713
9. Hamanaka, M., Nakatsuka, T., Morishima, S.: Melody slot machine. In: ACM SIGGRAPH 2019 Emerging Technologies ET-245 (2019)
10. Hirata, K., Tojo, S., Hamanaka, M.: Music, Mathematics and Language - The New Horizon of Computational Musicology Opened by Information Science. Springer, Singapore (2022). https://doi.org/10.1007/978-981-19-5166-4
11. Lerdahl, F., Jackendoff, R.: A Generative Theory of Tonal Music. The MIT Press, Cambridge (1983)
12. Nakatsuka, T., Hamanaka, M., Morishima, S.: Audio-guided video interpolation via human pose features. In: Proceedings of the 15th International Joint Conference on Computer Vision, Imaging and Computer Graphics Theory and Applications (VISIGRAPP 2020), pp. 27–35 (2020)
13. Nakatsuka, T., Tsuchiya, Y., Hamanaka, M., Morishima, S.: Audio-oriented video interpolation using key pose. Int. J. Pattern Recogn. Artif. Intell. 35(16) (2021)

# E2Evideo: End to End Video and Image Pre-processing and Analysis Tool

Faiga Alawad$^{(\boxtimes)}$ , Pål Halvorsen , and Michael A. Riegler

Department of Holistic Systems, Simula Metropolitan Center for Digital Engineering,
Oslo, Norway
`faiga@simula.no`

**Abstract.** In this demonstration paper, we present "e2evideo" a versatile Python package composed of domain-independent modules. These modules can be seamlessly customised to suit specialised tasks by modifying specific attributes, allowing users to tailor functionality to meet the requirements of a targeted task. The package offers a variety of functionalities, such as interpolating missing video frames, background subtraction, image resizing, and extracting features utilising state-of-the-art machine learning techniques. With its comprehensive set of features, "e2evideo" stands as a facilitating tool for developers in the creation of image and video processing applications, serving diverse needs across various fields of computer vision.

**Keywords:** Computer vision · Machine learning · Python · Feature extraction · Video pre-processing · Image pre-processing

## 1 Introduction

Video and image pre-processing tools play a crucial role in computer vision tasks, as they help mitigate the complexity inherent in subsequent stages of the video analysis pipeline [7]. By refining the input data at the pre-processing stage, these tools facilitate more efficient and streamlined processing, allowing enhanced accuracy and reliability in the resulting analyses.

Currently, there are several video and image pre-processing tools that perform one task at a time but do not offer a complete pipeline. For example, Imageio [6] provides a solution for reading and writing videos and images in different formats. On the other hand, there are some advanced tools available that, due to their extensive feature sets and voluminous reference manuals and books, can be overly complex for simple pre-processing tasks, one of them is OpenCV[1]. The work described in [2] delves into the functionality and associated complexity of OpenCV. In addition, certain tools are narrowly tailored to a specific domain or task, limiting their versatility. For example, MIScnn [8], built for medical image segmentation.

---

[1] https://opencv.org/.

© The Author(s), under exclusive license to Springer Nature Switzerland AG 2024
S. Rudinac et al. (Eds.): MMM 2024, LNCS 14557, pp. 258–264, 2024.
https://doi.org/10.1007/978-3-031-53302-0_19

We present a tool that facilitates a general video pre-processing pipeline. For example, it starts with a video, passes it to a video pre-processing module, then to an image pre-processing module, and finally to a feature extractor module. Figure 1 illustrates the flow of this pipeline. The pipeline can be executed in its entirety or run modularly, by processing through individual modules one at a time. The objective of this pipeline is to simplify the preliminary stages for researchers, enabling them to focus on addressing new and challenging problems instead of allocating extensive time and resources to prepare videos and images for deep learning tasks. This tool is accessible as a Python package under the MIT licence and is continuously being developed to enhance its functionality by adding new features. Additionally, it welcomes contributions from the open-source community[2]. Getting started with the package is straightforward and user-friendly, due to the availability of documentation and tutorials. Users simply clone the Git repository, navigate to the directory where the package reside and run *'pip install .'* in the command line or terminal. Once the installation is complete, all the modules in the package will be accessible through a simple *'import'* statement.

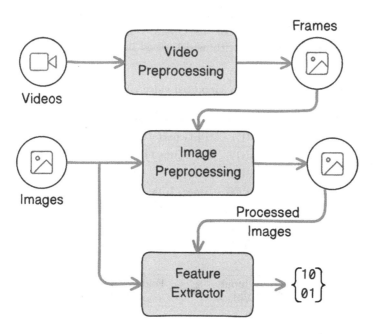

**Fig. 1.** E2evideo package pipeline.

---

[2] https://github.com/simulamet-host/video_analytics.

## 2    Video Pre-processing

This module is built using Python and uses functionalities from the OpenCV package. Figure 2, demonstrates an overview of the functionalities of this module. Therefore, it supports all the video and image formats supported by OpenCV, Table 1 lists examples of popular supported formats for both videos and images.

The input to this module is the system path where a folder containing supported videos is located. The module's main functionality is to extract frames from those videos and save them in a user selected image format, inside a user specified folder. Within the folder, images corresponding to the same video are saved in a folder with the same video file name. Users can choose to save all the frames "*every_ frame*" option, to save one frame per second "*per_ second*" option, or the third option "*fixed_ frames*". The third option allows the user to decide on a fixed number of frames to be extracted from the video. If the number of frames selected is less than the specified fixed frame number, the tool interpolates the missing frames. The interpolated frame is calculated as a weighted sum of the previous frame and the next frame.

This module comes with an optional background subtraction configuration. The user can choose to activate either Gaussian mixture-based (MOG2) or K-nearest neighbours (KNN) background/foreground segmentation. These two classes are implemented as part of the OpenCV background subtractor class, following the corresponding algorithms described in [13].

**Table 1.** Examples of supported video and image formats.

| Video Format | AVI - Audio video interleave - *.avi |
|---|---|
|  | MP4 - MPEG part 14 -*.mp4 |
|  | MOV - QuickTime file format -*.mov |
|  | MKV - *.mkv |
| **Image Format** | JPEG - *.jpeg, *.jpg, *.jpe |
|  | JPEG 2000 - *.jp2 |
|  | PNG - Portable Network Graphics (*.png) |
|  | TIFF - *.tiff |

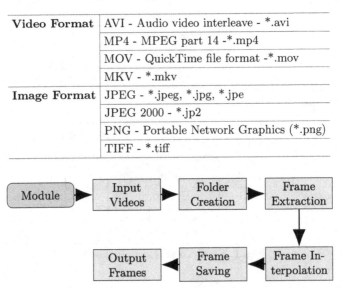

**Fig. 2.** An overview of the video pre-processing module.

## 3   Image Pre-processing

This Python module uses OpenCV and scikit-image [12]. Scikit-image is an image processing library. This module offers functionality for converting images to greyscale and resizing images to a given dimension. It structures the processed videos into a compressed NumPy array format (i.e., npz), which helps save disk space [3]. The module also aligns videos with different frame lengths by padding shorter videos with zeros. Additionally, it can create videos from the extracted frames and save them in a specified format. Figure 3, shows an overview of the functionalities of this module.

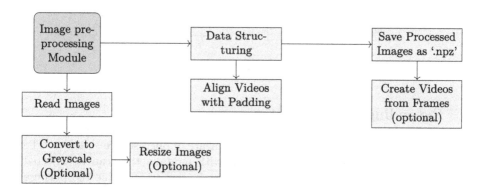

**Fig. 3.** Overview of the image pre-processing module.

## 4   Feature Extractor

Feature extraction methods transform the input space into a lower-dimensional subspace while retaining most of the relevant information from the original input. It is considered an important pre-processing task for dimensionality reduction and building robust machine learning models [5]. Figure 4, shows an overview of the feature extractor module. This module is highly adaptable and can be extended to integrate a wide range of machine learning feature extractors. Currently, it comes with a pre-trained ResNet18 [4] model and DINOv2 [10] to extract vectors of image embeddings. Users can choose between the two options, as well as choose which layer within ResNet18 to extract the embeddings. The extracted features can be represented as vectors, which are then saved as a CSV file in a specified output directory. These vectors can be used to train various machine learning models in downstream tasks.

Data visualisation and interpretation are important for feature extraction. The t-distributed stochastic neighbour embedding (t-SNE), stands out as a powerful tool in this domain, which is included in our implementation. This advanced method enables the visualisation of high dimensional feature vectors in 3D space.

It goes beyond mere presentation by offering an interactive platform to engage with the feature space. This interactivity is important for examining patterns within the data, which are often elusive in higher-dimensional spaces.

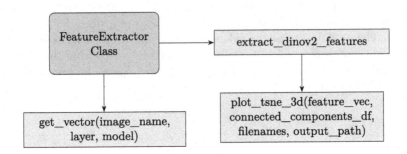

**Fig. 4.** Overview of the feature extractor module.

## 5    Extending Functionality

By considering the modular and extensible nature of open-source software. The functionality of this tool can be extended with modules that can be injected at any entry point of the pipeline. It is a good practise to define the purpose of the new module and specify its inputs, outputs, and possible interactions with the existing modules. Focusing on a single functionality and making it self-contained is recommended. Lastly, it is important to identify the correct entry point in the pipeline where the module should be invoked and handle the necessary inputs and outputs. Figure 5 outlines this process, highlighting the importance of updating the existing tests, documentation, and tutorials to ensure seamless integration of the new module.

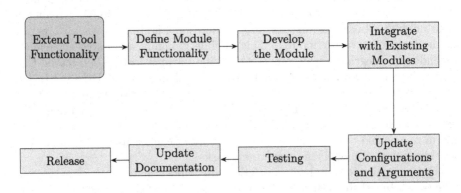

**Fig. 5.** Process outline for adding a new module to the existing e2evideo's modules.

# 6 Demonstration

For the demonstrations we will show how the framework works using three different datasets. The datasets will be from different domains and different technical specifications (videos, images, and time lapse).

Firstly, our demonstration begins with Njord dataset [9], comprised of video recordings from a fishing trawler. This dataset serves as a practical case for illustrating our framework's video pre-processing functionalities. Next, we will explore Toadstool dataset [11], primarily consisting of videos and images. This dataset is designed to assist in the training and development of emotionally intelligent agents. We will leverage Toadstool to showcase the strengths of our image pre-processing module. Lastly, we will explore the Cellular dataset [1], featuring time-lapse imagery of cell autophagy. Using Cellular, we aim to demonstrate the feature extraction module of our framework. This part of our demonstration underlines the capabilities of our framework in analysing and extracting meaningful information from complex biological sequences.

We will also demonstrate how the framework supports easy integration of current state-of-the-art foundational models and how they can be included into the pipeline to perform for example feature extraction. As an example for this we will use Meta's DINOv2 model.

# 7 Conclusion and Future Work

In this paper we described our video and image analysis framework that can easily be used to extract features from videos or images. We also explain how the framework can integrate new methods.

For future work we plan to integrate more state-of-the-art methods and extend it to other frames such as scikit-learn so that the extracted features can also be integrated into a follow up analysis system.

# References

1. Al Outa, A., et al.: Cellular, a cell autophagy imaging dataset. Sci. Data **10**(1), 806 (2023)
2. Culjak, I., Abram, D., Pribanic, T., Dzapo, H., Cifrek, M.: A brief introduction to OpenCV. In: 2012 Proceedings of the 35th International Convention MIPRO, pp. 1725–1730. IEEE (2012)
3. Harris, C.R., et al.: Array programming with NumPy. Nature **585**(7825), 357–362 (2020). https://doi.org/10.1038/s41586-020-2649-2
4. He, K., Zhang, X., Ren, S., Sun, J.: Deep residual learning for image recognition. In: Proceedings of the IEEE Conference on Computer Vision and Pattern Recognition, pp. 770–778 (2016)
5. Khalid, S., Khalil, T., Nasreen, S.: A survey of feature selection and feature extraction techniques in machine learning. In: 2014 Science and Information Conference, pp. 372–378. IEEE (2014)

6. Klein, A., et al.: imageio/imageio: v2.31.3 (2023). https://doi.org/10.5281/zenodo.8313513
7. Lu, T., Palaiahnakote, S., Tan, C.L., Liu, W.: Video preprocessing. In: Video Text Detection. ACVPR, pp. 19–47. Springer, London (2014). https://doi.org/10.1007/978-1-4471-6515-6_2
8. Müller, D., Kramer, F.: MIScnn: a framework for medical image segmentation with convolutional neural networks and deep learning. BMC Med. Imaging **21**, 1–11 (2021). https://doi.org/10.1186/s12880-020-00543-7
9. Nordmo, T.A.S., et al.: Njord: a fishing trawler dataset. In: Proceedings of the 13th ACM Multimedia Systems Conference, pp. 197–202 (2022)
10. Oquab, M., et al.: Dinov2: learning robust visual features without supervision. arXiv preprint: arXiv:2304.07193 (2023)
11. Svoren, H., et al.: Toadstool: a dataset for training emotional intelligent machines playing super mario bros. In: Proceedings of the 11th ACM Multimedia Systems Conference, pp. 309–314 (2020)
12. Van der Walt, S., et al.: scikit-image: image processing in python. PeerJ **2**, e453 (2014)
13. Zivkovic, Z., Van Der Heijden, F.: Efficient adaptive density estimation per image pixel for the task of background subtraction. Pattern Recogn. Lett. **27**(7), 773–780 (2006)

# Augmented Reality Photo Presentation and Content-Based Image Retrieval on Mobile Devices with *AR-Explorer*

Loris Sauter[1]([✉]) [ID], Tim Bachmann[1] [ID], Heiko Schuldt[1] [ID], and Luca Rossetto[2] [ID]

[1] University of Basel, Basel, Switzerland
{loris.sauter,t.bachmann,heiko.schuldt}@unibas.ch
[2] University of Zurich, Zurich, Switzerland
rossetto@ifi.uzh.ch

**Abstract.** Mobile devices are increasingly being used not only to take photos but also to display and present them to their users in an easily accessible and attractive way. Especially for spatially referenced objects, Augmented Reality (AR) offers new and innovative ways to show them in their actual, real-world context. In this demo, we present the *AR-Explorer* system, and particularly its use 'in-the-field' with local visual material. A demo video is available on Youtube: https://youtu.be/13f5vL9ZU3I.

**Keywords:** multimedia retrieval · spatial-retrieval · augmented reality · cultural heritage · mobile application

## 1 Introduction

Multimedia collections are growing at an accelerating rate, especially due to the ubiquity of smartphones and their improved sensor readings in terms of high image quality, as well as orientation and location sensors. However, the browsing and searching capabilities along the spatio-temporal metadata are not widely available to users. Spatial metadata, in particular, can greatly enhance an image's communicative power. Coupled with the ever-increasing quality of Augmented Reality (AR) technologies, spatio-temporal metadata can be used to enhance the way we revisit images based on an on-the-spot AR presentation.

Based on prior research [11,12], we introduce *AR-Explorer*, a mobile AR-based browser for contemporary and historical images at their real-world location, enriched with content-based search. The system builds on the open-source content-based multimodal multimedia retrieval stack *vitrivr* [10] and is implemented as an iOS application. The major novelty is the ability to capture photos and re-experience them in their real-world location through an AR image presentation combined with content-based search. The presented work builds on preliminary work targeting Android, which did not include contemporary imagery [12]. Especially for cultural heritage, there is existing work that exploits

S. Rudinac et al. (Eds.): MMM 2024, LNCS 14557, pp. 265–270, 2024.
https://doi.org/10.1007/978-3-031-53302-0_20

**Fig. 1.** Screenshots of the system in use with historical content (home screen on the left, search view on the right). In Basel, Switzerland, the depicted square used to be a parking lot in the 1930s. Nowadays, the old town part of the city is a pedestrian-only zone.

the capabilities of AR [4,6,8,9], see also a survey in [5], while recent work in this domain limits itself to traditional desktop applications [7].

The contribution of the paper is twofold: first, we present the architecture and methods of *AR-Explorer*, and second, we introduce a demonstration (together with an accompanying video) that showcases the capabilities of our system.

The remainder of this paper is structured as follows: in Sect. 2, we discuss the architecture of our system, and in Sect. 3, its implementation. The demonstration is presented in Sect. 4 and Sect. 6 concludes.

## 2  *AR-Explorer*

As capturing and presenting photos on smartphones is a well-integrated and well-established process, care must be given to an intuitive user experience when designing additional functionality. Challenges include the look and feel, query modes, and an immersive AR presentation mode. The *AR-Explorer* implementation is capable of capturing photos and annotating them with necessary metadata in order to recreate the exact same perspective in AR at a later time. Additionally, content-based retrieval methods provided by *vitrivr* [10], such as Query-by-Example, Query-by-Location, Query-by-TimePeriod and Query-by-Keyword, are available to users. The latter particularly employs deep learning-based visual text co-embedding to enable enhanced search [13]. On the data production side, a typical workflow is as follows: First, the user captures an image with the in-app camera. Then, the app sends the image and its metadata to the server that extracts content-based retrieval features. Finally, after a short while, the image is available for search. On the consumer side, typical interactions are as follows: First, the user starts the app in the map mode and uses one of the query modes to get results from the back-end. Then, either by using the gallery or the map view, results are presented. Finally, the AR presentation is started by either selecting one specific image or physically moving to such a location and starting the AR view by reacting to the notification.

When using *AR-Explorer*, users have multiple options for expressing their interests. The following provides an overview of the supported query modes which can be used individually or in combination:

**Query-by-Location** is the possibility to query for relevant content close to the user's current position or to an arbitrary location on a map.

**Query-by-Time-Period** is the ability to select all content that was created within a certain time interval.

**Query-by-Example** makes use of the content-based retrieval capabilities of the back-end by enabling the user to query using a picture taken by the mobile phone.

**Query-by-Text** makes use of deep-learning backed visual-text co-embedding [13] in order to textually search within the images.

## 3  Architecture

*AR-Explorer* consists of two parts: The iOS application as user interface and the back-end, which is an extension of the open-source content-based multimodal multimedia retrieval stack vitrivr,[1] as illustrated in the form of an architecture diagram in Fig. 2. The communication to the back-end is facilitated through a REST API, of which we leverage the spatio-temporal query support [2]. The user interface consists of multiple views, corresponding to the user interaction shown in Fig. 1 and a client for the retrieval tasks. Since iOS is the target platform, the app is written in Swift[2] and the AR functionality is provided by Apple's AR Kit.[3]

---

[1]  https://vitrivr.org.

[2]  https://www.apple.com/swift/.

[3]  https://developer.apple.com/augmented-reality/.

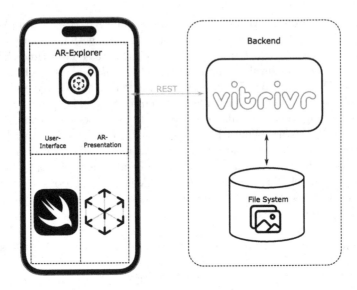

**Fig. 2.** The architecture diagram of the system. Communication with the back-end is facilitated via REST. The iOS app builds on Swift and ARKit.

## 4    Demonstration

*AR-Explorer* enables users to re-experience contemporary and historic photo material at their real-world location using AR. Thus, the demonstration scenarios are highly location-dependent. Particularly, historical images covering cultural, architectural, and urban heritage cannot be transferred from one place to another. Hence, the demonstration consists of two parts, an online (local) and an offline part:

**Offline:** This part is covered by means of the accompanying video, which will also be shown at the demo booth at the conference: Using the aforementioned query capabilities and AR presentation mode, we demonstrate *AR-Explorer*'s application for cultural heritage with images of Basel, the largest city in north-western Switzerland, with a well-preserved medieval old town (and therefore many interesting historic photos – most of which come without any copyright restrictions and thus can be easily and safely shared and re-used). A video showing the application on location is available on Youtube via: https://youtu.be/13f5vL9ZU3I.

**Online:** During the conference, we will provide opt-in beta testing for iPhone or iPad users who want to try out *AR-Explorer*. We will take conference participants on a physical journey through contemporary (and, if possible, also historic) Amsterdam. Users will be able to experience the city (surroundings of the conference venue) on their own according to the process outlined earlier, from capturing a moment to re-experiencing city history at its real-world location.

## 5    Evaluation

We conducted an initial user study employing Brooke's System Usability Scale (SUS) [3]. The participants in the user studies during the summers of 2022 and 2023 numbered 20 and 8, respectively, and were primarily associated with the Department of Mathematics and Computer Science at the University of Basel. A range of iOS devices were utilized with minimal restrictions, with the oldest device being an iPhone 6. The System Usability Scale (SUS) is a ten-item survey designed to evaluate the usability of a system in a general manner, resulting in a score. In 2022, the mean SUS score was 77 with a standard deviation of 19, which is considered a *good* rating according to [1]. The more recent user study in 2023 resulted in a mean SUS score of 80 with a standard deviation of 13, which falls into the same *good* category as the 2022 study. We recognize the limited expressiveness of the SUS scores and regard the user studies as a brief assessment of this work.

## 6    Conclusion and Outlook

We have presented *AR-Explorer*, a prototype system consisting of an iOS app providing AR image presentation and capturing capabilities and content-based image search for Query-by-Location, Query-by-Time, Query-by-Example, and Query-by-Text through the well-established content-based multimedia retrieval stack vitrivr. Necessary metadata are captured with the in-app camera functionality, and retrieved images, contemporary or manually annotated historical ones, are presented at their real-world location in AR. Users are directed toward AR image presentation locations with the in-app AR navigation mode that displays routes toward points of interest in AR. When not actively using the app, notifications alert users about possible AR interactions. A preliminary study with users employing the System Usability Scale has yielded encouraging findings, with a SUS score that meets the criteria for being classified as *good*.

Limitations of the current prototype include minor stability issues with the frameworks in use, such as the AR framework causing a slight drift or jumping caused by inaccuracies in GPS localization. One approach to negate this issue would be to employ landmark detection in order to achieve better alignment and positioning of AR images.

**Acknowledgements.** This work was partly funded by the Hasler Foundation in the context of the project "City-Stories" (contract no. 17055) and the Swiss National Science Foundation through project "MediaGraph" (contract no. 202125).

## References

1. Bangor, A., Kortum, P.T., Miller, J.T.: An empirical evaluation of the system usability scale. Int. J. Hum.-Comput. Interact. **24**(6), 574–594 (2008). https://doi.org/10.1080/10447310802205776

2. Beck, L., Schuldt, H.: City-stories: a spatio-temporal mobile multimedia search system. In: 2016 IEEE International Symposium on Multimedia (ISM), pp. 193–196. IEEE (2016)
3. Brooke, J.: SUS - a quick and dirty usability scale **57**, 187–191 (2013)
4. Cavallo, M., Rhodes, G.A., Forbes, A.G.: Riverwalk: incorporating historical photographs in public outdoor augmented reality experiences. In: International Symposium on Mixed and Augmented Reality (IEEE ISMAR), Merida, Mexico, pp. 160–165 (2016). https://doi.org/10.1109/ISMAR-Adjunct.2016.0068
5. Čopič Pucihar, K., Kljun, M.: ART for art: augmented reality taxonomy for art and cultural heritage. In: Geroimenko, V. (ed.) Augmented Reality Art. SSCC, pp. 73–94. Springer, Cham (2018). https://doi.org/10.1007/978-3-319-69932-5_3
6. Fond, A., Berger, M., Simon, G.: Facade proposals for urban augmented reality. In: International Symposium on Mixed and Augmented Reality (IEEE ISMAR), Nantes, France, pp. 32–41 (2017). https://doi.org/10.1109/ISMAR.2017.20
7. Geniet, F., Gouet-Brunet, V., Brédif, M.: ALEGORIA: joint multimodal search and spatial navigation into the geographic iconographic heritage. In: Proceedings of the 30th ACM International Conference on Multimedia, pp. 6982–6984. ACM, Lisboa, Portugal (2022). https://doi.org/10.1145/3503161.3547746
8. Lee, G.A., Dünser, A., Kim, S., Billinghurst, M.: CityViewAR: a mobile outdoor ar application for city visualization. In: Mixed and Augmented Reality (ISMAR-AMH), 2012 IEEE International Symposium on, pp. 57–64. IEEE (2012)
9. Lochrie, M., Čopič Pucihar, K., Gradinar, A., Coulton, P.: Designing seamless mobile augmented reality location based game interfaces. In: International Conference on Advances in Mobile Computing & Multimedia, p. 412. ACM (2013)
10. Rossetto, L., Giangreco, I., Tanase, C., Schuldt, H.: vitrivr: a flexible retrieval stack supporting multiple query modes for searching in multimedia collections. In: Proceedings of the 2016 ACM on Multimedia Conference, pp. 1183–1186. ACM (2016)
11. Sauter, L., Bachmann, T., Rossetto, L., Schuldt, H.: Spatially localised immsercive contemporary and historic photo presentation on mobile devices in augmented reality. In: Proceedings of the 5th Workshop on the analySis, Understanding and proMotion of heritAge Contents (SUMAC). to be published by ACM, Ottowa, ON, Canada (2023). https://doi.org/10.1145/3607542.3617358
12. Sauter, L., Rossetto, L., Schuldt, H.: Exploring cultural heritage in augmented reality with GoFind! In: 2018 IEEE International Conference on Artificial Intelligence and Virtual Reality (AIVR), pp. 187–188. IEEE, Taichung (2018). https://doi.org/10.1109/AIVR.2018.00041
13. Spiess, F., Gasser, R., Heller, S., Rossetto, L., Sauter, L., Schuldt, H.: Competitive interactive video retrieval in virtual reality with vitrivr-VR. In: Lokoč, J., et al. (eds.) MMM 2021. LNCS, vol. 12573, pp. 441–447. Springer, Cham (2021). https://doi.org/10.1007/978-3-030-67835-7_42

# Facilitating the Production
# of Well-Tailored Video Summaries
# for Sharing on Social Media

Evlampios Apostolidis$^{(\boxtimes)}$ ⓘ, Konstantinos Apostolidis ⓘ,
and Vasileios Mezaris ⓘ

Information Technologies Institute - Centre for Research and Technology, Hellas,
6th km Charilaou - Thermi Road, 57001 Thessaloniki, Greece
{apostolid,kapost,bmezaris}@iti.gr

**Abstract.** This paper presents a web-based tool that facilitates the
production of tailored summaries for online sharing on social media.
Through an interactive user interface, it supports a "one-click" video
summarization process. Based on the integrated AI models for video
summarization and aspect ratio transformation, it facilitates the generation of multiple summaries of a full-length video according to the needs
of target platforms with regard to the video's length and aspect ratio.

**Keywords:** Video summarization · Video aspect ratio transformation · Saliency prediction · Artificial Intelligence · Social media

## 1 Introduction

Social media users crave short videos that attract the viewers' attention and
can be ingested quickly. Therefore, for sharing on social media platforms, video
creators often need a trimmed-down version of their original full-length video.
However, different platforms impose different restrictions on the duration and
aspect ratio of the video that they accept, e.g., on Facebook's feed videos up to
2 min. appear in a 16:9 ratio, whereas Instagram and Facebook stories usually
allow for 20 sec. and are shown in a 9:16 ratio. This makes the generation of
tailored versions of video content for sharing on multiple platforms, a tedious
task. In this paper, we introduce a web-based tool that harnesses the power
of AI (Artificial Intelligence) to automatically generate video summaries that
encapsulate the flow of the story and the essential parts of the full-length video,
and are already adapted to the needs of different social media platforms in terms
of video length and aspect ratio.

---

This work was supported by the EU Horizon 2020 programme under grant agreement
H2020-951911 AI4Media.

S. Rudinac et al. (Eds.): MMM 2024, LNCS 14557, pp. 271–278, 2024.
https://doi.org/10.1007/978-3-031-53302-0_21

## 2    Related Work

Several video summarization tools can be found online, that are based on AI models. However, most of them produce a textual summary of the video, by analyzing the available [6,7] or automatically-extracted transcripts [1,8] using NLP (Natural Language Processing) models. Similar solutions are currently provided by various browser plugins [4,9]. Going one step further, another tool creates a video summary by stitching in chronological order the parts of the video that correspond to the selected transcripts for inclusion in the textual summary [5]. Focusing on the visual content, Cloudinary [2] released a tool that allows users to upload a video and receive a summarized version of it based on a user-defined summary length (max 30 s). In addition, Cognitive Mill [3] released a paid AI-based platform which, among other media content management tasks, supports the semi-automatic production of movie trailers and video summaries. The proposed solution in this paper is most closely related to the tool of Cloudinary. However, contrary to this tool, our solution offers various options for the video summary duration and applies video aspect ratio transformation techniques to fully meet the specifications of the target video-sharing platform.

## 3    Proposed Solution

The proposed solution (available at https://idt.iti.gr/summarizer) is an extension of the web-based service for video summarization, presented in [17]. It is composed of a front-end user interface (UI) that allows interaction with the user (presented in Sec. 3.1), and a back-end component that analyses the video and produces the video summary (discussed in Sec. 3.2). The front-end follows the HTML5 standard and integrates technologies of the JQuery JavaScript library (version 3.3.1) for supporting interaction with the user, and tools of the PHP library (version 7.4.26) for server-side processing. The back-end is a RESTful web service implemented in Python 3.7 using the Bottle framework (version 0.12). It hosts all the different AI-based video analysis methods that have been developed in Python 3.11 using the PyTorch framework, and it is exposed to the front-end UI through an API. The front-end and back-end communication is carried out via REST calls that initiate the analysis, periodically request its status, and, after completion, retrieve the video summary for presentation to the user. Our solution extends [17] by i) using an advanced AI-based method for video summarization, ii) integrating an AI-based approach for spatially cropping the video given a target aspect ratio, and iii) supporting customized values for the target duration and aspect ratio of the generated video summary.

Concerning scalability, our solution supports faster than real-time processing, as approximately 30 hours of video content can be analyzed per day on a modest PC (i5 CPU, GTX 1080Ti GPU). This can be easily scaled with the use of a more powerful PC or even multiple PCs and an established data load sharing mechanism. Moreover, the customized selection of the video summary size and duration, allows quick and easy adaptation to new video presentation formats and durations, that might appear as optimal choices in social media platforms in the future. With respect to data privacy and security, the submitted videos and the produced summaries are automatically removed from our server after 24 hours; during these 24 hours the summaries are accessible through a link that appears only on the user's browser and/or is sent to a user-provided e-mail. The temporarily-stored data in our processing server are protected by applying the commonly used security measures for preventing unauthorized server access.

## 3.1  Front-End UI

The UI of the proposed solution (see the top part of Fig. 1) allows the user to submit a video (that is either available online or locally stored in the user's device) for summarization, and choose the duration and aspect ratio of the produced summary. This choice can be made either by selecting among presets for various social media channels, or in a fully-custom manner. After initiating the analysis, the user can monitor its progress (see the middle part of Fig. 1) and submit additional requests while the previous ones are being analyzed. When the analysis is completed, the original video and the produced summary are shown to the user through an interactive page containing two video players that support all standard functionalities (see the bottom part of Fig. 1); through the same page, the user is able to download the produced video summary. Further details about the supported online sources, the permitted file types, and the management of the submitted and produced data can be found in [17].

## 3.2  Back-End Component

The submitted video is initially fragmented to shots using a pre-trained model of the method from [29]. This method relies on a 3D-CNN network architecture with two prediction heads; one predicting the middle frame of a shot transition and another one predicting all transition frames and used during training to improve the network's understanding of what constitutes a shot transition and how long it is. The used model has been trained using synthetically-created data from the TRECVID IACC.3 dataset [15] and the ground-truth data of the ClipShots dataset [30], and exhibits 11% improved performance on the RAI dataset [16] compared to the previous (used in [17]) approach. Following, **video summarization** is performed using a pre-trained model of AC-SUM-GAN [11], a top-performing unsupervised video summarization method [12]. This method embeds an Actor-Critic model into a Generative Adversarial Network and formulates the selection of important video fragments as a sequence generation task. At training time, the Actor-Critic model utilizes the Discriminator's feedback as a reward, to progressively explore a space of states and actions, and learn a value function (Critic) and a policy (Actor) for key-fragment selection. As shown in Table 1, AC-SUM-GAN performs much better than SUM-GAN-AAE [10] (used in [17]), on the SumMe [19] and TVSum [28] benchmark datasets for video summarization. Both methods learn the task using a summary-to-video reconstruction mechanism and the received feedback from an adversarially-trained Discriminator. We argue that the advanced performance of AC-SUM-GAN relates to the use of this feedback as a reward for training an Actor-Critic model and

274   E. Apostolidis et al.

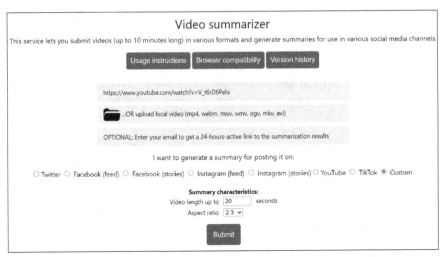

(a) The landing page of the UI.

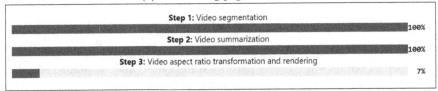

(b) The progress-reporting bars.

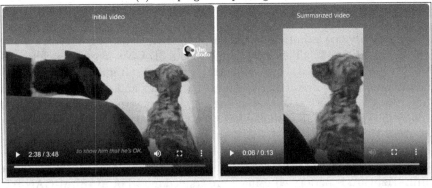

(c) The video players of the page showing the analysis results.

**Fig. 1.** Instances of the updated and extended UI.

learning a good policy for key-fragment selection, rather than using it as part of a loss function to train a bi-directional LSTM for frame importance estimation.

The proposed solution uses a model of AC-SUM-GAN that has been trained using augmented data. Following the typical approach in the literature [12], we extended the pool of training samples of the SumMe and TVSum datasets, by including videos of the OVP and YouTube [18] datasets. As presented in Table 2, the utilized data include videos from various categories and thus facili-

**Table 1.** Performance (F-Score (%)) of SUM-GAN-AEE and AC-SUM-GAN on SumMe and TVSum; the last row reports AC-SUM-GAN's performance for augmented training data.

| Method | SumMe | TVSum |
|---|---|---|
| SUM-GAN-AAE [10] (used in [17]) | 48.9 | 58.3 |
| AC-SUM-GAN [11] | 50.8 | 60.6 |
| AC-SUM-GAN$_{aug}$ (used now) | **52.0** | **61.0** |

**Table 2.** Used datasets for training and evaluating the AC-SUM-GAN model.

| Dataset | Videos | Duration (min) | Content |
|---|---|---|---|
| SumMe [19] | 25 | 1–6 | holidays, events, sports, travelling, cooking |
| TVSum [28] | 50 | 1–11 | news, how-to's, user-generated, documentaries |
| OVP [18] | 50 | 1–4 | documentary, educational, ephemeral, historical, lecture |
| YouTube [18] | 50 | 1–10 | cartoons, sports, tv-shows, commercial, home videos |

tate the training of a general-purpose video summarization model. Nevertheless, we anticipate a better summarization performance on videos from the different categories found in the used datasets, such as tutorials, "how-to", product demos, and event videos (e.g., birthday parties) that are commonly shared on social media, compared to the expected performance on videos from completely unseen categories, such as movies, football games, and music shows. This data augmentation process resulted in improvements on both benchmarking datasets (see the last row of Table 1) and to a very competitive performance against several state-of-the-art unsupervised methods from the literature that have been assessed under the same evaluation settings (see Table 3).

To minimize the possibility of losing semantically-important visual content or resulting in visually unpleasant results during **video aspect ratio transformation** (that would be highly possible when using naive approaches, such as fixed cropping of a central area of the video frames, or padding of black borders to reach the target aspect ratio) the proposed solution integrates an extension of the smart video cropping (SVC) method of [14]. The latter starts by computing the saliency map for each chosen frame for inclusion in the video summary. Then, to select the main part of the viewers' focus, the integrated method applies a filtering-through-clustering procedure on the pixel values of each predicted saliency map. Finally, it infers a single point as the center of the viewer's attention and computes a crop window for each frame based on the displacement of this point. The applied extension on [14], relates to the use of a state-of-the-art saliency prediction method [21], which resulted in improved

**Table 3.** Performance comparison (F-Score (%)) with state-of-the-art unsupervised approaches after using augmented training data. The reported scores for the listed methods are from the corresponding papers.

| Method | SumMe | TVSum |
|---|---|---|
| ACGAN [20] | 47.0 | 58.9 |
| RSGN$_{unsup}$ [31] | 43.6 | 59.1 |
| 3DST-UNet [24] | 49.5 | 58.4 |
| DSR-RL-GRU [26] | 48.5 | 59.2 |
| ST-LSTM [25] | **52.0** | 58.1 |
| CAAN [23] | 50.9 | 59.8 |
| SUM-GDA$_{unsup}$ [22] | 50.2 | 60.5 |
| SUM-FCN$_{unsup}$ [27] | 51.1 | 59.2 |
| AC-SUM-GAN$_{aug}$ | **52.0** | **61.0** |

**Table 4.** Video aspect ratio transformation performance (IoU (%)) on the RetargetVid dataset.

| | Method | Worst | Best | Mean |
|---|---|---|---|---|
| 1:3 target aspect ratio | SVC (used in [14]) | 51.7 | 53.8 | 52.9 |
| | SVC$_{ext}$ (used now) | **53.8** | **57.6** | **55.6** |
| 3:1 target aspect ratio | SVC (used in [14]) | 74.4 | 77.0 | 75.3 |
| | SVC$_{ext}$ (used now) | **76.3** | **78.0** | **77.6** |

performance on the RetargetVid dataset [13]. As shown in Table 4, the averaged Intersection-over-Union (IoU) scores for all video frames have been increased by more than 2 percentage points.

## 4   Conclusions and Future Work

In this paper, we presented a web-based tool which facilitates the generation of video summaries that are tailored to the specifications of various social media platforms, in terms of video length and aspect ratio. After reporting on the applied extensions to a previous instance of the tool, we provided more details about the front-end user interface and the back-end component of this technology, and we documented the advanced performance of the newly integrated AI models for video summarization and aspect ratio transformation. This tool will be demonstrated in MMM2024, while the participants in the relevant demonstration session will have the opportunity to test it in real-time. In the future, we will explore different options for including audio into the generated video summaries, while a relevant functionality will be made available via a user-defined parameter in the front-end UI of the tool. In addition, we will investigate various approaches for exploiting textual data associated with the submitted video (such as the video title and description) or user-defined queries, in order to perform text/query-driven summarization and produce video summaries that are more tailored to the main subject of the video or the users' demands.

# References

1. Brevify: Video Summarizer. https://devpost.com/software/brevify-video-summar izer. Accessed 29 Sept 2023
2. Cloudinary: Easily create engaging video summaries. https://smart-ai-transform ations.cloudinary.com. Accessed Sept 2023
3. Cognitive Mill: Cognitive Computing Cloud Platform For Media And Entertainment. https://cognitivemill.com. Accessed Sept 2023
4. Eightify: Youtube Summary with ChatGPT. https://chrome.google.com/webs tore/detail/eightify-youtube-summary/cdcpabkolgalpgeingbdcebojebfelgb. Accessed Sept 2023
5. Pictory: Automatically summarize long videos. https://pictory.ai/pictory-features/auto-summarize-long-videos. Accessed 29 Sept 2023
6. summarize.tech: AI-powered video summaries. https://www.summarize.tech. Accessed 29 Sept 2023
7. Video Highlight: the fastest way to summarize and take notes from videos. https://videohighlight.com. Accessed 29 Sept 2023
8. Video Summarizer - Summarize YouTube Videos. https://mindgrasp.ai/video-summarizer. Accessed 29 Sept 2023
9. VidSummize - AI YouTube Summary with Chat GPT. https://chrome.goo gle.com/webstore/detail/vidsummize-ai-youtube-sum/gidcfccogfdmkfdfmhfdmfni bafoopic. Accessed 29 Sept 2023
10. Apostolidis, E., Adamantidou, E., Metsai, A.I., Mezaris, V., Patras, I.: Unsupervised video summarization via attention-driven adversarial learning. In: Ro, Y.M., et al. (eds.) MMM 2020. LNCS, vol. 11961, pp. 492–504. Springer, Cham (2020). https://doi.org/10.1007/978-3-030-37731-1_40
11. Apostolidis, E., Adamantidou, E., Metsai, A.I., Mezaris, V., Patras, I.: AC-SUM-GAN: connecting actor-critic and generative adversarial networks for unsupervised video summarization. IEEE Trans. Circ. Syst. Video Technol. **31**(8), 3278–3292 (2021). https://doi.org/10.1109/TCSVT.2020.3037883
12. Apostolidis, E., Adamantidou, E., Metsai, A.I., Mezaris, V., Patras, I.: Video summarization using deep neural networks: a survey. Proc. IEEE **109**(11), 1838–1863 (2021). https://doi.org/10.1109/JPROC.2021.3117472
13. Apostolidis, K., Mezaris, V.: A fast smart-cropping method and dataset for video retargeting. In: Proceedings of the IEEE International Conference on Image Processing (ICIP), pp. 1956–1960 (2021)
14. Apostolidis, K., Mezaris, V.: A web service for video smart-cropping. In: 2021 IEEE International Symposium on Multimedia (ISM), pp. 25–26. IEEE (2021)
15. Awad, G., et al.: TRECVID 2017: evaluating ad-hoc and instance video search, events detection, video captioning and hyperlinking. In: 2017 TREC Video Retrieval Evaluation, TRECVID 2017, Gaithersburg, MD, USA, 13–15 November 2017. National Institute of Standards and Technology (NIST) (2017)
16. Baraldi, L., Grana, C., Cucchiara, R.: Shot and scene detection via hierarchical clustering for re-using broadcast video. In: Azzopardi, G., Petkov, N. (eds.) CAIP 2015. LNCS, vol. 9256, pp. 801–811. Springer, Cham (2015). https://doi.org/10. 1007/978-3-319-23192-1_67

17. Collyda, C., Apostolidis, K., Apostolidis, E., Adamantidou, E., Metsai, A.I., Mezaris, V.: A web service for video summarization. In: ACM International Conference on Interactive Media Experiences (IMX), pp. 148–153 (2020)

18. De Avila, S.E.F., Lopes, A.P.B., da Luz Jr, A., de Albuquerque Araújo, A.: VSUMM: a mechanism designed to produce static video summaries and a novel evaluation method. Pattern Recogn. Lett. **32**(1), 56–68 (2011)

19. Gygli, M., Grabner, H., Riemenschneider, H., Van Gool, L.: Creating summaries from user videos. In: Fleet, D., Pajdla, T., Schiele, B., Tuytelaars, T. (eds.) ECCV 2014. LNCS, vol. 8695, pp. 505–520. Springer, Cham (2014). https://doi.org/10.1007/978-3-319-10584-0_33

20. He, X., et al.: Unsupervised video summarization with attentive conditional generative adversarial networks. In: Proceedings of the 27th ACM International Conference on Multimedia (MM 2019), pp. 2296–2304. ACM, New York, NY, USA (2019)

21. Hu, F., et al.: TinyHD: efficient video saliency prediction with heterogeneous decoders using hierarchical maps distillation. In: Proceedings of the IEEE/CVF Winter Conference on Applications of Computer Vision, pp. 2051–2060 (2023)

22. Li, P., Ye, Q., Zhang, L., Yuan, L., Xu, X., Shao, L.: Exploring global diverse attention via pairwise temporal relation for video summarization. Pattern Recogn. **111**, 107677 (2021)

23. Liang, G., Lv, Y., Li, S., Zhang, S., Zhang, Y.: Video summarization with a convolutional attentive adversarial network. Pattern Recogn. **131**, 108840 (2022)

24. Liu, T., Meng, Q., Huang, J.J., Vlontzos, A., Rueckert, D., Kainz, B.: Video summarization through reinforcement learning with a 3D spatio-temporal U-Net. Trans. Image Proc. **31**, 1573–1586 (2022)

25. Min, H., Ruimin, H., Zhongyuan, W., Zixiang, X., Rui, Z.: Spatiotemporal two-stream LSTM network for unsupervised video summarization. Multimed. Tools Appl. **81**, 40489–40510 (2022)

26. Phaphuangwittayakul, A., Guo, Y., Ying, F., Xu, W., Zheng, Z.: Self-attention recurrent summarization network with reinforcement learning for video summarization task. In: Proceedings of the 2021 IEEE International Conference on Multimedia and Expo (ICME), pp. 1–6 (2021). https://doi.org/10.1109/ICME51207.2021.9428142

27. Rochan, M., Ye, L., Wang, Y.: Video summarization using fully convolutional sequence networks. In: Ferrari, V., Hebert, M., Sminchisescu, C., Weiss, Y. (eds.) ECCV 2018. LNCS, vol. 11216, pp. 358–374. Springer, Cham (2018). https://doi.org/10.1007/978-3-030-01258-8_22

28. Song, Y., Vallmitjana, J., Stent, A., Jaimes, A.: TVSum: summarizing web videos using titles. In: IEEE Conference on Computer Vision and Pattern Recognition (CVPR), pp. 5179–5187 (2015). https://doi.org/10.1109/CVPR.2015.7299154

29. Souček, T., Lokoč, J.: Transnet V2: an effective deep network architecture for fast shot transition detection. arXiv preprint arXiv:2008.04838 (2020)

30. Tang, S., Feng, L., Kuang, Z., Chen, Y., Zhang, W.: Fast video shot transition localization with deep structured models. In: Jawahar, C.V., Li, H., Mori, G., Schindler, K. (eds.) ACCV 2018. LNCS, vol. 11361, pp. 577–592. Springer, Cham (2019). https://doi.org/10.1007/978-3-030-20887-5_36

31. Zhao, B., Li, H., Lu, X., Li, X.: Reconstructive sequence-graph network for video summarization. IEEE Trans. Pattern Anal. Mach. Intell. **44**, 2793–2801 (2021). https://doi.org/10.1109/TPAMI.2021.3072117

# AI-Based Cropping of Soccer Videos for Different Social Media Representations

Mehdi Houshmand Sarkhoosh[2,3]([✉]), Sayed Mohammad Majidi Dorcheh[2,3],
Cise Midoglu[1,3], Saeed Shafiee Sabet[1,3], Tomas Kupka[3], Dag Johansen[4],
Michael A. Riegler[1,2], and Pål Halvorsen[1,2,3]

[1] SimulaMet, Oslo, Norway
[2] Oslo Metropolitan University, Oslo, Norway
mehdihou@oslomet.no
[3] Forzasys, Oslo, Norway
[4] UIT The Arctic University of Norway, Tromsø, Norway

**Abstract.** The process of re-publishing soccer videos on social media often involves labor-intensive and tedious manual adjustments, particularly when altering aspect ratios while trying to maintain key visual elements. To address this issue, we have developed an AI-based automated cropping tool called SmartCrop which uses object detection, scene detection, outlier detection, and interpolation. This innovative tool is designed to identify and track important objects within the video, such as the soccer ball, and adjusts for any tracking loss. It dynamically calculates the cropping center, ensuring the most relevant parts of the video remain in the frame. Our initial assessments have shown that the tool is not only practical and efficient but also enhances accuracy in maintaining the essence of the original content. A user study confirms that our automated cropping approach significantly improves user experience compared to static methods. We aim to demonstrate the full functionality of SmartCrop, including visual output and processing times, highlighting its efficiency, support of various configurations, and effectiveness in preserving the integrity of soccer content during aspect ratio adjustments.

**Keywords:** AI · aspect ratio · cropping · GUI · soccer · social media · video

## 1 Introduction

Media consumption has expanded beyond traditional platforms, and content must be curated and prepared in various formats [32]. Today, people are spending a huge amount of time to manually generate content to be posted in various distribution channels, targeting different user groups. With the rise of artificial intelligence (AI), such processes can be automated [7,21]. Using soccer as a case study, we have earlier researched and demonstrated tools to automatically detect events [23–25], clip events [34,35], select thumbnails [10,11] and summarize games [8,9]. The next step is to prepare the content for specific platforms,

S. Rudinac et al. (Eds.): MMM 2024, LNCS 14557, pp. 279–287, 2024.
https://doi.org/10.1007/978-3-031-53302-0_22

**Fig. 1.** Smart cropping using object detection. Red dot marks the calculated cropping-center point in the frame (the ball), red square marks the cropping area. (Color figure online)

ranging from large television screens to compact smartphones. Each platform and view mode might require different aspect ratios, necessitating an adaptive presentation that remains consistent in its delivery of content to audiences, irrespective of the viewing settings, i.e., cropping the original content to the target platform while preserving its essence.

To capture the saliency of soccer videos, the cropping operation must consider dynamic objects such as the ball and players at the same time as the camera itself is moving. Traditionally, video cropping is performed manually using commercial software tools, selecting the area of interest in a frame-by-frame manner, which is a tedious and time-consuming operation. With the sheer volume of content and the demand for timely publishing, manual approaches are unsustainable. Hence, the domain has witnessed a shift towards automated solutions using AI.

There are several areas of research related to cropping, such as changing the video aspect ratio, e.g., content-adaptive reshaping (warping) [17,22], segment-based exclusion (cropping) [5,12,19,27], seam extraction [14,37] and hybrids of these techniques [1,16,38], as well as detecting the location of important objects in a frame, e.g., single shot multibox detector [20], focal loss for dense object detection [18], faster R-CNN [29], and YOLO [28]. In the context of our work, YOLOv8 has a specialized ball detector. However, challenges inherent in soccer broadcasts, such as swift player movements, sudden changes in camera perspectives, and frequent occlusions, necessitate adjustments to YOLOv8's neural network layers. Furthermore, ensuring the consistent visibility of the soccer ball is non-trivial, and challenges such as inconsistent detection, rapid lighting changes, player occlusions, and camera angles can hinder its accurate detection.

The goal of our work is to develop an AI-based smart cropping pipeline tailored for soccer highlights to be published on social media. We have developed SMARTCROP [6] for delivering various media representations using a fine-tuned version of YOLOv8 for object detection, and tracking the ball through an extended logic including outlier detection and interpolation, for calculating an appropriate cropping-center for video frames. This is shown in Fig. 1. Initial results from both objective and subjective experiments show that SMARTCROP increases end-user Quality of Experience (QoE). We will demonstrate our pipeline step-by-step in interactive fashion, where participants can configure settings such as the target aspect ratio, and chose among various methods for object detection, outlier detection, and interpolation.

## 2    SMARTCROP

The guiding principle for the SMARTCROP pipeline [6] is to use the soccer ball as the Point of Interest (POI), i.e., the center of the cropping area, as illustrated by Fig. 1. When the ball is visible within the frame, it is the primary focal point for the cropping. If it is absent, we employ outlier detection and interpolation to select an appropriate alternative focal point. The pipeline is depicted in Fig. 2.

All video formats can be processed by SMARTCROP, in this demo we use an HTTP Live Streaming (HLS) playlist as **video input**. To reduce runtime, the pipeline searches inside the HLS manifest and selects the lowest quality stream available, which has the smallest resolution. After pre-processing, the first step is **scene detection**, which runs SceneDetect [3] and TransNetV2 [33]. These models jointly segment the video into distinct scenes, allowing each to serve as a separate unit for further steps.

The **object detection** module can employ one of the various supported alternative YOLOv8 models [13,36]. Here, we tested various alternatives as shown in Fig. 3 (Sc01 to Sc05 based on YOLOv8-nano, small, medium, large, xlarge respectively), and opted for the medium architecture as a baseline after considering several trade-offs. Larger architectures offer slightly better performance, but at a significantly increased computation time. On the other hand, smaller models provide quicker processing, but suffer from reduced accuracy. Starting with Sc06 (previously trained on 4 classes as detailed in [26]), we used the YOLOv8-medium model, fine-tuning it on a dataset that included 1,500 annotated images from Norwegian and Swedish soccer leagues, plus 250 images

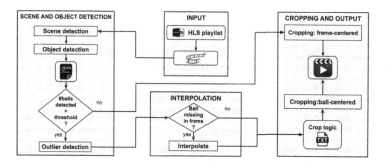

**Fig. 2.** SMARTCROP pipeline overview.

from a public soccer dataset [30]. The model went through multiple training scenarios with different hyperparameters, Sc11 representing the final and most effective scenario in this series. This specific configuration involved processing high-resolution images and utilized an increased batch size for more effective learning. The training approach for featured extended training epochs, a carefully calibrated learning rate for steady model development, and a tailored dropout rate to ensure robust generalization. This structured training regimen enabled Sc11 to achieve significant accuracy in detecting balls, players, and logos, establishing it as the optimal configuration for the object detection module.

The **outlier detection** module is designed to identify and exclude anomalous data points, known as outliers, from the detected positions of the soccer ball. This is crucial for enhancing the system's robustness. Outliers are data points that deviate significantly from the majority, lying beyond a pre-set threshold. To accurately determine these thresholds, we integrated three distinct methods: Interquartile Range (IQR), Z-score, and modified Z-score, all described in [31]. After outlier detection, we use **interpolation** to estimate the position of the ball for frames where the ball is not detected. The module can impute missing data points using various alternative interpolation techniques, including linear interpolation [2], polynomial interpolation [2], ease-in-out interpolation [15], and our own heuristic-based smoothed POI interpolation [6].

The **cropping** module is used to isolate regions of interest within the frames, reducing computational complexity and improving focus on key areas. It crops each frame using ffmpeg, based on the aspect ratio dictated by the relevant parameter in the user configuration. Overall, the pipeline undertakes either *ball-centered cropping*, where cropping dimensions are computed dynamically based on the coordinates created by the interpolation module (optimal operation), or *frame-centered cropping*, where cropping dimensions are computed statically based on the aspect ratio and frame dimensions, to point to the mathematical frame center (fallback in case of too few ball detections, or explicit user configuration *not* to use "smart" crop). As **output**, SMARTCROP returns an .mp4 file generated from the cropped frames. It also has additional functionality to prepare processed data for visualization, summarization, and further analysis.

## 3    Evaluation

To evaluate SMARTCROP, we performed both objective and subjective experiments. Figure 3 presents the **object detection** performance of various YOLO configurations. Notably, Sc11 significantly outperforms all other models. In this experiment, we monitored the precision and recall across training epochs to ensure effective learning without overfitting. Starting from epoch 1 with a precision of 0.71 and recall of 0.046, there was a gradual increase, achieving a precision of approximately 0.969 and a recall rate of 0.896 by epoch 109. This trend suggests that the model is effectively learning to identify more positive samples (high precision) and capturing a larger proportion of total positive samples (high

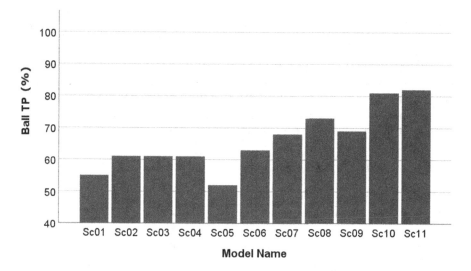

**Fig. 3.** The performance of various YOLO models in terms of ball true positive (TP) detections. Note that the y-axis starts at 40% to better highlight differences.

recall). To guard against overfitting, we implemented several strategies, including regular validation on a separate dataset, early stopping if the validation loss ceased to decrease, and a dropout rate.

To evaluate the **outlier detection** and **interpolation** performance, we assessed the Root Mean Square Error (RMSE) [4] of various methods against a meticulously annotated ground truth derived from a 30-second soccer video, detailing the ball's position in each frame. Following the validation, the IQR method demonstrated superior accuracy and has since been integrated as the default approach in our pipeline. Among alternative interpolation methods, heuristic interpolation [6] proved to perform best and has been adopted as the default.

To evaluate **system performance**, we tested a local deployment on an NVIDIA GeForce GTX 1050 GPU with 4 GB memory, focusing on the impact of the Skip Frame parameter. Most modules, such as scene detection and cropping, maintained consistent execution times across different configurations. However, by adjusting the Skip Frame parameter from 1 to 13, the object detection module's execution time decreased 70% due to processing fewer frames.

To assess the **end-to-end subjective performance** of the pipeline, we performed a user survey via Google Forms to retrieve feedback from participants in a crowd-sourced fashion on the final cropped video output. We compared 2 static and 4 dynamic approaches, as shown in Table 1. Two representative videos were selected to evaluate each approach, one of normal gameplay with fast motion and edge field features (video 1) and one of a goal with varying motion and ball occlusion features (video 2), each cropped to 1:1 and 9:16 target aspect ratios. Thus, four cases were constructed overall (two videos, each cropped to two aspect ratios) to compare the performance of six different cropping approaches (where the full SMARTCROP corresponds to approach 6). We recruited 23 participants (9 female, 14 male) with ages ranging from 18 to 63 (mean: 30.95), all of whom were active on social media. For each case, the participants first viewed the original video, and then each of the processed videos corresponding to cropping approaches 1–6, which they rated using a 5-point Absolute Category Rating (ACR) scale, as recommended by ITU-T P.800. As shown in Fig. 4, one-way repeated measures ANOVA results indicate an overall improved QoE for both videos in both aspect ratios, for SMARTCROP (approach 6).

**Table 1.** Cropping alternatives used in the subjective evaluation.

| No | Crop Centering | Description | Outlier detection | Interpolation |
|----|----------------|-------------|-------------------|---------------|
| 1 | frame-centered | static no padding | ✘ | ✘ |
| 2 | frame-centered | static w/black padding to 16:9 | ✘ | ✘ |
| 3 | ball-centered | use last detected ball position | ✘ | ✘ |
| 4 | ball-centered | w/interpolation | ✘ | ✔ |
| 5 | ball-centered | w/outlier detection | ✔ | ✘ |
| 6 | ball-centered | w/interpolation & outlier detection | ✔ | ✔ |

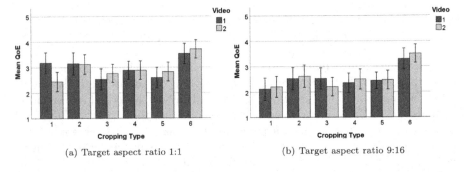

(a) Target aspect ratio 1:1             (b) Target aspect ratio 9:16

**Fig. 4.** QoE ratings for different target aspect ratios with 95% confidence intervals.

## 4    Demonstration

We demonstrate the SMARTCROP pipeline with a graphical user interface (GUI) as depicted in Fig. 5. First, the participants set up the cropping operation by selecting an HLS stream and setting configuration parameters such as output aspect ratio, skip frame, object detection model, outlier detection model, and interpolation model. Then, the pipeline can be started, which runs in three steps; i) scene and object detection; ii) outlier detection and interpolation; and iii) cropping and video output. For each of the steps, our GUI provides debugging information (e.g., execution time in the text output box), as well as intermediate results in terms of visual output (video output window), to demonstrate the individual contributions of various modules. Finally, the cropped video can be viewed. A video of the demo can be found here: https://youtu.be/aqqPWfrPmsE, with more details on the pipeline in [6].

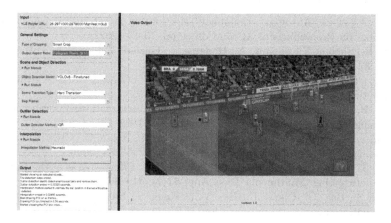

**Fig. 5.** SMARTCROP GUI - shown here is an intermediate result with POI markings.

**Acknowledgment.** This research was funded by the Research Council of Norway, project number 346671 (AI-storyteller).The authors would like to thank the Norwegian Professional Football League ("Norsk Toppfotball") for making videos available for the research.

# References

1. Apostolidis, K., Mezaris, V.: A fast smart-cropping method and dataset for video retargeting. In: Proceedings of IEEE ICIP, pp. 2618–2622 (2021)
2. Bourke, P.: Interpolation methods. Miscellaneous Projection Model. Rendering **1**(10), 1–9 (1999)
3. Castellano, B.: SceneDetect. https://github.com/Breakthrough/PySceneDetect/tree/main (2023)
4. Chai, T., Draxler, R.R.: Root mean square error (RMSE) or mean absolute error (MAE)? Arguments against avoiding RMSE in the literature. Geosci. Model Dev. **7**(3), 1247–1250 (2014)
5. Deselaers, T., Dreuw, P., Ney, H.: Pan zoom scan - time-coherent trained automatic video cropping. In: Proceedings of IEEE CVPR, pp. 1–8 (2008)
6. Dorcheh, S.M.M., et al.: SmartCrop: AI-based cropping of soccer videos. In: Proceedings of IEEE ISM (2023)
7. Gautam, S.: Bridging multimedia modalities: enhanced multimodal AI understanding and intelligent agents. In: Proceedings of ACM ICMI (2023)
8. Gautam, S., Midoglu, C., Shafiee Sabet, S., Kshatri, D.B., Halvorsen, P.: Assisting soccer game summarization via audio intensity analysis of game highlights. In: Proceedings of 12th IOE Graduate Conference, vol. 12, pp. 25–32. Institute of Engineering, Tribhuvan University, Nepal (2022)
9. Gautam, S., Midoglu, C., Shafiee Sabet, S., Kshatri, D.B., Halvorsen, P.: Soccer game summarization using audio commentary, metadata, and captions. In: Proceedings of of ACM MM NarSUM. pp. 13–22 (2022)
10. Husa, A., Midoglu, C., Hammou, M., Halvorsen, P., Riegler, M.A.: HOST-ATS: automatic thumbnail selection with dashboard-controlled ML pipeline and dynamic user survey. In: Proceedings of ACM MMSys, pp. 334–340 (2022)

11. Husa, A., et al.: Automatic thumbnail selection for soccer videos using machine learning. In: Proceedings of ACM MMSys, pp. 73–85 (2022)

12. Jain, E., Sheikh, Y., Shamir, A., Hodgins, J.: Gaze-driven video re-editing. ACM TOG **34**(2), 1–12 (2015)

13. Jocher, G., Chaurasia, A., Qiu, J.: Ultralytics YOLOv8. https://github.com/ultralytics/ultralytics (2023)

14. Kaur, H., Kour, S., Sen, D.: Video retargeting through spatio-temporal seam carving using kalman filter. IET Image Proc. **13**(11), 1862–1871 (2019)

15. Kemper, M., Rosso, G., Monnone, B., Kemper, M., Rosso, G.: Creating animated effects. In: Advanced Flash Interface Design, pp. 255–288 (2006)

16. Kopf, S., Haenselmann, T., Kiess, J., Guthier, B., Effelsberg, W.: Algorithms for video retargeting. Multimedia Tools Appl **51**(2), 819–861 (2011)

17. Lee, H.S., Bae, G., Cho, S.I., Kim, Y.H., Kang, S.: Smartgrid: video retargeting with spatiotemporal grid optimization. IEEE Access **7**, 127564–127579 (2019)

18. Lin, T.Y., Goyal, P., Girshick, R., He, K., Dollár, P.: Focal loss for dense object detection. In: Proceedings of IEEE ICCV, pp. 2980–2988 (2017)

19. Liu, F., Gleicher, M.: Video retargeting: automating pan and scan. In: Proceedings of ACM MM, pp. 241–250 (2006)

20. Liu, W., et al.: SSD: Single shot multibox detector. In: Proceedings of ECCV, pp. 21–37 (2016)

21. Midoglu, C., et al.: AI-based sports highlight generation for social media. In: Proceedings of ACM MHV (2024)

22. Nam, H., Park, D., Jeon, K.: Jitter-robust video re-targeting with kalman filter and attention saliency fusion network. In: Proceedings of IEEE ICIP, pp. 858–862 (2020)

23. Nergård Rongved, O.A., et al.: Using 3D convolutional neural networks (CNN) for real-time detection of soccer events. Int. J. Seman. Comput. **15**(2), 161–187 (2021)

24. Nergård Rongved, O.A., et al.: Real-time detection of events in soccer videos using 3D convolutional neural networks. In: Proceedings of IEEE ISM, pp. 135–144 (2020)

25. Nergård Rongved, O.A., et al.: Automated event detection and classification in soccer: the potential of using multiple modalities. Mach. Learn. Knowl. Extr. **3**(4), 1030–1054 (2021)

26. Noorkhokhar: YOLOv8-football: how to detect football players and ball in real-time using YOLOv8: a computer tutorial. https://github.com/noorkhokhar99/YOLOv8-football

27. Rachavarapu, K.K., Kumar, M., Gandhi, V., Subramanian, R.: Watch to edit: video retargeting using gaze. Comput. Graph. Forum **37**, 205–215 (2018)

28. Redmon, J., Divvala, S., Girshick, R., Farhadi, A.: You only look once (YOLO): unified, real-time object detection. In: Proceedings of IEEE CVPR, pp. 779–788 (2016)

29. Ren, S., He, K., Girshick, R., Sun, J.: Faster R-CNN: towards real-time object detection with region proposal networks. In: Advances in Neural Information Processing Systems, vol. 28 (2015)

30. Roboflow: football players detection dataset. https://universe.roboflow.com/roboflow-jvuqo/football-players-detection-3zvbc (2023)

31. Saleem, S., Aslam, M., Shaukat, M.R.: A review and empirical comparison of universe outlier detection methods. Pakistan J. Stat. **37**(4), 447–462 (2021)

32. Sarkhoosh, M.H., Dorcheh, S.M.M., Gautam, S., Midoglu, C., Sabet, S.S., Halvorsen, P.: Soccer on social media. arXiv preprint arXiv:2310.12328 (2023)

33. Soucek, T., Lokoc, J.: TransNet V2: an effective deep network architecture for fast shot transition detection. CoRR (2020)

34. Valand, J.O., et al.: Automated clipping of soccer events using machine learning. In: Proceedings of IEEE ISM, pp. 210–214 (2021)

35. Valand, J.O., Kadragic, H., Hicks, S.A., Thambawita, V.L., Midoglu, C., Kupka, T., Johansen, D., Riegler, M.A., Halvorsen, P.: AI-based video clipping of soccer events. Mach. Learn. Knowl. Extr. **3**(4), 990–1008 (2021)

36. Wang, C.Y., Bochkovskiy, A., Liao, H.Y.M.: YOLOv7: trainable bag-of-freebies sets new state-of-the-art for real-time object detectors. arXiv preprint arXiv:2207.02696 (2022)

37. Wang, S., Tang, Z., Dong, W., Yao, J.: Multi-operator video retargeting method based on improved seam carving. In: Proceedings of IEEE ITOEC, pp. 1609–1614 (2020)

38. Wang, Y.S., Lin, H.C., Sorkine, O., Lee, T.Y.: Motion-based video retargeting with optimized crop and warp. In: Proceedings of ACM SIGGRAPH, pp. 1–9 (2010)

# Few-Shot Object Detection as a Service: Facilitating Training and Deployment for Domain Experts

Werner Bailer[1]([✉])[ID], Mihai Dogariu[2][ID], Bogdan Ionescu[2][ID],
and Hannes Fassold[1][ID]

[1] Joanneum Research – Digital, Graz, Austria
{werner.bailer,hannes.fassold}@joanneum.at
[2] National University of Science and Technology POLITEHNICA Bucharest,
Bucharest, Romania
{mihai.dogariu,bogdan.ionescu}@upb.ro

**Abstract.** We propose a service-based approach for training few-shot object detectors and running inference with these models. This eliminates the need to write code or execute scripts, thus enabling domain experts to train their own detectors. The training service implements an efficient ensemble learning method in order to obtain more robust models without parameter search. The entire pipeline is deployed as a single container and can be controlled from a web user interface.

**Keywords:** few-shot learning · object detection · ensemble learning · data preparation

## 1 Introduction

Few-shot object detection is useful in order to extend object detection capabilities in archiving and sourcing of multimedia content with specific object classes of interest for a particular organization or production context. For example, a specific class of objects may be relevant for a media production project, but not among those commonly annotated. Current workflows and tools for training few-shot object detectors have been created for data scientists and developers, but are not usable by domain experts (for example documentalists or journalists), who can provide and annotate the training samples of interest. Data scientists are usually not part of a media production team, and involving them is not possible on short notice for a small task. In order to leverage the progress in research on few-shot object detection for practitioners, it is thus necessary to enable them to use these tools themselves.

We build on the work in [2], which proposed integrated training scripts and support in content annotation tools. In addition, we provide ensemble learning methods for more robust training processes. Both are built using the same FSOD framework [14] and can thus be integrated. [2] has addressed the issue of integrating all necessary steps into the training workflow, and providing an

S. Rudinac et al. (Eds.): MMM 2024, LNCS 14557, pp. 288–294, 2024.
https://doi.org/10.1007/978-3-031-53302-0_23

extension of an annotation tool to label the data for novel classes. However, it may still be hard for non-experts to select an appropriate model and find a good configuration for training. In addition, once trained, the result is a serialised model, requiring access (via remote shell or desktop) to the training server in order to use the model further.

This paper addresses the shortcomings that prevent domain experts from training their own few-shot object detectors. It demonstrates a service for few-shot object detection that (i) runs the entire training workflow for the provided data, (ii) includes ensemble learning methods in order to facilitate achieving good results without the need for lengthy parameter tuning and (iii) deploys the trained model automatically as a service. The code is available at https://github.com/wbailer/few-shot-object-detection.

The rest of this paper is organised as follows. Section 2 provides a brief overview of the related work on ensemble learning for few-shot object detection (FSOD) and tools for few-shot training. Section 3 presents the integrated ensemble learning method, the training service is described in Sect. 4, and Sect. 5 concludes the paper.

## 2 Related Work

### 2.1 Ensemble Learning for FSOD

Few-shot object detection is by itself a relatively new topic, being addressed by a modest, albeit growing, number of works. As a sub-topic, ensemble learning in few-shot object detection is even more arcane. Ensembling efforts have been mostly made in the image classification field [1,7,13], or for object detection [4, 5,9], in both cases approaching techniques such as cooperation, competition or voting schemes. Separately, different groups of researchers tackled the few-shot object detection problem [3,11,14,15], with focus on meta-learning, weight sharing or fine-tuning approaches. However, we could not find any work related to ensemble learning in the few-shot object detection scenario. Therefore, to our knowledge, our work is the first in this field.

### 2.2 Tooling for Few-Shot Training

Torchmeta [6] is a Python library for meta-learning, providing a common interface for working with different algorithms and datasets. However, no support for using own/aggregated datasets is provided. Argilla[1], a data framework for large language models, also provides some support for facilitating few-shot learning. Similarly, AutoML frameworks such as Amazon SageMaker[2] and AutoGluon[3] provide some support to set up few-shot learning tasks. However, all these approaches still require coding.

---

[1] https://argilla.io/.

[2] https://aws.amazon.com/de/sagemaker/.

[3] https://auto.gluon.ai/stable/index.html.

The approach in [2] focuses on using custom data for novel classes, possibly in combination with other datasets. An integration with the MakeSense[4] annotation tool is provided, in order to obtain a set of files for the training process, which can be fed into a single script running the two-stage finetuning approach [14]. Two recent works for action recognition [8] and medical image segmentation [12] propose interactive few-shot learning frameworks, aiming at getting required annotations, as well as guiding the human annotator towards creating discriminative annotations. However, these frameworks do not consider deployment of the final model, and require having the human user in the loop.

## 3  Integrated Ensemble Learning

We base our ensembling strategy on a mixture between the works of Wang et al. [14] and Dvornik et al. [7]. In the former, FSOD is introduced as a fine-tuning step on top of a pre-trained two-stage object detector. The authors argue that having a pre-trained detector, it is sufficient to freeze the entire network, except for the last two layers and perform fine-tuning solely on these layers in order to obtain improved performance. In the latter, the authors tackle the problem of image classification and argue that having several networks performing the same classification task together yields better results due to having as little as possible different random weight initialization. The authors study the impact brought by having several almost identical classifiers perform the same job, with the only difference between them being the random values used to initialize the weights in the training process.

Two-stage object detectors generally consist of two fundamental sections: the region (or object) proposal network (RPN) and the feature extractor, together with a classifier, working on top of it. Ensembling strategies usually deploy several networks and process their set of outcomes, as depicted in Fig. 1a. However, this bears the cost of training N different networks with N usually being greater than 5. From the resource point of view, this type of processing is very costly, especially GPU-wise. Furthermore, the ensemble is usually distilled in order to reduce inference time, possibly at the cost of also reducing the system's performance.

Our ensemble learning paradigm takes advantage of the fact that the framework presented in [14] freezes almost the entire two-stage object detection network. This leaves the object classifier part open for fine-tuning. Following [7], we apply a set of classifiers on top of the features extracted from the RPN's proposed boxes and generate N classification decisions for each proposed box, as depicted in Fig. 1b, thus approximately simulating an ensemble of N complete networks. Then, a regular non-maximum suppression (NMS) algorithm is applied for the resulting proposals. From this point on, the system approximates a single object detector, with enhanced detection capabilities.

The main difference between this method and the usual ensembling strategy is that we reduce the use of resources N-fold. One could argue that our proposed system's RPN does not behave in the same manner as in the original case,

---

[4] https://www.makesense.ai/.

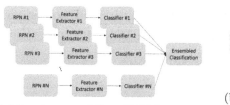

(a) Regular ensembling of FSOD systems.

(b) Proposed ensembling of FSOD systems.

**Fig. 1.** Comparison between regular and proposed ensembling architectures for FSOD systems.

having N times less proposed boxes to work on, but allowing the ensembled system's RPN to propose a large amount of possible objects (>2000) reaches the same performance as a combination of several RPNs, since the vast majority of the proposed regions are, in fact, not objects, and are therefore redundant. Another significant advantage of our method is that it is almost free to scale. Adding another network to the ensemble is reduced to adding another classification head to the architecture, which has insignificant impact from a memory standpoint. This method adds flexibility in the sense that it can be applied to a large number of network architectures that follow this working environment. Both the classifiers and the ensembling algorithm can remain unchanged from the regular ensembling setup. Performance-wise, our ensembling method adds a slight improvement to the detection performance of the original system [14]. To the best of our knowledge, this type of approach has not been tried before. Therefore, we compared our proposed system on the MS COCO [10] dataset with the original work of Wang et al. [14], while keeping the evaluation protocol unchanged. We obtained an AP@0.5 of 10.1 and 13.6 on 10 and 30 shots, respectively, compared to the original results of 10.0 and 13.4, respectively. Thus, our method essentially adds a marginal improvement with virtually no additional cost incurred.

## 4 Few-Shot Learning as a Service

We aim to provide few-shot learning as a service usable by domain experts. The extension of the MakeSense annotation tool described in our earlier work [2] allows annotating the new images and downloading the configuration files for the few-shot training task. We extend this by providing a RESTful API for starting and monitoring the training process, together with a TorchServe[5] instance that provides a RESTful interface for running inference using both pretrained and new models. A simple web page for testing the whole process is provided, and neither installation on the client side nor direct access to the remote server is required.

---

[5] https://pytorch.org/serve/.

All components are deployed as a Docker[6] container to facilitate deployment. The container is built to be deployed on an environment with GPU acceleration. The scripts to generate and run the container are provided on the Github repository.

## 4.1  Training Workflow and Service

The training workflow builds on the script described in [2], which prepares the training and validation datasets and related configurations on the fly from the submitted data, runs the training on the novel classes, merges the base and novel class weights in the classification head of the model, and runs the finetuning over all classes. This script has now been extended to support the ensemble training methods, and integrated into the backend of a server exposing a RESTful interface, implement using Flask[7].

The training endpoint of the service receives two ZIP files, one containing the configuration and annotation files generated by the annotation tool and the other the set of images. The configuration file specifies the name of the model to be trained, base model used and the number of novel classes. These files are stored on the training server, and the training procedure is invoked.

In order to monitor the training process, a logging endpoint has been implemented, which can be invoked to access the entire log or a number of tail lines.

A simple HTML page is provided to invoke the training, view logs and test inference. The page is hosted by the Flask server providing the training endpoint. Figure 2 shows screenshots of different stages of this page.

## 4.2  Auto-deployment

TorchServe is a framework for deploying PyTorch models for inference. A model is packaged as a model archive, containing a handler class, necessary configuration files and the trained weights of the model. We have built a custom handler based on TorchServe's object detection handler, which can generically serve the base models as well as any novel/combined models trained on the server.

Once few-shot training is completed, the required configuration files for the running inference in TorchServe are generated, and model archive is built. Using the TorchServe management API, the model is deployed to the server. To the best of our knowledge, this is the first work proposing a fully automated chain from training to deployment for few-shot object detection.

The TorchServe instance provided in the container starts already with base models trained on MS COCO, either on all or 60 of the classes (which is a common setting for evaluating few-shot object detection). Additional models are deployed and reachable via an endpoint that is named with the label provided for the model in the configuration file.

---

[6] https://www.docker.com/.

[7] https://flask.palletsprojects.com.

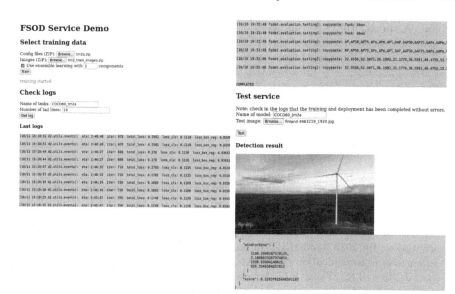

**Fig. 2.** Screenshot of the demo HTML page when configuring training and viewing logs (left) and testing inference with the newly trained model (right).

## 5 Conclusion

Building on previous work for facilitating training of few-shot object detectors, we provide the entire training and testing workflow of detectors as a service, usable for domain experts without the need to write code or install any software on the client device. The training service also integrates an efficient ensemble learning method.

**Acknowledgements.** The research leading to these results has been funded partially by the European Union's Horizon 2020 research and innovation programme under grant agreement n° 951911 AI4Media (https://ai4media.eu).

**Disclosure of Interests.** The authors have no competing interests to declare that are relevant to the content of this article.

## References

1. Afrasiyabi, A., Larochelle, H., Lalonde, J.F., Gagné, C.: Matching feature sets for few-shot image classification. In: Proceedings of the IEEE/CVF Conference on Computer Vision and Pattern Recognition, pp. 9014–9024 (2022)
2. Bailer, W.: Making few-shot object detection simpler and less frustrating. In: THornór Jónsson, B., et al. (eds.) MMM 2022. LNCS, vol. 13142, pp. 445–451. Springer, Cham (2022). https://doi.org/10.1007/978-3-030-98355-0_37

3. Bar, A., et al.: DETReg: unsupervised pretraining with region priors for object detection. In: Proceedings of the IEEE/CVF Conference on Computer Vision and Pattern Recognition, pp. 14605–14615 (2022)
4. Carranza-García, M., Lara-Benítez, P., García-Gutiérrez, J., Riquelme, J.C.: Enhancing object detection for autonomous driving by optimizing anchor generation and addressing class imbalance. Neurocomputing **449**, 229–244 (2021)
5. Casado-García, Á., Heras, J.: Ensemble methods for object detection. In: ECAI 2020, pp. 2688–2695. IOS Press (2020)
6. Deleu, T., Würfl, T., Samiei, M., Cohen, J.P., Bengio, Y.: TorchMeta: a meta-learning library for PyTorch (2019). https://arxiv.org/abs/1909.06576, https://github.com/tristandeleu/pytorch-meta
7. Dvornik, N., Schmid, C., Mairal, J.: Diversity with cooperation: ensemble methods for few-shot classification. In: Proceedings of the IEEE/CVF International Conference on Computer Vision, pp. 3723–3731 (2019)
8. Gassen, M., et al.: I3: interactive iterative improvement for few-shot action segmentation. In: 2023 32nd IEEE International Conference on Robot and Human Interactive Communication (RO-MAN), Busan, South Korea (2023)
9. Lee, J., Lee, S.K., Yang, S.I.: An ensemble method of CNN models for object detection. In: 2018 International Conference on Information and Communication Technology Convergence (ICTC), pp. 898–901. IEEE (2018)
10. Lin, T.Y., et al.: Microsoft COCO: common objects in context. In: Fleet, D., Pajdla, T., Schiele, B., Tuytelaars, T. (eds.) Computer Vision-ECCV 2014: 13th European Conference, Zurich, Switzerland, 6–12 September 2014, Proceedings, Part V 13, pp. 740–755. Springer, Cham (2014). https://doi.org/10.1007/978-3-319-10602-1_48
11. Liu, F., et al.: Integrally migrating pre-trained transformer encoder-decoders for visual object detection. In: Proceedings of the IEEE/CVF International Conference on Computer Vision, pp. 6825–6834 (2023)
12. Miyata, S., Chang, C.M., Igarashi, T.: Trafne: a training framework for non-expert annotators with auto validation and expert feedback. In: Degen, H., Ntoa, S. (eds.) International Conference on Human-Computer Interaction, pp. 475–494. Springer, Cham (2022). https://doi.org/10.1007/978-3-031-05643-7_31
13. Tian, Y., Wang, Y., Krishnan, D., Tenenbaum, J.B., Isola, P.: Rethinking few-shot image classification: a good embedding is all you need? In: Vedaldi, A., Bischof, H., Brox, T., Frahm, J.M. (eds.) Computer Vision-ECCV 2020: 16th European Conference, Glasgow, UK, 23–28 August 2020, Proceedings, Part XIV 16, pp. 266–282. Springer, Cham (2020). https://doi.org/10.1007/978-3-030-58568-6_16
14. Wang, X., Huang, T., Gonzalez, J., Darrell, T., Yu, F.: Frustratingly simple few-shot object detection. In: International Conference on Machine Learning, pp. 9919–9928. PMLR (2020)
15. Xiao, Y., Lepetit, V., Marlet, R.: Few-shot object detection and viewpoint estimation for objects in the wild. IEEE Trans. Pattern Anal. Mach. Intell. **45**(3), 3090–3106 (2022)

# DatAR: Supporting Neuroscience Literature Exploration by Finding Relations Between Topics in Augmented Reality

Boyu Xu[1]([✉])[iD], Ghazaleh Tanhaei[1][iD], Lynda Hardman[1,2][iD], and Wolfgang Hürst[1][iD]

[1] Utrecht University, Princetonplein 5, 3584 CC Utrecht, Netherlands
b.xu@uu.nl
[2] Centrum Wiskunde & Informatica, Science Park 123,
1098 XG Amsterdam, Netherlands

**Abstract.** We present DatAR, an Augmented Reality prototype designed to support neuroscientists in finding fruitful directions to explore in their own research. DatAR provides an immersive analytics environment for exploring relations between topics published in the neuroscience literature. Neuroscientists need to analyse large numbers of publications in order to understand whether a potential experiment is likely to yield a valuable contribution. Using a user-centred design approach, we have identified useful tasks in collaboration with neuroscientists and implemented corresponding functionalities in DatAR. This facilitates querying and visualising relations between topics. Participating neuroscientists have stated that the DatAR prototype assists them in exploring and visualising seldom-mentioned direct relations and also indirect relations between brain regions and brain diseases. We present the latest incarnation of DatAR and illustrate the use of the prototype to carry out two realistic tasks to identify fruitful experiments.

**Keywords:** Topic-based Literature Exploration · Data Visualisation · Augmented Reality · Human-centered Computing

## 1 Motivation and Related Work

An important task for neuroscientists is analysing large numbers of publications to identify potentially fruitful experiments. This process is necessary before undertaking any costly practical experiments; however, it is time-consuming and challenging [7]. Informed by communications with neuroscientist Cunqing Huangfu[1] [5,8–10], utilising topic-based literature exploration, such as finding relations between brain regions and brain diseases, can aid in analysing numerous publications to identify fruitful experiments. Immersive analytics (IA) allows

---

[1] Dr Cunqing Huangfu, a neuroscientist at the Institute of Automation of the Chinese Academy of Sciences.

© The Author(s), under exclusive license to Springer Nature Switzerland AG 2024
S. Rudinac et al. (Eds.): MMM 2024, LNCS 14557, pp. 295–300, 2024.
https://doi.org/10.1007/978-3-031-53302-0_24

researchers to examine 3D structures, in our case brain regions, allowing neuroscientists to find relations between topics such as brain diseases, brain regions and genes [6]. Augmented Reality (AR) allows researchers to have an immersive and interactive experience during their literature exploration [2].

Four neuroscientists (Cunqing Huangfu, Yu Mu[2], Danyang Li and Yu Qian) proposed several explorative tasks that they felt would be useful to illustrate topic-based literature exploration for identifying fruitful experiments during interviews [5,10]. These tasks include finding seldom-mentioned direct relations between topics. For example, the brain disease *Bipolar Disorder* and the brain region *Cerebral Hemisphere Structure* co-occur only twice among all the sentences of the 414,224[3] titles or abstracts in the repository. These twice direct relations are unusual, indicating that there is as yet no common knowledge about the relation between the topics *Bipolar Disorder* and *Cerebral Hemisphere Structure*. These seldom-mentioned direct relations between topics could offer fruitful directions for further investigation.

Indirect relations occur when two topics do not appear together in the same sentence but co-occur with the same intermediate topic [3,4]. For example, the brain disease *Bipolar Disorder* and the brain region *Cerebellar vermis* lack any direct relation but co-occurs with the intermediate topic *GFAP Gene*, establishing an indirect relation between them. In addition to finding seldom-mentioned direct relations, neuroscientists also proposed that finding unknown or indirect relations between topics (through an intermediate topic) would provide additional opportunities to identify fruitful experiments. We present the DatAR prototype [5,8–10], demonstrating how it can assist neuroscientists in identifying potentially fruitful experiments by explaining a walk-through of two scenarios[4].

## 2    DatAR Prototype

The DatAR prototype incorporates various functionalities in the form of AR widgets, see Table 1. The widgets provide selection, querying and visualisation mechanisms that allow users to explore the titles and abstracts in the PubMed neuroscience repository through topics. We briefly explain the functionalities of the widgets for finding seldom-mentioned direct relations and also indirect relations between topics.

## 3    Task Description

We collaborated with the neuroscientist Danyang Li to define representative tasks. Her research topic was the brain disease *Bipolar Disorder*. When she

---

[2] Dr Yu Mu, Danyang Li, and Yu Qian are neuroscientists at Institute of Neuroscience of the Chinese Academy of Sciences.

[3] Neuroscience publications in PubMed https://pubmed.ncbi.nlm.nih.gov/ as of February 3, 2022.

[4] See also the video https://drive.google.com/file/d/12sMvClw1zcytNPk7zl3J5NTe g22dsm2b/view.

**Table 1.** An overview of relevant widgets for finding relations between topics.

| Name of Widget | Description |
|---|---|
| **Brain Regions Visualisation**, Figs. 1 and 2, Label 1 | Visualises 274 brain regions sourced from the Scalable Brain Atlas[a] (SBA) as spheres in 3D AR [1]. The position of each sphere is determined by the 3D coordinates of brain regions taken from SBA [10] |
| **Co-occurrences Widget**, Figs. 1 and 2, Label 2 | Queries the direct co-occurrences between a single brain region and multiple brain diseases or a single brain disease and multiple brain regions in the PubMed repository [10]. The visualisation of direct relations is represented as pink spheres in 3D AR |
| **Max-Min Co-occurrences Filter Widget**, Fig. 1, Label 3 | Selects few direct co-occurrences by the max-min filter [10]. The visualisation of seldom-mentioned direct relations is represented as yellow spheres in 3D AR |
| **Sentences Extractor Widget**, Fig. 1, Label 4 | Queries which sentences indicate the direct relation between the specific brain region and brain disease [10] |
| **Indirect Relations Querier**, Fig. 2, Label 5 | Selects an intermediate topic, such as genes, to find indirect relations between brain regions and brain diseases in the PubMed repository. The visualisation of indirect relations is represented as green spheres in 3D AR |

[a] Scalable Brain Atlas (SBA) https://scalablebrainatlas.incf.org/index.php

explored *Bipolar Disorder*, her objective was to uncover seldom-mentioned brain regions, that are not yet common topics related to *Bipolar Disorder*. These seldom-mentioned regions could offer useful inspirations for further investigation. Additionally, she explained that brain diseases can also be connected indirectly to brain regions, through genes. She wanted to find which brain regions are indirectly related to *Bipolar Disorder* to provide additional options for identifying fruitful experiments.

To *find seldom-mentioned direct relations between topics*, Danyang carries out the following steps in the DatAR prototype.

**Step 1:** She wants to understand the known relations between the brain disease *Bipolar Disorder* and all brain regions, Fig. 1. She queries co-occurrences between topics via the Co-occurrences Widget.

**Step 2:** To find seldom-explored and lesser-known relations between topics, she filters the brain regions by the Max-Min Co-occurrences Filter Widget. The brain disease *Bipolar Disorder* co-occurs twice with the brain region *Cerebral Hemisphere Structure*, Fig. 1. The seldom-mentioned region *Cerebral Hemisphere Structure* is visualised as a yellow sphere.

**Step 3:** Finding few co-occurrence relations, such as the relation between *Bipolar Disorder* and *Cerebral Hemisphere Structure*, is not sufficient to convince her to start an experiment. She needs additional contextual information, including sentences and publication dates, that indicate these seldom-mentioned direct relations are indeed useful for her research. She observes the co-occurrence between the brain disease *Bipolar Disorder* and the brain region *Cerebral Hemisphere Structure* in the sentence: "We evaluated the structural changes of the cerebrum in patients with bipolar disorder (BD) by these dMRI techniques," published in May 2019[5] Fig. 1. This sentence provides useful context for the co-occurrence found.

To *find indirect relations between topics*, Danyang navigates the Indirect Relations Querier to find and visualise brain regions indirectly related to *Bipolar Disorder* via genes.

**Step 4:** She also wants to find unknown indirect relations between topics to provide more options for potentially fruitful experiments. She chooses genes as the intermediate topic, using the Indirect Relations Querier, to navigate indirect relations between topics. The brain disease *Bipolar Disorder* has an indirect relation with the brain region *Cerebellar vermis* via genes, Fig. 2.

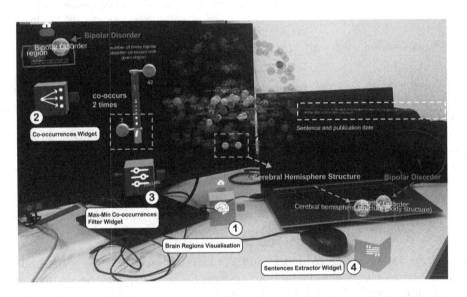

**Fig. 1.** The brain disease *Bipolar Disorder* co-occurs twice with the brain region *Cerebral Hemisphere Structure*.

---

[5] DOI: https://doi.org/10.1016/j.jad.2019.03.068.

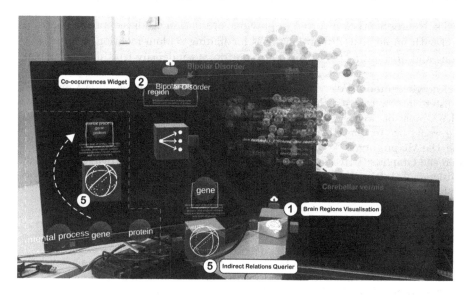

**Fig. 2.** The brain disease *Bipolar Disorder* has an indirect relation with the brain region *Cerebellar vermis* via genes.

## 4  Implementation

The topic co-occurrences used in the DatAR prototype are stored in the Knowledge Graphs of Brain Science[6] (KGBS) in Triply. KGBS contains an analysis of sentences in the titles and abstracts of 414,224 neuroscience publications in PubMed[7]. Topics include brain regions, brain diseases, genes, proteins and mental processes. The DatAR prototype is constructed using Unity[8] (v2020.3.15f2) and the Microsoft Mixed Reality Toolkit[9] (MRTK) (v2.7.0) and is deployed offline on a HoloLens 2[10] head-mounted display.

## 5  Conclusion

We have demonstrated the potential of the immersive AR-based 3D DatAR prototype in assisting neuroscientists in conducting and visualising relation-finding

---

[6] Knowledge Graphs of Brain Science in Triply
https://krr.triply.cc/BrainScienceKG/-/queries/Brain-Region---Brain-Disease/1.
[7] 414,224 neuroscience publications in PubMed https://pubmed.ncbi.nlm.nih.gov/ as of February 3, 2022.
[8] Unity https://unity.com/.
[9] Microsoft Mixed Reality Toolkit (MRTK) https://learn.microsoft.com/en-us/windows/mixed-reality/mrtk-unity/.
[10] Microsoft HoloLens 2 is an Augmented Reality headset developed and manufactured by Microsoft, https://www.microsoft.com/en-us/hololens.

tasks. Neuroscientists involved in previous evaluation studies expressed satisfaction with implemented functionalities for finding seldom-mentioned direct relations and also indirect relations between topics to identify fruitful experiments.

**Acknowledgments.** We would like to thank Zhisheng Huang at Vrije Universiteit, Amsterdam, for creating the repository of co-occurrences between topics in Knowledge Graphs of Brain Science and Triply (Triply https://triply.cc/) for hosting these. We would like to thank Ivar Troost for implementing the Co-occurrences Widget in the DatAR prototype. We are grateful to our colleagues in the DatAR, Visualisation and Graphics (Utrecht University) and Human-Centred Data Analytics (Centrum Wiskunde & Informatica) groups for their support. This work is supported by a scholarship from the China Scholarship Council (CSC): No. 202108440064.

# References

1. Bakker, R., Tiesinga, P., Kötter, R.: The Scalable Brain Atlas: instant web-based access to public brain atlases and related content. Neuroinformatics **13**, 353–366 (2015). https://link.springer.com/article/10.1007/s12021-014-9258-x
2. Bravo, A., Maier, A.M.: Immersive visualisations in design: using augmented reality (AR) for information presentation. In: Proceedings of the Design Society: DESIGN Conference, vol. 1, pp. 1215–1224. Cambridge University Press (2020). https://doi.org/10.1017/dsd.2020.33
3. Gopalakrishnan, V., Jha, K., Jin, W., Zhang, A.: A survey on literature based discovery approaches in biomedical domain. J. Biomed. Inform. **93**, 103141 (2019). https://doi.org/10.1016/j.jbi.2019.103141
4. Henry, S., McInnes, B.T.: Literature based discovery: models, methods, and trends. J. Biomed. Inform. **74**, 20–32 (2017). https://doi.org/10.1016/j.jbi.2017.08.011
5. Kun, C.: Assisting neuroscientists in exploring and understanding brain topic relations using augmented reality. Master's thesis (2022). https://studenttheses.uu.nl/handle/20.500.12932/41512
6. Marriott, K., et al.: Immersive Analytics, vol. 11190. Springer (2018). https://doi.org/10.1007/978-3-030-01388-2
7. Shardlow, M., et al.: A text mining pipeline using active and deep learning aimed at curating information in computational neuroscience. Neuroinformatics **17**, 391–406 (2019). https://link.springer.com/article/10.1007/s12021-018-9404-y
8. Tanhaei, G.: Exploring neuroscience literature and understanding relations between brain-related topics-using augmented reality. In: Proceedings of the 2022 Conference on Human Information Interaction and Retrieval, pp. 387–390 (2022). https://doi.org/10.1145/3498366.3505806
9. Tanhaei, G., Troost, I., Hardman, L., Hürst, W.: Designing a topic-based literature exploration tool in AR-an exploratory study for neuroscience. In: 2022 IEEE International Symposium on Mixed and Augmented Reality Adjunct (ISMAR-Adjunct), pp. 471–476. IEEE (2022). https://doi.org/10.1109/ISMAR-Adjunct57072.2022.00099
10. Troost, I., Tanhaei, G., Hardman, L., Hürst, W.: Exploring relations in neuroscientific literature using augmented reality: a design study. In: Designing Interactive Systems Conference 2021, pp. 266–274 (2021). https://doi.org/10.1145/3461778.3462053

# EmoAda: A Multimodal Emotion Interaction and Psychological Adaptation System

Tengteng Dong[1], Fangyuan Liu[1], Xinke Wang[2,3], Yishun Jiang[3,4], Xiwei Zhang[4], and Xiao Sun[1,3(✉)]

[1] School of Computer Science and Information Engineering, Hefei University of Technology, Hefei, China
{2022111070,2020212378}@mail.hfut.edu.cn, sunx@hfut.edu.cn
[2] AHU-IAI AI Joint Laboratory, Anhui University, Hefei, China
wa22101009@stu.ahu.edu.cn
[3] Institute of Artificial Intelligence, Hefei Comprehensive National Science Center, Hefei, China
{ysjiang,zhangxw}@iai.ustc.edu.cn
[4] Hefei Zhongjuyuan Intelligent Technology Co., Ltd, Hefei, China

**Abstract.** In recent years, anxiety and depression have placed a significant burden on society. However, the supply of psychological services is inadequate and costly. With advancements in multimedia computing and large language model technologies, there is hope for improving the current shortage of psychological resources. In this demo paper, we proposed a multimodal emotional interaction large language model (MEILLM) and develop EmoAda, A Multimodal Emotion Interaction and Psychological Adaptation System, providing users with cost-effective psychological support. EmoAda possesses multimodal emotional perception, personalized emotional support dialogue, and multimodal emotional interaction capabilities, helping users alleviate psychological stress and enhance psychological adaptation.

**Keywords:** Mental Health · Large Language Model · Multimodal Emotion Interaction

## 1 Introduction

The global prevalence of psychological disorders such as depression and anxiety has significantly increased after the COVID-19 pandemic. [5] However, there is a substantial gap in the availability of professional psychological services. Psychological research has shown that timely social support can effectively mitigate the impact of daily stress on mental health [9].

Mental health researchers believe that virtual psychological assistants can offer patients more timely, user-friendly, and privacy-conscious support. In recent years, multimodal emotion interaction has become a hot research topic and has significant results. For example, Feifei Li et al. introduced a method to measure

S. Rudinac et al. (Eds.): MMM 2024, LNCS 14557, pp. 301–307, 2024.
https://doi.org/10.1007/978-3-031-53302-0_25

the severity of depression symptoms from spoken language and 3D facial expressions [2]. Xiao Sun et al. proposed a multimodal attention network based on category-based mean square error (CBMSE) for personality assessment [7]. Minlie Huang et al. defined the task of emotion-support dialogue and presented an emotion-support dialogue framework [4]. Many research organizations are attempting to provide richer psychological support services with artificial intelligence. In May 2022, Google's LaMDA demonstrated emotion handling capabilities. OpenAI introduced ChatGPT in early 2023, which achieved some emotional recognition and inference. Tsinghua University's CoAI group developed Emohaa, an emotion-healing AI robot that offers emotion support.

However, despite technological progress laying the foundation for mental health support tools, existing emotion interaction, systems still face several challenges:

1) They rely on text as a single modality for emotion recognition, lacking the ability to perceive emotions through multiple modalities.
2) They have limited and relatively fixed dialogue response strategies, lacking personalized intervention strategy planning.
3) They predominantly rely on text and voice for interaction, lacking multimodal emotion interaction methods.

To address these challenges, we introduced a multimodal emotional interaction large language model based on open-source LLM. We also have developed EmoAda, a multimodal emotion interaction and psychological adaptation system.

EmoAda can analyze real-time multimedia data, including video, audio, and transcribed text, to create a comprehensive psychological profile for user. Based on the user's psychological profile and the intervention knowledge base, EmoAda customizes personalized intervention strategies and generates more natural and empathetic emotional support dialogue. EmoAda enables multimodal emotion interaction through a digital psychological avatar and offers various psychological adjustment methods such as meditation training and music for stress relief. Specifically, in the second section, we describe the system architecture; in the third section, we detail the implementation of the system.

## 2    System Overview

This section will introduced our developed multimodal emotion interaction and psychological adaptation system - EmoAda. EmoAda aims to assist individuals with limited access to mental health services to obtain timely, 24-hour psychological support.

Figure 1 shows the system structure,and the system consists of four main modules:

1) Real-time Multimodal Data Collection: This module collects audio and video data from users through cameras and microphones. Automatic Speech recognition(ASR) technology is used to convert speech into text.

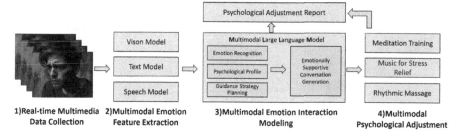

**Fig. 1.** EmoAda System Architecture Diagram

2) Multimodal Emotion Feature Extraction: This module extracts emotional features from the raw data of video, audio, and text using visual models, speech models, and text models. It obtains visual features such as facial key points, eye gaze tracks, body movements, and acoustic features like rhythm and MFCC. Sequence neural networks extract deep features, producing a multimodal emotional vector.

3) Multimodal Emotion Interaction Modeling: This module analyzes the multimodal emotional vectors using a multimodal large language model, enabling real-time emotion recognition, psychological profiling, guidance strategy planning, and emotional dialogue generation.

4) Multimodal Psychological Adaptation: This module synthesizes psychological digital avatars driven by multimodal emotions, which can express emotions more naturally and human-likely. Based on the user's psychological profile, the system recommends personalized multimedia content, such as meditation guidance music and emotion-stimulating videos.

The system consists of both hardware and software components. The hardware includes input devices such as microphones and cameras, output devices like speakers and screens. The software comprises a visual interactive front-end, core algorithms, and application system interfaces.

The system's core technology is derived from the research accomplishments of our team [1,3,8,10] and the open-source community. The multimodal emotion interaction large language model(MEILLM) is built upon the baichuan-13B-chat, an open-source project by Baichuan Intelligence. The psychological digital avatars are created using Silicon Intelligence's silicon-based AI digital humans.

In the next section, we will explain the multimodal emotion sensing algorithm and the technical details of the MEILLM.

## 3   Model

### 3.1   Multimodal Emotion Perception Model

Human emotions are complex and rich attributes. Emotions can be recognised from various means, such as a segment of a social media comment, an audio

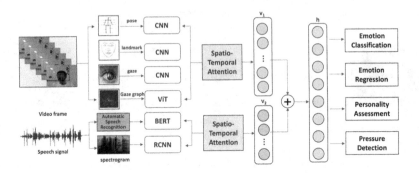

**Fig. 2.** Multimodal Emotion Perception Model Architecture Diagram

recording, facial expressions, and more. In human-computer interaction, accurately recognizing users' emotions can significantly enhance the interaction experience.

In this paper, we introduced a multimodal emotion perception approach that integrates the analysis of audio, video, and text data to accomplish various emotion perception tasks, such as emotion classification, emotion regression, personality assessment, and stress detection. Figure 2 shows the architecture of the multimodal emotion perception model. First, we segment video data into frames and extract multimodal features, including body posture, facial landmark, eye gaze, and gaze heatmaps. We then utilize Convolutional Neural Network (CNN) and Vision Transformers (ViT) to obtain deep feature vectors. Semantic fusion of these deep feature vectors is achieved through spatio-temporal attention mechanisms, creating visual emotion feature vectors $v_1$.

We perform speech recognition for audio data to obtain the users' dialogue text and transform the original speech signal into a spectrogram. We encode the text data using BERT and analyze the spectrogram with Recurrent Convolutional Neural Networks (RCNN) to extract deep feature vectors for audio. As with the visual data, we use spatiotemporal attention mechanisms to fuse these deep feature vectors semantically, generating audio emotion feature vectors $v_2$.

By concatenating $v_1$ and $v_2$, we obtain a multimodal emotion feature vector $h$. Different deep models are then employed for emotion classification, regression, personality assessment, and stress detection, leveraging this combined feature vector to enhance the accuracy and granularity of emotion perception.

## 3.2 Multimodal Emotion Interaction Large Language Model

Researchers have defined the Multimodal Large Language Model (MLLM) as an extension of the Large Language Model (LLM) with the ability to receive and reason about multimodal information. Compared to unimodal LLMs, MMLMs offer many advantages such as handling multimodal information, a more user-friendly interaction interface, and broader task support [11].

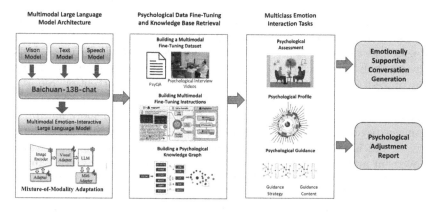

**Fig. 3.** Multimodal Emotion Interaction Large Language Model Architecture Diagram

In this paper, we introduced a Multimodal Emotion Interaction Large Language Model (MEILLM) capable of performing various emotion interaction tasks, including psychological assessment, psychological profiling, and psychological guidance. Figure 3 shows the architecture of the Multimodal Emotion Interaction Large Language Model. We used the open-source model baichuan13B-chat as the foundation, integrating deep feature vectors extracted from visual, text, and audio models through an MLP layer into baichuan13B-chat. We employed a Mixture-of-Modality Adaptation technique to achieve multimodal semantic alignment, enabling the LLM model to process multimodal information. We also constructed a multimodal emotion fine-tuning dataset, including open-source PsyQA dataset and a team-collected dataset of psychological interview videos. Using HuggingGPT [6], we developed a multimodal fine-tuning instruction set to enhance the model's multimodal interaction capabilities. We are creating a psychological knowledge graph to improve the model's accuracy in responding to psychological knowledge and reduce model hallucinations.

By combining these various techniques, MEILLM can perform psychological assessments, conduct psychological interviews using psychological scales, and generate psychological assessment reports for users. MEILLM can also create comprehensive psychological profiles for users, including emotions, moods, and personality, and provides personalized psychological guidance based on the user's psychological profile. MEILLM offers users a more natural and humanized emotional support dialogue and a personalized psychological adaptation report.

## 4   Conclusion and Future Work

In this demo paper, we constructed a multimodal emotion interaction large language model and developed EmoAda, a system designed for multimodal emotion interaction and psychological adaptation. EmoAda enables real-time analysis of users' multimodal data, creating a comprehensive psychological profile encompassing emotions, moods, and personality. It provides personalized intervention strategies and a variety of psychological adaptation methods for users. Importantly, we analyzed the changes in users' emotional states throughout the interaction process and generated a detailed psychological intervention report for users.

Through user testing, our system has been shown to offer more natural and humanized psychological support, effectively helping users relieve psychological stress and improve negative emotions. Nevertheless, there is still room for improvement and optimization in several areas: a)Multimodal Emotion Interaction Large Language Model: The model may generate model hallucinations that produce incorrect information. Enhancing the model's safety through constraints is necessary. b)High Inference Costs of Multimodal Large Models: Efficient model compression techniques are needed to improve model inference performance and reduce inference costs. c)Integration of Psychological Expert Knowledge Base: Incorporating a knowledge base of psychological experts will enhance the system's reliability and professionalism. These areas represent ongoing challenges that require further research and development to refine and enhance the capabilities of the EmoAda system.

**Acknowledgments.** This work was supported by the National Key R&D Programme of China (2022YFC3803202), Major Project of Anhui Province under Grant 202203a05020011 and General Programmer of the National Natural Science Foundation of China (62376084).

# References

1. Fan, C., Wang, Z., Li, J., Wang, S., Sun, X.: Robust facial expression recognition with global-local joint representation learning. Multimed. Syst. 1–11 (2022)
2. Haque, A., Guo, M., Miner, A.S., Fei-Fei, L.: Measuring depression symptom severity from spoken language and 3D facial expressions. arXiv preprint arXiv:1811.08592 (2018)
3. Huang, Y., Zhai, D., Song, J., Rao, X., Sun, X., Tang, J.: Mental states and personality based on real-time physical activity and facial expression recognition. Front. Psych. **13**, 1019043 (2023)
4. Liu, S., et al.: Towards emotional support dialog systems. arXiv preprint arXiv:2106.01144 (2021)
5. Santomauro, D.F., et al.: Global prevalence and burden of depressive and anxiety disorders in 204 countries and territories in 2020 due to the COVID-19 pandemic. Lancet **398**, 1700–1712 (2021)
6. Shen, Y., Song, K., Tan, X., Li, D., Lu, W., Zhuang, Y.: HuggingGPT: solving AI tasks with ChatGPT and its friends in huggingface. arXiv preprint arXiv:2303.17580 (2023)
7. Sun, X., Huang, J., Zheng, S., Rao, X., Wang, M.: Personality assessment based on multimodal attention network learning with category-based mean square error. IEEE Trans. Image Process. **31**, 2162–2174 (2022)
8. Sun, X., Song, Y., Wang, M.: Toward sensing emotions with deep visual analysis: a long-term psychological modeling approach. IEEE Multimed. **27**(4), 18–27 (2020)

9. Turner, R.J., Brown, R.L.: Social support and mental health. Handb. Study Mental Health Soc. Contexts Theor. Syst. **2**, 200–212 (2010)

10. Wang, J., Sun, X., Wang, M.: Emotional conversation generation with bilingual interactive decoding. IEEE Trans. Comput. Soc. Syst. **9**(3), 818–829 (2021)

11. Yin, S., et al.: A survey on multimodal large language models. arXiv preprint arXiv:2306.13549 (2023)

# Video Browser Showdown

# Waseda_Meisei_SoftBank at Video Browser Showdown 2024

Takayuki Hori[2,3]([✉]) [ID], Kazuya Ueki[1] [ID], Yuma Suzuki[2], Hiroki Takushima[2],
Hayato Tanoue[2], Haruki Sato[2], Takumi Takada[2],
and Aiswariya Manoj Kumar[2]

[1] Meisei University, 2-1-1 Hodokubo, Hino, Tokyo 191-8506, Japan
kazuya.ueki@meisei-u.ac.jp
[2] Softbank Corp, Kaigan 1-7-1, Minato-ku, Tokyo 105-7529, Japan
{takayuki.hori,yuma.suzuki,hiroki.takushima,hayato.tanoue,haruki.sato,
takumi.takada,aiswariya.kumar}@g.softbank.co.jp
[3] Waseda University, 3-4-1, Okubo, Shinjuku-ku, Tokyo 169-8555, Japan
hori@akane.waseda.jp

**Abstract.** This paper presents our first interactive video browser system "System 4 Vision". Because our system is based on the system that achieved the highest video retrieval accuracy in the AVS task of the TRECVID benchmark 2022, high retrieval accuracy can be expected in the Video Browser Showdown competition. Our system is characterized by the availability of rich text input, including complicated multiple conditions as queries, because our system uses the visual-semantic embedding method represented by Contrastive Language-Image Pre-training (CLIP).

**Keywords:** Video retrieval · Visual-semantic embedding · Pre-trained models · CLIP

## 1 Introduction

With the massive volume of videos uploaded to the Internet and recorded by surveillance cameras, locating specific video clips within extensive collections presents a considerable challenge for users. To address this issue, the Video Browser Showdown (VBS) competition was established, encouraging participants to create interactive video browsing systems for more practical and realistic tasks. This competition offers valuable insights into interactive video search tasks and lays the groundwork for innovative methodologies and systems. In the VBS2024 competition, participants utilize the V3C1 and V3C2 datasets from the Vimeo Creative Commons Collection, along with marine video and LapGyn100 datasets. The V3C1 and V3C2 datasets contain 7,475 and 9,760 video files, which translate to 1,082,659 and 1,425,454 predefined segments, respectively. Given the substantial size of these video datasets, finding the target video immediately still remains a very challenging task at the moment.

The search target should not only include people and objects, but also a combination of detailed information such as where, when, and what they are

S. Rudinac et al. (Eds.): MMM 2024, LNCS 14557, pp. 311–316, 2024.
https://doi.org/10.1007/978-3-031-53302-0_26

doing. Therefore, we thought that it is necessary to provide a system that allows users to search for the videos they need by inputting free sentences, rather than simply inputting queries as words.

## 2    "System 4 Vision": Video Retrieval System

**Fig. 1.** Examples of top 20 retrieved results. Left: Retrieved results for the query "A woman wearing dark framed glasses." Right: Retrieved results for the query "At least two persons are working on their laptops together in the same room indoors."

Our video retrieval system is based on a highly accurate retrieval method that won first place in the AVS task of TRECVID 2022. As shown in Fig. 1, it excels at precisely locating the videos users need, even when inputting complicated queries involving people, objects, locations, actions, and etc. The system utilizes the latest multimodal models to achieve high precision video retrieval.

**Fig. 2.** System Overview: User Interface and Search Functionality

The image shown in Fig. 2, is an external view of our demo system. This system includes a search bar where users enter their search queries. Concerning the search text, when the system conducts a video search, the search results are presented as thumbnails at the bottom of the screen, and these search results are presented to the user. Videos with similar search scores are displayed, starting from the top left. The search function implemented here utilizes the video search technology introduced at TRECVID 2022.

## 2.1 System Architecture

In Figs. 3 and 4, we introduce the architecture of the video retrieval technology implemented in our system. For the videos under consideration, we utilize the algorithm described in the following section. This algorithm is also employed by TRECVID to precompute the image feature vectors. These feature vectors are subsequently compared with the feature vectors of search queries to determine their similarity. Since videos consist of multiple frames, we sample keyframes at regular intervals as part of the video retrieval process. The results of comparing the feature vectors of the sampled keyframes with those of the search queries are aggregated to produce the final result. It's important to note that the ranking of these results is closely tied to the degree of similarity between the query and the image feature vectors, ensuring that the most relevant matches are presented at the top.

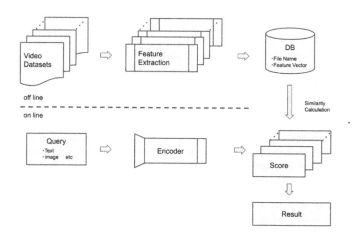

**Fig. 3.** Architecture Overview

## 2.2 Pre-trained Multimodal Models

An early method for mapping images and language into a common space is deep visual-semantic embedding (DeViSE) [1], proposed by Frome et al. Since then, many approaches such as VSE++ [3], SCAN [4], GSMN [5], CLIP [6], and SLIP

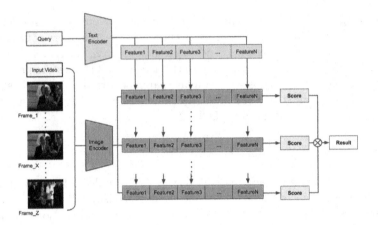

**Fig. 4.** Detailed Architectural Process for Video Retrieval

[7] have been proposed and their accuracy in image and text retrieval tasks has been reported to improve year by year. We have also implemented these approaches in the TRECVID benchmark and have seen a significant improvement in video retrieval accuracy. Among these models, CLIP stands out. It was pre-trained with an extensive collection of text/image pairs, surpassing 400 million in number. CLIP has made substantial contributions to the field, achieving state-of-the-art accuracy in various vision and language multimodal tasks. These tasks include image/text retrieval, image caption generation, and image generation. In recent years, multiple models that have been pre-trained on large image datasets with captions, such as LAION-5B [8] (5 billion images), have been released in OpenCLIP[1], making it possible to significantly improve video retrieval accuracy by utilizing these models. In our system, we select and utilize several of these models with high accuracy.

### 2.3 Preprocessing

Once the pre-trained multimodal models have been selected, features are extracted from the video and stored in advance to output results in real time as soon as a search query is entered. To use all frames in a video would require a significant amount of processing time. Therefore, we perform frame decimation, such as using only keyframes within the video or extracting frames at intervals of 10 frames, in order to reduce computational load.

### 2.4 Video Retrieval by Query Sentences

By pre-extracting and storing image features in advance, it becomes possible to perform video searches based on the similarity between text query features

---

[1] https://github.com/mlfoundations/open_clip.

and image features. Users input their query sentences, and the system extracts their text features. These features are then compared to the pre-extracted image features. The system calculates the similarity between the query features and the image features, ranking the results from high to low similarity. This approach allows for efficient video searches, presenting the most relevant results at the top based on the calculated similarity scores.

Moreover, since this method ranks results based on similarity scores, it is straightforward to integrate multiple models, leading to an enhancement in video retrieval performance. Specifically, after calculating similarity scores with each model, integration can be achieved through simple methods such as taking their average. These models exhibit complementarity, and our previous research [9] has shown that integrating them leads to improved search accuracy. By combining complementary models and taking advantage of their respective strengths, we enhance the overall precision of the search results, making it a powerful approach for video retrieval tasks.

In our approach, we calculate scores using cosine similarity, and these scores are further computed through a weighted harmonic mean.

## 2.5   Function to Enter Multiple Conditions

As mentioned earlier, the ability to compute similarity between text and image features allows for the calculation of multiple similarities for multiple sentences as well. By calculating these similarities and integrating them, it becomes possible to narrow down videos that meet multiple criteria. This approach enables the filtering of videos that satisfy various conditions by considering the combined similarity scores, providing a powerful tool for more detailed content retrieval.

Furthermore, because score-based integration is possible, it allows not only positive prompts but also negative prompts to be input, enabling the exclusion of specific conditions. For instance, in the context of a query provided in TRECVID 2023 AVS task, such as "A man carrying a bag on one of his shoulders excluding backpacks," user can employ "A man carrying a bag on one of his shoulders" as a positive prompt for calculating similarity scores and employ "backpacks" as a negative prompt for score subtraction. By applying this approach, you have the capability to exclude results that match the negative prompts, effectively refining search results.

## 2.6   Video Retrieval by Image Input

Moreover, as this system utilizes a multimodal model for both images and text, it is also possible to input actual frames from successful videos as queries. By doing so, users can find videos that are visually similar to the correct one, which opens up the possibility of discovering new videos that may not have been found through text queries alone. By combining both text and visual cues, the system allows users to explore a wider range of video content.

## 3   Summary and Future Work

Our system leverages the power of multimodal processing, utilizing both text and image data, to provide a robust and versatile video capabilities. Users have the freedom to input multiple, unrestricted queries, allowing for a highly flexible and tailored search process. With the ability to include negative prompts, our system enables users to refine their search results by excluding specific criteria, enhancing result precision. Furthermore, our system allows users to input actual images as queries, making it possible to find visually similar content, potentially uncovering new videos that may have eluded text-based searches alone.

In future work, we aim to continually improve our system's capabilities, exploring advanced techniques in multimodal processing and user experience to deliver even more effective and intuitive video retrieval solutions.

**Acknowledgments.** This work was partially supported by the Telecommunications Advancement Foundation.

## References

1. Frome, A., et al.: DeViSE: a deep visual-semantic embedding model. In: Proceedings of Advances in Neural Information Processing Systems (NIPS), vol. 26 (2013)
2. Schoeffmann, K., Lokoč, J., Bailer, W.: 10 years of video browser showdown. In: MMAsia 2020: ACM Multimedia Asia (2022)
3. Faghri, F., Fleet, D.J., Kiros, R., Fidler, S.: VSE++: improved visual-semantic embeddings. arXiv:1707.05612 (2017)
4. Lee, K.-H., Chen, X., Hua, G., Hu, H., He, X.: Stacked cross attention for image-text matching. In: Proceedings of European Conference on Computer Vision (ECCV) (2018)
5. Liu, C., Mao, Z., Zhang, T., Xie, H., Wang, B., Zhang, Y.: Graph structured network for image-text matching. In: Proceedings of IEEE Conference on Computer Vision and Pattern Recognition (CVPR) (2020)
6. Radford, A., et al.: Learning transferable visual models from natural language supervision. arXiv:2103.00020 (2021)
7. Mu, N., Kirillov, A., Wagner, D., Xie, S.: SLIP: self-supervision meets language-image pre-training. arXiv:2112.12750 (2021)
8. Schuhmann, C., et al.: LAION-5B: an open large-scale dataset for training next generation image-text models. In: 36th Conference on Neural Information Processing Systems (NeurIPS) (2022)
9. Ueki, K., Suzuki, Y., Takushima, H., Okamoto, H., Tanoue, H., Hori, T.: Waseda_Meisei_SoftBank at TRECVID 2022 ad-hoc video search. In: Notebook paper of the TRECVID 2022 Workshop (2022)

# Exploring Multimedia Vector Spaces with vitrivr-VR

Florian Spiess[1]([✉]) [iD], Luca Rossetto[2] [iD], and Heiko Schuldt[1] [iD]

[1] Department of Mathematics and Computer Science, University of Basel,
Basel, Switzerland
{florian.spiess,heiko.schuldt}@unibas.ch
[2] Department of Informatics, University of Zurich, Zurich, Switzerland
rossetto@ifi.uzh.ch

**Abstract.** Virtual reality (VR) interfaces are becoming more common-place as the number of capable and affordable devices increases. How-ever, VR user interfaces for common computing tasks often fail to take full advantage of the affordances provided by this immersive interface modality. New interfaces designed for VR must be developed in order to fully explore the possibilities for user interaction.

In this paper, we present vitrivr-VR and the improvements made for its participation in the Video Browser Showdown (VBS) 2024. We describe the current state of vitrivr-VR, with a focus on a novel point cloud browsing interface and improvements made to its text input methods.

**Keywords:** Video Browser Showdown · Virtual Reality · Interactive Video Retrieval · Content-based Retrieval

## 1 Introduction

Virtual reality (VR) is becoming increasingly commonplace as a user interface modality as more VR hardware is brought to market. While many VR-only applications, such as games and interactive entertainment, implement and explore ways to fully utilize the immersion and interactive possibilities afforded by virtual reality, interfaces for tasks also performed on conventional interfaces, such as search and browsing, are often direct translations of their conventional interface counterparts.

vitrivr-VR is a virtual reality multimedia retrieval system research prototype that aims to explore how virtual reality can be utilized to improve multimedia retrieval and in particular user interaction with immersive query and results browsing interfaces. To evaluate how the methods and interfaces implemented in vitrivr-VR compare to other state-of-the-art systems, vitrivr-VR participates in multimedia retrieval campaigns such as the Video Browser Showdown (VBS) [2]. The VBS is a valuable source of insights, which has led to multiple important improvements of vitrivr-VR through previous participations [9–11].

© The Author(s), under exclusive license to Springer Nature Switzerland AG 2024
S. Rudinac et al. (Eds.): MMM 2024, LNCS 14557, pp. 317–323, 2024.
https://doi.org/10.1007/978-3-031-53302-0_27

In this paper, we describe the version of the vitrivr-VR system participating in the VBS 2024, with a focus on our novel point cloud results browsing interface and improvements made to text input. We describe the system and its components in Sect. 2, introduce our novel point cloud display in Sect. 3, describe our text input improvements in Sect. 4, and conclude in Sect. 5.

## 2   vitrivr-VR

vitrivr-VR is an experimental virtual reality multimedia retrieval and analytics system. Based on components of the vitrivr stack [8], the system architecture of vitrivr-VR consists of three parts: the vector database *Cottontail DB* [1], the feature extraction and retrieval engine *Cineast* [7], and the *vitrivr-VR*[1] virtual reality user interface. The vitrivr-VR user interface is developed in Unity with C# and is capable of running on any platform compatible with OpenXR. All parts of the vitrivr-VR stack are open-source and available on GitHub.[2]

The goal of vitrivr-VR is to explore methods of multimedia retrieval and analytics with virtual reality user interfaces. To achieve this goal, vitrivr-VR implements query formulation, results browsing, and multimedia interaction methods developed with virtual reality user interfaces in mind, enabling an immersive user experience.

vitrivr-VR implements a number of query formulation methods for queries of different kinds. Due to the prevalence of semantic multimodal embedding features such as CLIP [5], text is an important query modality. Text input for VR interfaces is still subject to active research. vitrivr-VR implements a combination of speech-to-text and a word-gesture keyboard [12], which allows the input of entire sentences at a time with speech-to-text and fast word-level corrections using the word-gesture keyboard. Other query modalities supported by vitrivr-VR include query-by-concept, -pose, -sketch, and -example (frame). Queries can consist of individual query terms, combinations, or even temporally ordered query terms for temporal queries.

To browse query results, vitrivr-VR implements a number of results display and browsing methods. The main results display of vitrivr-VR is a cylindrical grid, an example of which can be seen in Fig. 1a. This display method is a direct translation of conventional grid results displays into VR, with the only modification being that the results curve around the user, allowing more intuitive browsing through head movement. The cylindrical results display can be rotated to reveal more results, replacing already viewed results. Results can be selected from this display to be viewed in greater detail.

To allow easier browsing within a video vitrivr-VR implements a multimedia drawer view, as depicted in Fig. 1b. This interactive display allows the temporally ordered browsing within the key frames of a video. By hovering a hand over a keyframe, the frame is selected and hovers above the drawer for easier viewing. In this way, a user can move their hand through the drawer to view the key

---

[1] https://github.com/vitrivr/vitrivr-vr.

[2] https://github.com/vitrivr.

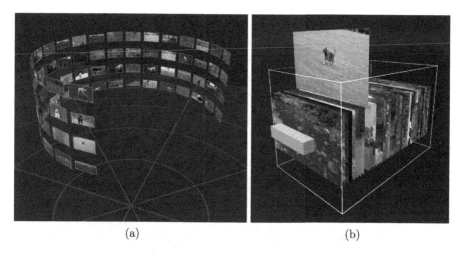

(a)                                    (b)

**Fig. 1.** Screenshots of the vitrivr-VR user interface: (a) cylindrical grid results display and (b) multimedia drawer segment view.

frames of a video in a fashion similar to a flip book. Keyframes can be selected to skip to that part of the video and view the respective video segment in greater detail. The handle at the front of the drawer allows it to be extended, increasing the spacing between keyframes and allowing easier frame selection for videos with many keyframes.

## 3   Point Cloud Browsing

Many of the theoretical advantages of VR interfaces over conventional interfaces in multimedia browsing and retrieval are enabled by the additional usable space for displaying and interacting with multimedia content. While vitrivr-VR already implements interfaces to explore these advantages for intra-video browsing, inter-video browsing in vitrivr-VR, such as browsing of query results, was previously only possible through interfaces heavily inspired by interfaces used on conventional displays.

Conventional grid-based displays for similarity search results have two properties that limit their effectiveness for browsing: due to the human cognitive limit, they are only able to present a very limited number of results at any given time, and since they only take into account the similarity of result items to the search query, results can only be displayed and traversed as sorted by this similarity measure. Furthermore, even though grid-based displays are inherently two-dimensional, the similarity of each result to the query is the only meaningful dimension of the data displayed. While wrapping the results into a grid allows more results to be displayed within a user's perceptive field, these grid-based displays use both dimensions available to them to express only a single quality of the data.

**Fig. 2.** The point cloud display shows a preview frame corresponding to the video segment represented by the currently hovered point. Points are colored by their coordinates in the reduced feature space to make distinguishing and keeping track of them easier.

To overcome these limitations of grid-based displays and further explore the possibilities of multimedia browsing in virtual reality, we have implemented a new point cloud-based results browsing interface for vitrivr-VR. Utilizing all three spatial dimensions available in virtual reality, this new point cloud display shows the highest-ranked results of a query as a point cloud in virtual space. To make the most use of the available dimensions, points in the point cloud are placed according to a dimensionality reduction of a high-dimensional feature space, allowing the similarity between results presented by the points to be estimated at a glance. By only including the highest-ranked results of the similarity query in the dimensionality reduction, the three reduced dimensions are fully utilized for differences within the result set rather than needing to include information about the collection as a whole. To reduce the cognitive strain of viewing a large number of multimedia items at once, results in the point cloud display are represented as points in space, only showing a preview of the respective query result when hovered by a user's hand. An example of the point cloud display is shown in Fig. 2.

**Fig. 3.** The point cloud display for the query "A whale shark". The points are colored by their relative query score, where the best scoring result is colored green, and the worst scoring result is colored red. (Color figure online)

The point cloud display can be used with any feature space, dimensionality reduction method, and point coloring, irrespective of the similarity query used to generate the results. This parameterization allows the point cloud display to be tuned to a specific task or use case. In our implementation of the point cloud display for the VBS 2024, we use a semantic co-embedding feature, experiment with t-SNE [3] and UMAP [4] dimensionality reduction, and color the points configurably by the relevance to the query (as shown in Fig. 3) or the coordinates in the reduced feature space.

The point cloud display is not meant to replace conventional browsing displays, such as the cylindrical grid display implemented in vitriv-VR, but rather to complement them by providing alternative forms of user interaction and result display. While the point cloud display shows fewer result previews at once, it allows top-scoring results to be viewed in the context of their similarity to other results, making it easy to disregard entire clusters of irrelevant results and focus on areas of the point cloud containing more related results.

The most closely related browsing method previously used at a VBS is that implemented in EOLAS [13]. While this method also uses dimensionality reduction to show query results within the context of similar multimedia objects, it allows the viewing of only a very limited number of results at once and does not allow quick scanning of the result space.

## 4 Improved Text Input

In addition to the implementation of the point cloud display, this iteration of vitrivr-VR includes a number of further improvements. The two most notable of these improvements both concern how users of the system can input text. Analyses of the VBS 2023 indicate, once more, that the most successful systems rely on text queries and that text input is much slower in VR systems than in comparable systems using conventional interfaces [11].

The first improvement to vitrivr-VR text input is the replacement of our speech-to-text input method. Previously based on Mozilla DeepSpeech[3], our new speech-to-text input is now based on the much more capable OpenAI Whisper [6]. Whisper addresses some of the most common limitations of our implementation of DeepSpeech, namely the accuracy of transcription and support for languages other than English.

Our second improvement to text input is based on the observation that a query can be input by keyboard much more quickly than it can even be spoken for speech-to-text input. This puts VR systems at a disadvantage at the beginning of a search task when no results are available that can be scanned while refining the query. To give users the option to conveniently input their initial query using a physical keyboard, vitrivr-VR implements support for headsets with passthrough augmented reality capabilities, such as the Vive Focus 3.

## 5 Conclusion

This paper presents the state of the vitrivr-VR system in which we intend to participate in the VBS 2024. We focus particularly on our novel point cloud browsing interface, which presents results within the context of a similarity feature space, and on our improvements to text input.

**Acknowledgements.** This work was partly supported by the Swiss National Science Foundation through projects "Participatory Knowledge Practices in Analog and Digital Image Archives" (contract no. 193788) and "MediaGraph" (contract no. 2021-25).

## References

1. Gasser, R., Rossetto, L., Heller, S., Schuldt, H.: Cottontail DB: an open source database system for multimedia retrieval and analysis. In: International Conference on Multimedia, pp. 4465–4468. Association for Computing Machinery, New York, NY, USA (2020). https://doi.org/10.1145/3394171.3414538
2. Lokoč, J., et al.: Interactive video retrieval in the age of effective joint embedding deep models: lessons from the 11$^{th}$ VBS. Multimedia Syst. (2023). https://doi.org/10.1007/s00530-023-01143-5

---

[3] https://github.com/mozilla/DeepSpeech.

3. van der Maaten, L., Hinton, G.: Visualizing Data Using T-SNE. J. Mach. Learn. Res. **9**(86), 2579–2605 (2008). http://jmlr.org/papers/v9/vandermaaten08a.html

4. McInnes, L., Healy, J., Melville, J.: UMAP: uniform manifold approximation and projection for dimension reduction. arXiv (2018). https://doi.org/10.48550/ARXIV.1802.03426

5. Radford, A., et al.: Learning transferable visual models from natural language supervision. arXiv (2021). https://doi.org/10.48550/ARXIV.2103.00020

6. Radford, A., Kim, J.W., Xu, T., Brockman, G., Mcleavey, C., Sutskever, I.: Robust speech recognition via large-scale weak supervision. In: Proceedings of the 40th International Conference on Machine Learning, vol. 202, pp. 28492–28518. PMLR (2023). https://proceedings.mlr.press/v202/radford23a.html

7. Rossetto, L., Giangreco, I., Schuldt, H.: Cineast: a multi-feature sketch-based video retrieval engine. In: IEEE International Symposium on Multimedia (2014). https://doi.org/10.1109/ISM.2014.38

8. Rossetto, L., Giangreco, I., Tanase, C., Schuldt, H.: vitrivr: a flexible retrieval stack supporting multiple query modes for searching in multimedia collections. In: Proceedings of the 2016 ACM Conference on Multimedia Conference, MM 2016, Amsterdam, The Netherlands, October 15–19, 2016, pp. 1183–1186. ACM (2016). https://doi.org/10.1145/2964284.2973797

9. Spiess, F., et al.: Multi-modal video retrieval in virtual reality with vitrivr-VR. In: Þór Jónsson, B., et al. (eds.) MMM 2022. LNCS, vol. 13142, pp. 499–504. Springer, Cham (2022). https://doi.org/10.1007/978-3-030-98355-0_45

10. Spiess, F., Gasser, R., Heller, S., Rossetto, L., Sauter, L., Schuldt, H.: Competitive interactive video retrieval in virtual reality with vitrivr-VR. In: Lokoč, J., et al. (eds.) MMM 2021. LNCS, vol. 12573, pp. 441–447. Springer, Cham (2021). https://doi.org/10.1007/978-3-030-67835-7_42

11. Spiess, F., Gasser, R., Heller, S., Schuldt, H., Rossetto, L.: A comparison of video browsing performance between desktop and virtual reality interfaces. In: Proceedings of the 2023 ACM International Conference on Multimedia Retrieval, pp. 535–539. ACM (2023). https://doi.org/10.1145/3591106.3592292

12. Spiess, F., Weber, P., Schuldt, H.: Direct interaction word-gesture text input in virtual reality. In: 2022 IEEE International Conference on Artificial Intelligence and Virtual Reality (AIVR), pp. 140–143. IEEE, CA, USA (2022). https://doi.org/10.1109/AIVR56993.2022.00028

13. Tran, L.-D., et al.: A VR interface for browsing visual spaces at VBS2021. In: Lokoč, J., et al. (eds.) MMM 2021. LNCS, vol. 12573, pp. 490–495. Springer, Cham (2021). https://doi.org/10.1007/978-3-030-67835-7_50

# A New Retrieval Engine for Vitrivr

Ralph Gasser[1]([✉])[iD], Rahel Arnold[1][iD], Fynn Faber[1][iD], Heiko Schuldt[1][iD], Raphael Waltenspül[1][iD], and Luca Rossetto[2][iD]

[1] University of Basel, Basel, Switzerland
{ralph.gasser,rahel.arnold,fynn.faber,heiko.schuldt,
raphael.waltenspul}@unibas.ch
[2] University of Zurich, Zurich, Switzerland
rossetto@ifi.uzh.ch

**Abstract.** While the vitrivr stack has seen many changes in components over the years, its feature extraction and query processing engine traces its history back almost a decade. Some aspects of its architecture and operation are no longer current, limiting the entire stack's applicability in various use cases. In this paper, we present the first glimpse into vitrivr's next-generation retrieval engine and our plan to overcome previously identified limitations.

**Keywords:** Video Browser Showdown · Interactive Video Retrieval · Content-based Retrieval

## 1 Introduction

The content-based multimedia retrieval stack vitrivr [18] has been a long-time participant in the Video Browser Showdown (VBS). Including its predecessor, the iMotion System [17], from which vitrivr ultimately emerged, the system participates for the $10^{th}$ year in a row. Architecturally, the vitrivr stack consists of three primary components: a database, a retrieval engine, and a user interface. While both the database and the user interface have been replaced several times over the years, the retrieval engine *Cineast* traces its origins back almost a decade [16]. Despite much work and many contributions made to Cineast over the years, many insights gained through past activities and technological changes cannot easily be retrofitted to its underlying architecture, which is why an overhaul of the retrieval engine has been long overdue.

In this light, we, therefore, use our $10^{th}$ VBS participation as an opportunity to take the first steps towards Cineast's successor, a system we simply call the *vitrivr-engine*. The remainder of this paper is structured as follows: Sect. 2 starts with a brief history of the vitrivr stack and its components and capabilities, especially in the context of VBS. It then continues by looking forward and discussing some of our future plans with the new vitrivr-engine. Section 3 then goes into more detail about the exact state of the system that will participate in the VBS in 2024. Finally, Sect. 4 offers some concluding remarks.

S. Rudinac et al. (Eds.): MMM 2024, LNCS 14557, pp. 324–331, 2024.
https://doi.org/10.1007/978-3-031-53302-0_28

# 2   vitrivr

This section first provides a brief history of the development of the vitrivr stack before outlining the goals we are trying to achieve with the vitrivr-engine.

## 2.1   The "vitrivr-Story" Thus Far

The vitrivr stack grew out of the iMotion system, which participated in the VBS in 2015 [17] and 2016 [15]. The iMotion system used a database system called *ADAM* [4], which was an extension of the PostgreSQL [28] database system, augmented with the ability to perform vector space operations. The retrieval engine was Cineast, which used a tightly coupled browser-based user interface.

The *ADAM* database was replaced with its successor, $ADAM_{pro}$ [5], for the 2017 participation [19], which would be the last under the iMotion name and the first time the system scored highest among all participants [9]. In contrast to the purely relational *ADAM* database, $ADAM_{pro}$ took a polyglot approach to accommodate the different types of queries needed for multimedia retrieval.

2018 saw a substantial extension in the capabilities of Cineast, turning it from a video-only retrieval engine into one capable of processing several media types, including images, audio, and 3D models [3]. That year's VBS participation [14] happened under the name *vitrivr* for the first time, for which we also introduced a new user interface called vitrivr-ng. With the official transition to the new name, we made a consistent effort to open-source the entire stack. 2019 [12] saw various improvements throughout the system but no fundamental changes. Nevertheless, the system outperformed all other VBS participants [13].

In 2020 [23], we introduced the $3^{rd}$ database system to be used in the stack: Cottontail DB [2]. This was the last more significant architectural change to the vitrivr stack since then. Even so, in 2021 [7], vitrivr was the highest-scoring challenge participant for the third time [8]. That year also saw the first participation of a new user interaction approach, using virtual reality rather than a traditional desktop interface. vitrivr-VR [26] uses the same database and retrieval engine backend but introduces a completely new user interface with much more effective browsing capabilities. In 2022 [6,25] and 2023 [22,27], both vitrivr and vitrivr-VR participated without any fundamental architectural changes.

Throughout all this time, the only system component that, despite benefiting from various improvements, stayed the same was the feature extraction and query processing engine Cineast. While it served us well over the years in a broad range of applications, the implicit and explicit assumptions baked into its architecture are increasingly at odds with today's requirements. Especially in recent years, we have encountered more and more applications, e.g., in the context of the projects XReco[1] and Archipanion,[2] that could benefit from vitrivr's multimedia retrieval capabilities but that were not easily integrated due to Cineast's rather monolithic setup. The following section, therefore, provides a brief overview of some of our plans and gives a glimpse into Cineast's successor, the vitrivr-engine.

---

[1] See https://xreco.eu/.
[2] See https://dbis.dmi.unibas.ch/research/projects/archipanion/.

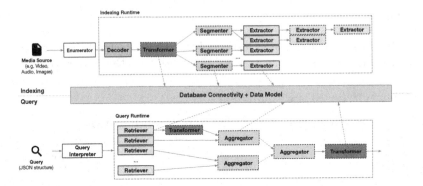

**Fig. 1.** A high-level depiction of vitrivr-engine's internals. The runtime environments for indexing and querying are strictly separated, and both are modeled as a pipeline of operators. The two modules share a common database connectivity and data model.

## 2.2   Where to Go from Here

With the lessons learned over the past years, we have identified several weaknesses in Cineast that we plan to address in the new vitrivr-engine to prepare vitrivr as a whole for a future in a wider range of different setups. An overview of vitrivr-engine's high-level architecture is provided in Fig. 1.

We plan to make vitrivr-engine more modular at different levels. At the highest level, there will be separate modules for facilities used for executing queries – the *query runtime* – and those used for indexing media collections – the *indexing runtime*. Both facilities will share a common core that defines the basic processing and data model. Within these facilities, the indexing and retrieval processes are explicitly modeled as configurable pipelines that consist of operators generating, processing, and transforming a stream of common entities. A plug-able database abstraction layer will back all this.

**Data Model.** Cineast's original data model is very strict in how information units are structured and connected. Firstly, it expects the main *media object* entity to always refer to individual files (with associated metadata) that are then temporally divided into *segments*, from which all features are derived. This is a remnant of Cineast's roots as a video retrieval engine. Consequently, some use cases could not easily be covered, e.g., having multiple independent or non-temporal segmentation schemes. Secondly, more complex relationships between parts of the media items could not be modeled.

The vitrivr-engine data model is more flexible in that regard. Fundamentally, it processes a stream of *content* from some source (e.g., a video file), from which one (or many) *retrievable(s)* are derived, e.g., through segmentation. The segmentation process can be temporal, spatial, or anything that can be applied to a particular type of content stream. In addition, vitrivr-engine allows for the modeling of explicit relationships between *retrievables*, which leads to a graph-like data structure with much more descriptive power than a static data model.

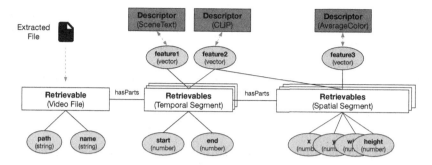

**Fig. 2.** An indexed video file is represented as three types of retrievables: One retrievable for the file itself, one for the temporal segments thereof, and one for the spatial segments generated for every temporal one. Each level of retrievable can hold its own fields with various types of attributes. A descriptor explicitly backs most fields.

*Fields* can then describe any *retrievable*, which can hold anything from a feature vector to a scalar value. This Entity-Attribute-Value (EAV) approach to describing *retrievables* gives us more flexibility in how instances of vitrivr can be tailored to a particular use case (with their own data model). Most fields are backed by some type of *descriptor*, an algorithm that extracts the *feature* from the *retrievable*. An example of the new data model's expressive power is provided in Fig. 2.

**Query Model.** Cineast's query model has always been tied to the functionality provided by the vitrivr-ng and (later) the vitrivr-VR user interfaces. The query model was extended whenever a new feature emerged, leading to a complex data structure.

In vitrivr-engine, we model queries as a pipeline of operators, as is illustrated in Fig. 1. vitrivr-engine's query model simply describes the shape and elements of such a pipeline in a declarative fashion, leading to a richer and more expressive query model. We see this as a necessary step to generalizing mechanisms for query formulation and execution and to decoupling them from particular implementations offered by some user interfaces.

**Extensibility and Modularity.** Cineast has always been extensible in that new feature descriptors (and other types of functionality) could be added by implementing a set of defined interfaces. However, such extensions were constantly added to the core codebase, leading to many feature implementations that are rarely ever used.

In vitrivr-engine, we have conceived a plugin architecture that facilitates adding functionality using a set of *extension points*. In addition to feature descriptors, these extension points involve all elements along the indexing and query pipeline, particularly all the operators depicted in Fig. 1, down to the database connectivity layer itself.

All the relevant feature descriptors existing in Cineast today will be bundled in a *plugin* that can be used to bootstrap a new installation. Consequently, the new vitrivr-engine can still be used as an off-the-shelf solution by using the default plugins if one wishes to do so. However, using the vitrivr-engine core facilities is also possible exclusively with custom implementations of decoders, extractors, and feature descriptors, leading to a leaner project better tailored to a particular application.

**Decoupling the Engine from the Interface.** Various use cases require different public programming interfaces that facilitate interaction with the extraction and retrieval engine. That is why Cineast featured three types of interfaces – gRPC, HTTP/REST, and WebSocket – that have seen major and breaking changes over the years to accommodate new use cases. In Cineast, the engine facilitating the execution of all these processes was strongly intertwined with these different interfaces.

In vitrivr-engine, we will decouple the processing facilities and the public interfaces completely. While vitrivr-engine will come with a default server that allows access via an OpenAPI REST interface, it will also be possible to use vitrivr-engine's processing components merely as libraries, without the overhead that comes with the default API infrastructure. This is especially useful in projects such as XReco, which rely on vitrivr's retrieval capabilities but come with their own design paradigm for public APIs.

## 3   Implementation

The vitrivr setup used at VBS 2024 will be similar to the one used in previous installments [6,7,22]. However, instead of using Cineast as the feature extraction and query engine, we will rely on the first version of the new vitrivr-engine, written in Kotlin. This middleware will be backed by the multimedia DBMS Cottontail DB [2] and will use a new, minimalistic UI that debuted at CBMI 2023 [24]. That UI is focused on a simple, efficient user interaction rather than a rich set of functionality.

Due to some architectural changes in vitrivr-engine, we will be able to serve all three independent collections[3] used during VBS within a common deployment of vitrivr-engine. For this year's installment of VBS and based on the experiences gathered in previous years, we restrict ourselves to a minimal set of features, which we know to be effective in this highly competitive setting: We use OpenClip [1] for text-based querying and DinoV2 [10] for similarity search queries (more-like-this). As these two features rely on neural networks, they will be hosted in an external feature server written in Python and are accessed via a RESTful API. In addition, we will index the V3C speech transcripts [20] generated using whisper [11] as well as OCR data already used in previous years.

---

[3] V3C (shards 1 and 2) [21], the Marine Video Kit [29], and LapGyn100 laparoscopic gynaecology video dataset.

# 4   Conclusion

In this paper, we have provided an overview of vitrivr's history, focusing on its retrieval engine, Cineast. While Cineast has served us well over the years, some of its architectural decisions, made about a decade ago, have become limiting. We therefore offer a first glimpse into its successor, the vitrivr-engine, and outline some of our plans towards vitrivr's future.

**Acknowledgements.** This work was partly supported by the Swiss National Science Foundation through project MediaGraph (contract no. 202125), the InnoSuisse project Archipanion (contract no. 2155012542), and the Horizon Europe project XR mEdia eCOsystem (XReco), based on a grant from the Swiss State Secretariat for Education, Research and Innovation (contract no. 22.00268).

# References

1. Cherti, M., et al.: Reproducible scaling laws for contrastive language-image learning. In: IEEE/CVF Conference on Computer Vision and Pattern Recognition, CVPR 2023, Vancouver, BC, Canada, 17–24 June 2023, pp. 2818–2829. IEEE (2023)
2. Gasser, R., Rossetto, L., Heller, S., Schuldt, H.: Cottontail DB: an open source database system for multimedia retrieval and analysis. In: The 28th ACM International Conference on Multimedia, Virtual Event/Seattle, MM 2020, WA, USA, 12–16 October 2020, pp. 4465–4468. ACM (2020)
3. Gasser, R., Rossetto, L., Schuldt, H.: Multimodal multimedia retrieval with vitrivr. In: Proceedings of the 2019 on International Conference on Multimedia Retrieval, ICMR 2019, Ottawa, ON, Canada, 10–13 June 2019, pp. 391–394. ACM (2019)
4. Giangreco, I., Al Kabary, I., Schuldt, H.: ADAM - A database and information retrieval system for big multimedia collections. In: 2014 IEEE International Congress on Big Data, Anchorage, AK, USA, 27 June–2 July 2014, pp. 406–413. IEEE Computer Society (2014)
5. Giangreco, I., Schuldt, H.: $ADAM_{pro}$: database support for big multimedia retrieval. Datenbank-Spektrum **16**(1), 17–26 (2016)
6. Heller, S., et al.: Multi-modal interactive video retrieval with temporal queries. In: Þór Jónsson, B., et al. (eds.) MMM 2022. LNCS, vol. 13142, pp. 493–498. Springer, Cham (2022). https://doi.org/10.1007/978-3-030-98355-0_44
7. Heller, S., et al.: Towards explainable interactive multi-modal video retrieval with vitrivr. In: Lokoč, J., et al. (eds.) MMM 2021. LNCS, vol. 12573, pp. 435–440. Springer, Cham (2021). https://doi.org/10.1007/978-3-030-67835-7_41
8. Heller, S., et al.: Interactive video retrieval evaluation at a distance: comparing sixteen interactive video search systems in a remote setting at the 10th video browser showdown. Int. J. Multimed. Inf. Retrieval **11**(1), 1–18 (2022)
9. Lokoc, J., Bailer, W., Schoeffmann, K., Münzer, B., Awad, G.: On influential trends in interactive video retrieval: video browser showdown 2015–2017. IEEE Trans. Multim. **20**(12), 3361–3376 (2018)
10. Oquab, M., et al.: DINOv2: learning robust visual features without supervision. CoRR abs/2304.07193 (2023)

11. Radford, A., Kim, J.W., Xu, T., Brockman, G., McLeavey, C., Sutskever, I.: Robust speech recognition via large-scale weak supervision. In: International Conference on Machine Learning, ICML 2023, 23–29 July 2023, Honolulu, Hawaii, USA. Proceedings of Machine Learning Research, vol. 202, pp. 28492–28518. PMLR (2023)
12. Rossetto, L., Amiri Parian, M., Gasser, R., Giangreco, I., Heller, S., Schuldt, H.: Deep learning-based concept detection in vitrivr. In: Kompatsiaris, I., Huet, B., Mezaris, V., Gurrin, C., Cheng, W.-H., Vrochidis, S. (eds.) MMM 2019. LNCS, vol. 11296, pp. 616–621. Springer, Cham (2019). https://doi.org/10.1007/978-3-030-05716-9_55
13. Rossetto, L., et al.: Interactive video retrieval in the age of deep learning - detailed evaluation of VBS 2019. IEEE Trans. Multim. **23**, 243–256 (2021)
14. Rossetto, L., Giangreco, I., Gasser, R., Schuldt, H.: Competitive video retrieval with vitrivr. In: Schoeffmann, K., et al. (eds.) MMM 2018. LNCS, vol. 10705, pp. 403–406. Springer, Cham (2018). https://doi.org/10.1007/978-3-319-73600-6_41
15. Rossetto, L., et al.: IMOTION – searching for video sequences using multi-shot sketch queries. In: Tian, Q., Sebe, N., Qi, G.-J., Huet, B., Hong, R., Liu, X. (eds.) MMM 2016. LNCS, vol. 9517, pp. 377–382. Springer, Cham (2016). https://doi.org/10.1007/978-3-319-27674-8_36
16. Rossetto, L., Giangreco, I., Schuldt, H.: Cineast: a multi-feature sketch-based video retrieval engine. In: 2014 IEEE International Symposium on Multimedia, ISM 2014, Taichung, Taiwan, 10–12 December 2014, pp. 18–23. IEEE Computer Society (2014)
17. Rossetto, L., et al.: IMOTION — a content-based video retrieval engine. In: He, X., Luo, S., Tao, D., Xu, C., Yang, J., Hasan, M.A. (eds.) MMM 2015. LNCS, vol. 8936, pp. 255–260. Springer, Cham (2015). https://doi.org/10.1007/978-3-319-14442-9_24
18. Rossetto, L., Giangreco, I., Tanase, C., Schuldt, H.: vitrivr: a flexible retrieval stack supporting multiple query modes for searching in multimedia collections. In: ACM Conference on Multimedia (2016)
19. Rossetto, L., Giangreco, I., Tănase, C., Schuldt, H., Dupont, S., Seddati, O.: Enhanced retrieval and browsing in the IMOTION system. In: Amsaleg, L., Guðmundsson, G.Þ, Gurrin, C., Jónsson, B.Þ, Satoh, S. (eds.) MMM 2017. LNCS, vol. 10133, pp. 469–474. Springer, Cham (2017). https://doi.org/10.1007/978-3-319-51814-5_43
20. Rossetto, L., Sauter, L.: Vimeo Creative Commons Collection (V3C) Whisper Transcripts (2022)
21. Rossetto, L., Schuldt, H., Awad, G., Butt, A.A.: V3C – a research video collection. In: Kompatsiaris, I., Huet, B., Mezaris, V., Gurrin, C., Cheng, W.-H., Vrochidis, S. (eds.) MMM 2019. LNCS, vol. 11295, pp. 349–360. Springer, Cham (2019). https://doi.org/10.1007/978-3-030-05710-7_29
22. Sauter, L., et al.: Exploring effective interactive text-based video search in vitrivr. In: Dang-Nguyen, D.T., et al. (eds.) MMM 2023. LNCS, vol. 13833, pp. 646–651. Springer, Cham (2023). https://doi.org/10.1007/978-3-031-27077-2_53
23. Sauter, L., Amiri Parian, M., Gasser, R., Heller, S., Rossetto, L., Schuldt, H.: Combining Boolean and multimedia retrieval in vitrivr for large-scale video search. In: Ro, Y.M., et al. (eds.) MMM 2020. LNCS, vol. 11962, pp. 760–765. Springer, Cham (2020). https://doi.org/10.1007/978-3-030-37734-2_66
24. Sauter, L., Schuldt, H., Waltenspül, R., Rossetto, L.: Novice-friendly text-based video search with vitrivr. In: 20th International Conference on Content-based Multimedia Indexing (CBMI 2023), 20–22 September 2023, Orléans, France. ACM (2023)

25. Spiess, F., et al.: Multi-modal video retrieval in virtual reality with vitrivr-VR. In: Þór Jónsson, B., et al. (eds.) MMM 2022. LNCS, vol. 13142, pp. 499–504. Springer, Cham (2022). https://doi.org/10.1007/978-3-030-98355-0_45

26. Spiess, F., Gasser, R., Heller, S., Rossetto, L., Sauter, L., Schuldt, H.: Competitive interactive video retrieval in virtual reality with vitrivr-VR. In: Lokoč, J., et al. (eds.) MMM 2021. LNCS, vol. 12573, pp. 441–447. Springer, Cham (2021). https://doi.org/10.1007/978-3-030-67835-7_42

27. Spiess, F., Heller, S., Rossetto, L., Sauter, L., Weber, P., Schuldt, H.: Traceable asynchronous workflows in video retrieval with vitrivr-VR. In: Dang-Nguyen, D.T., et al. (eds.) MMM 2023. LNCS, vol. 13833, pp. 622–627. Springer, Cham (2023). https://doi.org/10.1007/978-3-031-27077-2_49

28. Stonebraker, M., Rowe, L.A.: The design of postgres. In: Proceedings of the 1986 ACM SIGMOD International Conference on Management of Data, Washington, DC, USA, 28–30 May 1986, pp. 340–355. ACM Press (1986)

29. Truong, Q., et al.: Marine video kit: a new marine video dataset for content-based analysis and retrieval. In: Dang-Nguyen, D.T., et al. (eds.) MMM 2023. LNCS, vol. 13833, pp. 539–550. Springer, Cham (2023). https://doi.org/10.1007/978-3-031-27077-2_42

# VISIONE 5.0: Enhanced User Interface and AI Models for VBS2024

Giuseppe Amato⬝, Paolo Bolettieri⬝, Fabio Carrara⬝, Fabrizio Falchi⬝,
Claudio Gennaro⬝, Nicola Messina⬝, Lucia Vadicamo(✉)⬝,
and Claudio Vairo⬝

CNR -ISTI, Via G. Moruzzi 1, 56124 Pisa, Italy
{giuseppe.amato,paolo.bolettieri,fabio.carrara,fabrizio.falchi,
claudio.gennaro,nicola.messina,lucia.vadicamo,claudio.vairo}@isti.cnr.it

**Abstract.** In this paper, we introduce the fifth release of VISIONE, an advanced video retrieval system offering diverse search functionalities. The user can search for a target video using textual prompts, drawing objects and colors appearing in the target scenes in a canvas, or images as query examples to search for video keyframes with similar content. Compared to the previous version of our system, which was runner-up at VBS 2023, the forthcoming release, set to participate in VBS 2024, showcases a refined user interface that enhances its usability and updated AI models for more effective video content analysis.

**Keywords:** Information Search and Retrieval · Content-based video retrieval · Video search · Surrogate Text Representation · Multi-modal Retrieval · Cross-modal retrieval

## 1 Introduction

Video Browser Showdown (VBS) [13,15,17] is an important international competition for interactive video search [16], which is held annually at the International Conference on MultiMedia Modeling (MMM) since 2012. To date, it has comprised three different search tasks, namely *visual Known-Item Search* (KIS-V), *textual Known-Item Search* (KIS-T) and *Ad-hoc Video Search* (AVS). In KIS-V and KIS-T, the user must retrieve a specific segment of a video starting from, respectively, a visual or textual hint. The visual cue is given by playing the target video segment on a large screen visible to all competition participants (or on a browser using the DRES evaluation server [23], to which all teams' systems are connected). The textual hint is the description of the video segment, which is gradually extended during the task time with more details useful to identify the target video segment uniquely. In the AVS, the user has to retrieve as many as possible video segments matching a textual description provided.

The size or diversity of datasets used in the competition has increased over the years to make the competition more challenging. Last year, two datasets were used: a large dataset (V3C1+V3C2) [24], composed of 17,235 diverse videos for a total duration of 2,300 h, and the Marine Video Kit (MVK) dataset [29], which

S. Rudinac et al. (Eds.): MMM 2024, LNCS 14557, pp. 332–339, 2024.
https://doi.org/10.1007/978-3-031-53302-0_29

is composed of 1,372 highly redundant videos taken from moving cameras in underwater environments (scuba diving).

Usually, the competition is organized in expert and novice sessions. In the novice session, volunteers from the audience will be recruited to solve tasks with the search system after a brief tutorial on how to use it. Therefore, one of the focuses when implementing such systems is also to make them easy for a non-expert user to use.

There will be two main novelties in the 2024 edition of the competition. First, another dataset has been added as a target dataset for queries, the VBSLHE dataset. It is composed of 75 videos capturing laparoscopic gynecology surgeries. This dataset is particularly challenging because videos are very similar to each other, and it is difficult for non-medic users to understand what is happening in the scene and to submit more precise queries. The other novelty is the introduction of the Question Answering (Q/A) task, where the user is asked to submit textual answers to questions regarding the collection (e.g., "What airline name appears most frequently in the dataset?") instead of submitting video segments.

Our video retrieval system VISIONE [1–3,7] has participated in four editions of the VBS competition since 2019, skipping the 2020 edition. In the 2023 competition, with 13 systems participating, our system ranked first in the KIS visual task and second in the entire competition, behind Vibro [25] and ahead of other very effective systems, including Vireo [19], vitrivrVR [28], and CVHunter [18]. This paper presents the fifth version of VISIONE, which includes some changes compared to the previous versions. These changes are mainly focused on a new simplified and easy-to-use interface described in Sect. 3. In addition, it is worth noting that recently, we have made available the features we extracted from all the AI models we exploit in our system to the scientific community through several Zenodo repositories [4–6].

## 2   System Overview

VISIONE provides a set of search functionalities that empower the user to search specific video segments through text and visual queries, which can also be combined into a temporal search. The system provides *free text search, spatial color and object search,* and *visual-semantic similarity search.* For a comprehensive understanding of the system's architecture and an overview of the user interface, please refer to our previous publications, which describe these details [1,3].

In pursuit of robust and efficient free text and semantic similarity search, we leverage three cross-modal feature extractors, each driven by pre-trained models. OpenCLIP ViT-L/14 [14,22] model pre-trained on LAION-2B dataset [27] (in the following we refer to this model as ClipLAION), CLIP2Video [12], and ALADIN [20]. By default, a combination of all three models is used when performing a free text search, which is achieved by merging the results with a late fusion approach. However, if needed, it is possible to select one of the aforementioned models by selecting the desired model in the dedicated radio button in the UI. Three models are also used for the object detection: VfNet

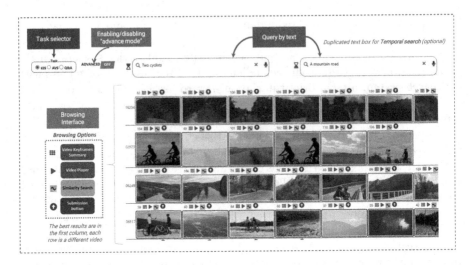

**Fig. 1.** New simplified user interface with textual, temporal and similarity search. When clicking on "Advanced Mode", the user can access advanced search options, including object search and the capability to choose a specific AI model for cross-modal searches.

trained on the COCO dataset, Mask R-CNN trained on the LVIS dataset, and a Faster R-CNN+Inception ResNet V2 trained on Open Images V4 (see [3]). The visual similarity search leverages the comparative analysis of the extracted DINOv2 [21] features that have been proven to be effective for several image-level visual tasks (image classification, instance retrieval, video understanding). Finally, the spatial color-based search is enabled by the annotations extracted using the two chip-based color naming techniques (see [3]).

For our indexing strategy, we leverage two different techniques: the Facebook FAISS library[1] is used to store and access both CLIP2Video and ClipLAION features. The second index is tailored to store all other descriptors and can be effortlessly queried using Apache Lucene[2]. Notably, for indexing all the extracted descriptors with Lucene, we devised specialized text encodings based on the Surrogate Text Representations (STRs) approach described in [1,8–10].

## 3   Recent Changes to the VISIONE System

In this section, we present the changes that we introduced in VISIONE with respect to the system version participating in the last VBS competition.

*Simplified Interface.* We added a simplified version of the user interface to help novice users to perform searches. The new simplified interface, now the default

---

[1] https://github.com/facebookresearch/faiss.
[2] https://lucene.apache.org/.

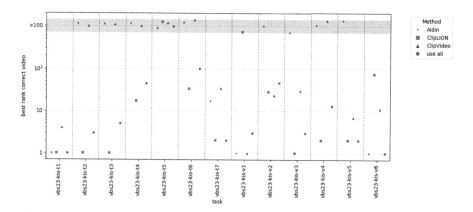

**Fig. 2.** Distributions of the best ranks of correct video per task and method. We used all the textual queries issued by the two users of the VISIONE system during the last VBS2023 competition. Note that only text queries made on the V3C dataset were considered, excluding the combination with other search features, e.g., temporal search.

one, shows basic search functionalities (see Fig. 1). In the text boxes, it is possible to submit a query by text that exploits a combination of the three networks we used to extract features from the videos, namely ALADIN, Clip2Video, and ClipLAION. The second text box (the green one on the right in Fig. 1) is optional and can be used to perform a temporal query, in which the user can describe a scene occurring after the scene described in the primary red text box on the left. On top of each returned result, there is the ID of the shot that the frame refers to and four buttons that allow some browsing options. From left to right, the grid button allows to show all the keyframes of that video, the play button starts the playback of the video, the tilde button performs an aggregated visual similarity search to find frames visually similar to the selected one, and the white arrow in the green circle button submits the corresponding frame to DRES.

Regarding the similarity search, in the previous versions of VISIONE, there was the possibility to perform three different visual similarity searches, one for each network used to extract visual features, namely DINOv2, ALADIN, and Clip2Video. In this version, we use an aggregated method of similarity search that relies on late fusion to combine the ranked results obtained from the single approaches. The late fusion employed is the Reciprocal Rank Fusion [11], with an additional enhancement on the top-5 results from each ranked list. This is the same heuristic used to combine results in the free text search. Based on our preliminary experiments using the VISIONE logs collected during VBS2023 (see Fig. 2), while ClipLION exhibits the overall best performance, there are instances in which other methods correctly identify the video with a better ranking (e.g., tasks vbs23-kis-v1 and vbs23-kis-v6). Our late fusion strategy doesn't consistently achieve the best rankings, however, it enables us to strike a favorable compromise among the various methods employed, allowing the user not to be forced to choose one of the search methods.

An "Advanced mode" button in the top-left corner of the interface allows enabling an advanced version of the interface with all the search functionalities available. In this modality, the user can select a specific model to be used for text search. In the new interface, we also added the possibility to click on a selected frame to get a zoomed view of that particular frame.

We presented this new interface to the "Interactive Video Retrieval for Beginners" special session at the 20th International Conference on Content-Based (CBMI 2023) [7]. This occasion has been very useful in collecting valuable feedback from the novice users who actually utilized our system. We will consider this feedback as we work to enhance the usability of VISIONE further.

Finally, we observe that in the top-left corner of the interface, it is possible to switch between the different tasks of the competition (KIS, AVS, and Q&A), and between the different datasets (V3C, MVK, VBSLHE). This triggers some ad-hoc modifications to the user interface to address functionalities tailored for that particular task or dataset, for example, a different object palette for different datasets, the addition of a text box for the Q&A task, and a different browsing behavior for the AVS task. To search the VBSLHE dataset, we used BiomedCLIP [30], a biomedical vision-language foundation model that is pretrained on PMC-15M [30], a dataset of 15 million figure-caption pairs extracted from biomedical research articles in PubMed Central, using contrastive learning.

*Keyboard Shortcuts.* AVS is one of the tasks where VISIONE performed worst in the previous competitions. Inspired by lifeXplore system [26], we tried to improve the usability of the system in this task by adding some keyboard shortcuts. In particular, it is possible to browse the frames resulting from a search using the arrow keys and to submit the current selected frame by pressing the "s" key. This makes selecting and submitting frames for the AVS task quicker than using the mouse.

*Features of all Datasets Available in Zenodo.* We extracted and made publicly available on Zenodo [4–6] the features and annotations extracted from all the dataset used in the completion (V3C, MVK, VBSLHE) for all the three multimodal models used in our system (ALADIN, Clip2Video and ClipLAION) and for the object detectors (VfNet, Mask R-CNN, and Faster R-CNN+Inception ResNet V2). We also provide a Python script to extract the keyframes used by the VISIONE system for which features and annotations are provided.

*Question and Answering Task.* To date, VISIONE does not allow automatic processing for the new question and answering task. We simply added a dedicated textbox to the UI by which the user can manually submit the answer when found. To do that, the user should rephrase the questions into text queries suitable for finding a set of relevant results, which she/he should then review to locate the desired answer.

## 4  Conclusions and Future Works

The paper introduces the fifth version of the VISIONE, a video retrieval system that offers advanced functionalities to search for specific video segments using free text search, spatial color and object search, and visual-semantic similarity search. Compared to the previous version, the new release for VBS 2024 features a simplified user interface, which could be helpful for novice users, and enhanced usability through keyboard shortcuts. It also includes the public release of features and annotations extracted from the datasets used for the competition. The 2024 VBS competition introduces new challenges, such as integrating the VBSLHE dataset and a Question Answering (Q/A) task. VISIONE addresses these challenges by improving its capabilities, incorporating new AI models, and enhancing its natural language processing capabilities.

We plan to release the code of our system as open source soon. This release will encompass not only the core search engine but also the user interface (UI). By making our system open source, we aim to encourage broader usage and invite collaboration for further system development. An example of a potential direction we envision is creating a Virtual Reality (VR) user interface for the system.

In our future work, we aim to automate the execution of the Q&A task by implementing natural language processing and question-answering techniques, enabling the user to directly input her/his questions and obtain a pool of potential textual answers.

**Acknowledgements.** This work was partially funded by AI4Media - A European Excellence Centre for Media, Society and Democracy (EC, H2020 n. 951911), the PNRR-National Centre for HPC, Big Data and Quantum Computing project CUP B93C22000620006, and by the Horizon Europe Research & Innovation Programme under Grant agreement N. 101092612 (Social and hUman ceNtered XR - SUN project). Views and opinions expressed in this paper are those of the authors only and do not necessarily reflect those of the European Union. Neither the European Union nor the European Commission can be held responsible for them.

## References

1. Amato, G., et al.: The VISIONE video search system: exploiting off-the-shelf text search engines for large-scale video retrieval. J. Imag. **7**(5), 76 (2021)
2. Amato, G., et al.: Visione: a large-scale video retrieval system with advanced search functionalities. In: Proceedings of the 2023 ACM International Conference on Multimedia Retrieval,D pp. 649–653 (2023)
3. Amato, G., et al.: VISIONE at video browser showdown 2023. In: Dang-Nguyen, D.-T., et al. (eds.) MultiMedia Modeling: 29th International Conference, MMM 2023, Bergen, Norway, January 9–12, 2023, Proceedings, Part I, pp. 615–621. Springer International Publishing, Cham (2023). https://doi.org/10.1007/978-3-031-27077-2_48
4. Amato, G., et al.: VISIONE Feature Repository for VBS: Multi-Modal Features and Detected Objects from LapGyn100 Dataset, October 2023. https://doi.org/10.5281/zenodo.10013328

5. Amato, G., et al.: VISIONE Feature Repository for VBS: Multi-Modal Features and Detected Objects from MVK Dataset (2023). https://doi.org/10.5281/zenodo.8355037

6. Amato, G.,et al.: VISIONE feature repository for VBS: multi-modal features and detected objects from V3C1+V3C2 dataset (Jul 2023). https://doi.org/10.5281/zenodo.8188570

7. Amato, G., et al.: VISIONE for newbies: an easier-to-use video retrieval system. In: Proceedings of the 20th International Conference on Content-based Multimedia Indexing. Association for Computing Machinery (2023)

8. Amato, G., Carrara, F., Falchi, F., Gennaro, C., Vadicamo, L.: Large-scale instance-level image retrieval. Inform. Process. Manage. **57**(6), 102100 (2020)

9. Carrara, F., Gennaro, C., Vadicamo, L., Amato, G.: Vec2Doc: transforming dense vectors into sparse representations for efficient information retrieval. In: Pedreira, O., Estivill-Castro, V. (eds.) Similarity Search and Applications: 16th International Conference, SISAP 2023, A Coruña, Spain, October 9–11, 2023, Proceedings, pp. 215–222. Springer Nature Switzerland, Cham (2023). https://doi.org/10.1007/978-3-031-46994-7_18

10. Carrara, F., Vadicamo, L., Gennaro, C., Amato, G.: Approximate nearest neighbor search on standard search engines. In: Skopal, T., Falchi, F., Lokoč, J., Sapino, M.L., Bartolini, I., Patella, M. (eds.) Similarity Search and Applications: 15th International Conference, SISAP 2022, Bologna, Italy, October 5–7, 2022, Proceedings, pp. 214–221. Springer International Publishing, Cham (2022). https://doi.org/10.1007/978-3-031-17849-8_17

11. Cormack, G.V., Clarke, C.L., Buettcher, S.: Reciprocal rank fusion outperforms condorcet and individual rank learning methods. In: Proceedings of the 32nd International ACM SIGIR Conference on Research and Development in Information Retrieval, pp. 758–759 (2009)

12. Fang, H., Xiong, P., Xu, L., Chen, Y.: Clip2video: Mastering video-text retrieval via image clip. arXiv preprint arXiv:2106.11097 (2021)

13. Heller, S., Gsteiger, V., Bailer, W., Gurrin, C., Jónsson, B.Þ, Lokoč, J., et al.: Interactive video retrieval evaluation at a distance: comparing sixteen interactive video search systems in a remote setting at the 10th video browser showdown. Int. J. Multimed. Inform. Retrieval **11**(1), 1–18 (2022)

14. Ilharco, G., et al.: Openclip (2021). https://doi.org/10.5281/zenodo.5143773

15. Lokoč, J., et al.: Interactive video retrieval in the age of effective joint embedding deep models: lessons from the 11th vbs. Multimedia Systems, pp. 1–24 (2023)

16. Lokoč, J., et al.: A Task Category Space for User-Centric Comparative Multimedia Search Evaluations. In: Þór Jónsson, B., Gurrin, C., Tran, M.-T., Dang-Nguyen, D.-T., Hu, A.M.-C., Huynh Thi Thanh, B., Huet, B. (eds.) MultiMedia Modeling: 28th International Conference, MMM 2022, Phu Quoc, Vietnam, June 6–10, 2022, Proceedings, Part I, pp. 193–204. Springer International Publishing, Cham (2022). https://doi.org/10.1007/978-3-030-98358-1_16

17. Lokoč, J., et al.: Is the reign of interactive search eternal? findings from the video browser showdown 2020. ACM Trans. Multimed. Comput. Commun. Appl. **17**(3), 1–26 (2021)

18. Lokoč, J., Vopálková, Z., Dokoupil, P., Peška, L.: Video search with CLIP and interactive text query reformulation. In: Dang-Nguyen, D.-T., Gurrin, C., Larson, M., Smeaton, A.F., Rudinac, S., Dao, M.-S., Trattner, C., Chen, P. (eds.) MultiMedia Modeling: 29th International Conference, MMM 2023, Bergen, Norway, January 9–12, 2023, Proceedings, Part I, pp. 628–633. Springer International Publishing, Cham (2023). https://doi.org/10.1007/978-3-031-27077-2_50

19. Ma, Z., Wu, J., Loo, W., Ngo, C.W.: Reinforcement learning enhanced pichunter for interactive search. In: MultiMedia Modeling (2023)
20. Messina, N., et al.: Aladin: distilling fine-grained alignment scores for efficient image-text matching and retrieval. In: Proceedings of the 19th International Conference on Content-based Multimedia Indexing, pp. 64–70 (2022)
21. Oquab, M., et al.: Dinov2: learning robust visual features without supervision. arXiv preprint arXiv:2304.07193 (2023)
22. Radford, A., et al.: Learning transferable visual models from natural language supervision. In: Proceedings of the 38th International Conference on Machine Learning, ICML 2021, pp. 8748–8763. PMLR (2021)
23. Rossetto, L., Gasser, R., Sauter, L., Bernstein, A., Schuldt, H.: A system for interactive multimedia retrieval evaluations. In: Lokoč, J., Skopal, T., Schoeffmann, K., Mezaris, V., Li, X., Vrochidis, S., Patras, I. (eds.) MultiMedia Modeling: 27th International Conference, MMM 2021, Prague, Czech Republic, June 22–24, 2021, Proceedings, Part II, pp. 385–390. Springer International Publishing, Cham (2021). https://doi.org/10.1007/978-3-030-67835-7_33
24. Rossetto, L., Schuldt, H., Awad, G., Butt, A.A.: V3C – a research video collection. In: Kompatsiaris, I., Huet, B., Mezaris, V., Gurrin, C., Cheng, W.-H., Vrochidis, S. (eds.) MultiMedia Modeling: 25th International Conference, MMM 2019, Thessaloniki, Greece, January 8–11, 2019, Proceedings, Part I, pp. 349–360. Springer International Publishing, Cham (2019). https://doi.org/10.1007/978-3-030-05710-7_29
25. Schall, K., Hezel, N., Jung, K., Barthel, K.U.: Vibro: video browsing with semantic and visual image embeddings. In: Dang-Nguyen, D.-T., Gurrin, C., Larson, M., Smeaton, A.F., Rudinac, S., Dao, M.-S., Trattner, C., Chen, P. (eds.) MultiMedia Modeling: 29th International Conference, MMM 2023, Bergen, Norway, January 9–12, 2023, Proceedings, Part I, pp. 665–670. Springer International Publishing, Cham (2023). https://doi.org/10.1007/978-3-031-27077-2_56
26. Schoeffmann, K.: lifexplore at the lifelog search challenge 2023. In: Proceedings of the 6th Annual ACM Lifelog Search Challenge, pp. 53–58 (2023)
27. Schuhmann, C., et al.: Laion-5b: an open large-scale dataset for training next generation image-text models. Adv. Neural. Inf. Process. Syst. **35**, 25278–25294 (2022)
28. Spiess, F., Heller, S., Rossetto, L., Sauter, L., Weber, P., Schuldt, H.: Traceable asynchronous workflows in video retrieval with vitrivr-VR. In: Dang-Nguyen, D.-T., Gurrin, C., Larson, M., Smeaton, A.F., Rudinac, S., Dao, M.-S., Trattner, C., Chen, P. (eds.) MultiMedia Modeling: 29th International Conference, MMM 2023, Bergen, Norway, January 9–12, 2023, Proceedings, Part I, pp. 622–627. Springer International Publishing, Cham (2023). https://doi.org/10.1007/978-3-031-27077-2_49
29. Truong, Q.-T., et al.: Marine Video Kit: a new marine video dataset for content-based analysis and retrieval. In: Dang-Nguyen, D.-T., Gurrin, C., Larson, M., Smeaton, A.F., Rudinac, S., Dao, M.-S., Trattner, C., Chen, P. (eds.) MultiMedia Modeling: 29th International Conference, MMM 2023, Bergen, Norway, January 9–12, 2023, Proceedings, Part I, pp. 539–550. Springer International Publishing, Cham (2023). https://doi.org/10.1007/978-3-031-27077-2_42
30. Zhang, S., et al. Large-scale domain-specific pretraining for biomedical vision-language processing (2023). https://doi.org/10.48550/ARXIV.2303.00915

# PraK Tool: An Interactive Search Tool Based on Video Data Services

Jakub Lokoč[1](✉), Zuzana Vopálková[1], Michael Stroh[2], Raphael Buchmueller[3], and Udo Schlegel[3]

[1] SIRET Research Group, Department of Software Engineering Faculty of Mathematics and Physics, Charles University, Prague, Czech Republic
jakub.lokoc@matfyz.cuni.cz
[2] Visual Computing, Department of Computer Science Faculty of Natural Sciences, Konstanz University, Konstanz, Germany
michael.stroh@uni-konstanz.de
[3] Data Analysis and Visualization, Department of Computer Science Faculty of Natural Sciences, Konstanz University, Konstanz, Germany
{raphael.buchmueller,u.schlegel}@uni-konstanz.de

**Abstract.** This paper presents a tool relying on data service architecture, where technical details of all VBS datasets are completely hidden behind an abstract stateless data layer. The data services allow independent development of interactive search interfaces and refinement techniques, which is demonstrated by a smart front-end component. The component supports common search features and allows users to exploit content-based statistics for effective filtering. We believe that video data services might be a valuable addition to the open-source VBS toolkit, especially when available for the competition on a shared server with all VBS datasets, extracted features, and meta-data behind.

**Keywords:** interactive video retrieval · deep features · text-image retrieval

## 1 Introduction

The Video Browser Showdown (VBS) [6,14] is a well-established event for comparison of interactive video search systems. Every year, new findings and insights are revealed at the competition, which affects the research and implementation of new video search prototypes. For example, joint embedding approaches based on Open CLIP trained with LAION-2B [7,13] turned out to be highly competitive at the VBS competition in 2023.

For VBS 2024, we present a new system created in cooperation between Konstanz University and Charles University, Prague. The system currently relies on extracted keyframes and features from the VISIONE [1] system but can be easily extended with novel feature extraction approaches. On top of the directory with the provided meta-data, we created a stateless application component called

S. Rudinac et al. (Eds.): MMM 2024, LNCS 14557, pp. 340–346, 2024.
https://doi.org/10.1007/978-3-031-53302-0_30

video data services. Our ambition is to propagate the idea that teams can participate at VBS even without the need to download, process, and maintain huge volumes of data from VBS datasets (V3C [15], MVK [18]).

In the following, we review available data access options of the front-ends of selected top VBS systems. The VIREO system [11] performs ranking operations in their back-end engine. However, the front-end application still requires indexed data (videos, thumbnails) locally to start searching. The Vibro system [17] is a standalone java application where all data has to be loaded into RAM for each instance. Similarly, both VIRET [10] and CVHunter [9] are WPF .NET applications that require to load all meta-data to RAM, while thumbnail images are stored and accessed locally from disk. Only text to joint-embedding vector transformation requires an online service for CVHunter and later versions of VIRET. The SOMHunter system [8] used Node.js containing SOMHunter core library that performed computations and maintained state. The latest version of SOMHunter core used an HTTP API. However, the core library maintained its state and did not return feature vectors to the front-end. The vitrivr system [16] relies on a three-tier architecture, where the data layer is hidden behind the retrieval engine accessible by a REST API or a WebSocket. The engine allows to run queries and returns items with some meta-data but does not return features to the front-end. The VISIONE system [1] has a retrieval engine accessible via a REST API as well; however, CLIP features cannot be directly attached to ranked result sets. The VERGE system [12] uses a web service for obtaining a ranked list of image names, but no features are attached to this list. We conclude that many systems use HTTP calls for ranked result sets; however, the option to obtain a complete set of video frames and their features for a more complex front-end application is indeed not common in regularly participating VBS systems.

## 2   System Description

Our video search system consists of video data services providing access to subsets of VBS video datasets and a smart front-end component for refinement of the subsets. We emphasize that the services are stateless, meaning it is up to front-end components to maintain the state in more complex search scenarios over data subsets. The video data services provide an abstract interface independent of used video data sources, and thus, front-end applications can remain unchanged (except configuration or parameter tuning) in case new joint embedding models or datasets are prepared for VBS. The current version of the services uses meta-data provided by the VISIONE system [1–3], where selected representative keyframes are accompanied by extracted features from CLIP based models [7,13] trained on LAION-2B. However, the meta-data feature set can be easily extended by the most recent joint-embedding models released before VBS 2024. The only change in the front-end application would be a new value of an existing parameter of the video data service interface. The overall architecture of the presented video search system is depicted in Fig. 1.

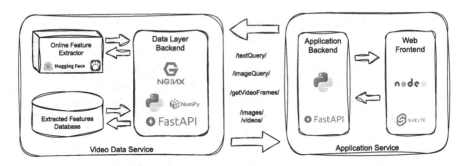

**Fig. 1.** Architecture of the presented video search system based on video data services. In the current version, the Application Backend provides just a web server functionality.

## 2.1   Video Data Services

The idea behind video data services is to provide access to video data, either one particular video or selected video frames with some properties (e.g., frames related to a text query). From this perspective, video data services combine a general data-sharing platform and retrieval engine available on the internet. Based on findings from previous VBS events, we identified four essential video data services for VBS search scenarios:

- *textQuery(text, k, dataset, model, addFeatures)* returning a list of the k most relevant representative video frames to a text query, cosine distance is assumed for vectors of a specified joint embedding model, temporal query option is allowed as well using > symbol in the text parameter
- *imageQuery(image, k, dataset, model, addFeatures)* returning a list of the k most relevant representative video frames to an example image query (any image can be used), cosine distance is assumed for vectors of a specified joint embedding model
- *getVideoFrames(itemID, k, dataset, model, addFeatures)* returning a list of k temporally preceding/subsequent representative frames to itemID from the same video, for $k = 0$ all representative frames are returned for the video
- *getVideo(itemID, dataset)* returning URI of a given video

The list of returned video frames is a JSON file with a sorted array of items. The items are sorted based on rank or time information for the list of video items. Each item contains a video frame thumbnail URI, item rank/time, cosine distance with respect to the query, item ID for the VBS server (for item submission), and optionally, video frame feature vector and detected classes. Since the last two fields can significantly expand the volume of each item (from several bytes to kilobytes), these fields are returned only if parameter $addFeatures = true$ is specified. Please note that a lightweight option $addFeatures = false$ leads to limited front-end functionality (e.g., no relevance feedback actions). On the other hand, both options can be combined to obtain quickly list of images and then list of corresponding features in a background process. Since the number of supported features could vary in the future, this parameter could be changed to a binary based encoding of requested features (e.g., "1010 ..." for first and third feature set).

After each video services call, a data chunk of k items is returned to a front-end that allows users to refine the result set interactively. Although primarily designed for video search applications, this architecture can provide access to various types of data for interactive search refinement. For example, the system can be used for interactive search experiments utilizing data visualizations based on Chernoff faces [4].

For all teams using video data services, a result log json file can be created and stored in the video data services back-end as long as a userID parameter is added to the interface.

## 2.2    Front-End Search Options and Interface

The front-end component is responsible for the refinement of subsets obtained by video data services. We first recapitulate common features of successful systems and then describe a new feature helpful in specific search scenarios.

Assuming a ranked list of video frames obtained by a primary ranking model, a list of common search options [6,14] that our tool supports comprise:

- Two types of result set visualization techniques – grid of ranked images and per-line visualization of ranked items with their temporal video context
- Text query reformulation interface as frequent query reformulation is indeed a popular search option
- Example image query interface, including images from external sources (e.g., as was used by VIRET [10])
- Presentation filters allowing to show only given $k \in \{1, 2, ..., 9\}$ most relevant items from each video (e.g., as was used in CVHunter [9])
- Bayesian relevance feedback reformulation [5] (e.g., used by SOMHunter [8]), where the score is updated only for items available for the front-end, joint-embedding features are necessary

In addition to these popular features, we plan to add a novel feature relying on content-based statistics of the available subsets on front-end. Specifically, we assume that each data item has a list of detected classes (e.g., using zero-shot CLIP classification) or that the classes correspond to centroids obtained by an unsupervised clustering technique (e.g., k-means). With these classes and corresponding class confidence scores, it is possible to add interactive search features like informative statistic charts or label-selection-based filters/rankers. Mainly, label-based filters/rankers represent a novel approach at VBS (to the best of our knowledge) that combines automatically detected content-based statistics (e.g., classes that can equally divide the current set into two subsets) and user knowledge of the searched target, which affects filtering/re-ranking decisions.

The current interface of our tool is illustrated on Fig. 2.

**Fig. 2.** The Front-end of the presented video search system based on video data services. In the top panel, users can select dataset, feature representation model and submit custom text for question answering tasks. The left panel allows users to run text queries or use external image to initialize the search. Bayes update button is available for selected images as well. Bellow the button, statistics of detected class labels are presented for the current result set. Users can click on each class label to set it as filter.

## 3    Conclusions

In the paper, we present an example of a VBS system relying on stateless video data services and a smart front-end, allowing interactive refinement of actual result sets. The prototype supports common search features popular at VBS and investigates a novel approach based on detected classes in video frames. For future VBS events, it would be a convenient option to run such video data services for (new) VBS teams that would like to focus more on front-ends with result-set refinement and HCI techniques. The data processing burden would be significantly lower, while the available feature set could be sufficiently competitive. Especially if the joint-embedding features are up-to-date. At the same time, result logs could be created automatically for the teams.

**Acknowledgment.** This work has been supported by Czech Science Foundation (GAČR) project 22-21696S. We would like to thank the authors of the discussed VBS systems for clarification of architecture details.

# References

1. Amato, G.: VISIONE at video browser showdown 2023. In: Dang-Nguyen, D.T., et al. (eds.) MMM 2023. LNCS, vol. 13833, pp. 615–621. Springer, Cham (2023). https://doi.org/10.1007/978-3-031-27077-2_48
2. Amato, G., et al.: VISIONE Feature Repository for VBS: Multi-Modal Features and Detected Objects from MVK Dataset (2023). https://doi.org/10.5281/zenodo.8355037
3. Amato, G., et al.: VISIONE Feature Repository for VBS: Multi-Modal Features and Detected Objects from V3C1+V3C2 Dataset (2023). https://doi.org/10.5281/zenodo.8188570
4. Chernoff, H.: The use of faces to represent points in k-dimensional space graphically. J. Am. Stat. Assoc. **68**(342), 361–368 (1973). https://doi.org/10.1080/01621459.1973.10482434
5. Cox, I.J., Miller, M.L., Minka, T.P., Papathomas, T.V., Yianilos, P.N.: The Bayesian image retrieval system, PicHunter: theory, implementation, and psychophysical experiments. IEEE Trans. Image Process. **9**(1), 20–37 (2000)
6. Heller, S., et al.: Interactive video retrieval evaluation at a distance: comparing sixteen interactive video search systems in a remote setting at the 10th video browser showdown. Int. J. Multim. Inf. Retr. **11**(1), 1–18 (2022). https://doi.org/10.1007/s13735-021-00225-2
7. Ilharco, G., et al.: Openclip (2021). https://doi.org/10.5281/zenodo.5143773, if you use this software, please cite it as below
8. Kratochvíl, M., Mejzlík, F., Veselý, P., Souček, T., Lokoč, J.: SOMHunter: lightweight video search system with SOM-guided relevance feedback. In: Proceedings of the 28th ACM International Conference on Multimedia, MM 2020. ACM (2020, in press)
9. Lokoč, J., Mejzlík, F., Souček, T., Dokoupil, P., Peška, L.: Video search with context-aware ranker and relevance feedback. In: Þór Jónsson, B., et al. (eds.) MMM 2022. LNCS, vol. 13142, pp. 505–510. Springer, Cham (2022). https://doi.org/10.1007/978-3-030-98355-0_46
10. Lokoč, J., Kovalčík, G., Souček, T., Moravec, J., Čech, P.: A framework for effective known-item search in video. In: Proceedings of the 27th ACM International Conference on Multimedia (MM 2019), Nice, France, 21–25 October 2019, pp. 1–9 (2019). https://doi.org/10.1145/3343031.3351046
11. Ma, Z., Wu, J., Loo, W., Ngo, C.W.: Reinforcement learning enhanced PicHunter for interactive search. In: Conference on Multimedia Modeling (2023)
12. Pantelidis, N., et al.: VERGE in VBS 2023. In: Dang-Nguyen, D.T., et al. (eds.) MMM 2023. LNCS, vol. 13833, pp. 658–664. Springer, Cham (2023). https://doi.org/10.1007/978-3-031-27077-2_55
13. Radford, A., et al.: Learning transferable visual models from natural language supervision. CoRR abs/2103.00020 (2021). https://arxiv.org/abs/2103.00020
14. Rossetto, L., et al.: Interactive video retrieval in the age of deep learning–detailed evaluation of VBS 2019. IEEE Trans. Multimed. **23**, 243–256 (2020). https://doi.org/10.1109/TMM.2020.2980944
15. Rossetto, L., Schuldt, H., Awad, G., Butt, A.A.: V3C – a research video collection. In: Kompatsiaris, I., Huet, B., Mezaris, V., Gurrin, C., Cheng, W.-H., Vrochidis, S. (eds.) MMM 2019. LNCS, vol. 11295, pp. 349–360. Springer, Cham (2019). https://doi.org/10.1007/978-3-030-05710-7_29

16. Sauter, L., et al.: Exploring effective interactive text-based video search in vitrivr. In: Dang-Nguyen, D.T., et al. (eds.) MMM 2023. LNCS, vol. 13833, pp. 646–651. Springer, Cham (2023). https://doi.org/10.1007/978-3-031-27077-2_53

17. Schall, K., Hezel, N., Jung, K., Barthel, K.U.: Vibro: video browsing with semantic and visual image embeddings. In: Dang-Nguyen, D.T., et al. (eds.) MMM 2023. LNCS, pp. 665–670. Springer, Cham (2023). https://doi.org/10.1007/978-3-031-27077-2_56

18. Truong, Q.T., et al.: Marine video kit: a new marine video dataset for content-based analysis and retrieval. In: Dang-Nguyen, D.T., et al. (eds.) MMM 2023. LNCS, vol. 13833, pp. 539–550. Springer, Cham (2023). https://doi.org/10.1007/978-3-031-27077-2_42

# Exquisitor at the Video Browser Showdown 2024: Relevance Feedback Meets Conversational Search

Omar Shahbaz Khan[1(✉)], Hongyi Zhu[2], Ujjwal Sharma[2], Evangelos Kanoulas[2],
Stevan Rudinac[2], and Björn Þór Jónsson[1]

[1] Reykjavik University, Reykjavik, Iceland
omark@ru.is
[2] University of Amsterdam, Amsterdam, The Netherlands

**Abstract.** An important open problem in video retrieval and exploration concerns the generation and refinement of queries for complex tasks that standard methods are unable to solve, especially when the systems are used by novices. In conversational search, the proposed approach is to ask users to interactively refine the information provided to the search process, until results are satisfactory. In user relevance feedback (URF), the proposed approach is to ask users to interactively judge query results, which in turn refines the query used to retrieve the next set of suggestions. The question that we seek to answer in the long term is how to integrate these two approaches: can the query refinements of conversational search directly impact the URF model, and can URF judgments directly impact the conversational search? We extend the existing Exquisitor URF system with conversational search. In this version of Exquisitor, the two modes of interaction are separate, but the user interface has been completely redesigned to (a) prepare for future integration with conversational search and (b) better support novice users.

**Keywords:** Interactive learning · Video browsing · Multimedia retrieval · Multimodal representaion learning · Multimodal conversational search

## 1 Introduction

Video Browser Showdown (VBS) is an annual video retrieval challenge where researchers showcase their novel multimedia retrieval tools to solve complex interactive tasks. A plethora of retrieval paradigms have been seen at VBS, such as textual search [18,26], similarity search with content-based retrieval [2,25], sketch/object-based search [1,25], and relevance feedback [9,14,27]. VBS 2024 covers three types of interactive tasks. The objective of Known Item Search (KIS) tasks is to locate a frame of a segment in a specific video. For Ad-hoc Video Search (AVS) tasks, the objective is to find as many video shots as possible that match a given description. The third task type is the Question and

S. Rudinac et al. (Eds.): MMM 2024, LNCS 14557, pp. 347–355, 2024.
https://doi.org/10.1007/978-3-031-53302-0_31

Answer (Q&A) task, where the objective is to answer a question regarding the dataset, rather than submitting video clips. Another major aspect of VBS is the division into expert and novice sessions. In the expert session, each participating system's users are typically system developers. In the novice session, the users are volunteers from the audience who are completely unfamiliar with the retrieval system. For researchers working on multimedia information retrieval, VBS is an ideal venue for real-world testing with different types of users [16,17,19]. Given the nature of videos, the tasks in VBS often include temporal information, and systems that introduce techniques to support temporal querying tend to perform better [1,2,14,18,22,25]. However, in the last two editions of VBS, systems that use search with CLIP [23] features have dominated the competition, regardless of their temporal features [2,22,26]. There are still some tasks, however, that CLIP models struggle to solve, notably where the task does not contain much distinct information or is very domain-specific. We believe that an interactive process, where users can incrementally refine the information given to the retrieval system, will be beneficial in solving such tasks.

Exquisitor is an interactive learning system for large-scale multimedia exploration [9,10,13]. At its core, Exquisitor relies on user relevance feedback to build one or more classifiers using positive and negative labeled video shots determined by the user for their current information need. To support the relevance feedback process with search-oriented tasks, the user has the option to apply metadata filters that reduce the search space. Additionally, the user can search for shots using keywords (concepts, actions) or a textual query, to find better examples for the classifier. Exquisitor supports temporal querying by merging the output of two classifiers using relational operators [10] and in the most recent participation introduced a basic CLIP search implementation [13]. Based on recent VBS participation, we have observed that (a) the interface was too complex for novice users, and (b) the workflow process between search and relevance feedback was unintuitive, at times leaving users stranded in the search interface [11]. We have redesigned the Exquisitor user interface to address both aspects.

Conversational search is a method where the user starts a dialogue with a conversational agent to get answers for an information need. With advancements in image encoders and Large Language Models (LLM), it is now possible to have dialogues about multimedia content through joint visual-language models (VLM) [15]. For interactive retrieval, conversational search is a powerful tool to refine queries, by having the user provide clarifications or additional details to the replies of the agent. Unlike CLIP-based search or keyword search, where the user continuously reformulates queries, but similar to URF, conversational search works together with the user to solve the task.

In this paper, we extend the Exquisitor URF system with conversational search. Figure 1 outlines the intended workflow of the Exquisitor system, where the left side of the figure represents the flow of the conversational search and the right side represents the URF approach. The conversational flow utilizes textual input and image descriptions together with a VLM to reformulate the original query. The URF flow aims to build a classifier, using the keyframes' semantic

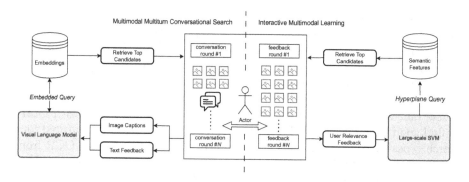

**Fig. 1.** Exquisitor workflow for interactive task refinement.

visual features, to acquire the top relevant keyframes from the high-dimensional index. The intention is to investigate how these two retrieval paradigms can be integrated: whether their different refinement methods can directly impact each other and what are the benefits and drawbacks of such an approach?

In the remainder of this paper, we first reevaluate Exquisitor's current features based on observed usage and present an entirely new user interface that aims to better support conversational search and novice usage, before outlining how conversational search is integrated into the current version of Exquisitor.

## 2    User Relevance Feedback

### 2.1    Exquisitor

Exquisitor is the state-of-the-art large-scale interactive learning approach with subsecond performance on collections with over 100 million images [12,24]. The core component of Exquisitor is an interactive learning server, which typically runs locally on a machine but can also be served for remote access. Exquisitor uses semantic features to represent multimedia items. These features are typically extracted using a deep neural network, where the resulting high-dimensional vector is compressed and stored in an approximate cluster-based high-dimensional index [6,33]. The server has three main functions for the general user relevance feedback process: train an interactive classifier, provide suggestions, and apply filters. The interactive classifier is linear SVM; one for each modality used. Once a classifier is trained, it is used to retrieve new suggestions from the index. This retrieval process consists of traversing the index to obtain a list of clusters to consider. The items from these clusters are then ranked and the top items are returned to the user. When retrieving from clusters with active filters, query optimization policies are used to incrementally consider additional clusters until enough top items are found. If multiple modalities are used, a fused ranked list is obtained through rank aggregation [12].

The original Exqusitor UI consisted of a grid view, showing the suggestions from the underlying model, which the user could label as positive or negative.

Two containers, displaying the user's positive and negative labeled items, were placed on the right and left sides of the suggestions. Below the suggestions was a set of buttons, which the user could use to get new sets of suggestions, view previously seen items, or reset the current model. Above the suggestions, the user could apply filters using text fields.

To perform a search, the user had to click on a search button that opened a new tab, where they could perform keyword searches, or search with CLIP, to find good examples to label as positives. The original interface featured four search types: ImageNet concepts [21], actions from Kinetics-700 [4], OCR from Tesseract and EasyOCR [8], and CLIP [23]. From participation in VBS and other similar demonstrations, it became evident that users strongly preferred CLIP search over the others. Furthermore, opening the search in a new tab disrupted the URF process. It frequently disconnected the user from the previous results, which were presented to them in the grid view. This abrupt transition often caused novice users to forget the primary purpose of the search, which was to find good examples for the classifier. Admittedly, CLIP search often returned task-relevant shots and occasionally directly solved the task at hand, but in cases where CLIP would not succeed, users would repeatedly search in vain, instead of using the returned items for relevance feedback.

## 2.2   Exquisitor's User Interface

While the typical workflow was often performed well by an expert, the performance of novices varied heavily. One reason is their inexperience with interactive learning approaches, and another is the interface not being intuitive towards the workflow. In this section, we review the new Exquisitor interface, which attempts to make the interactive learning approach more intuitive, while also decluttering the screen by reducing options. In addition, the UI is made with future development in mind for the conversational search component.

Figure 2 depicts the Exquisitor interface. The top orange bar is where the user can add or remove models. A quality of life improvement for this is the addition of a dialog prompt when removing models, as in the previous interface users could accidentally hit the "x" and lose all information from a working model. The items always show the available button options for the user below it. Instead of "+" and "−" for positive and negative, we now use "Thumbs up" and "Thumbs down" buttons, which resonates better with today's users. To submit a shot the ">" button is used, instead of the previous "S" button. At the bottom center of the grid is a hovering panel of buttons, currently only having "Update" and "Clear". We have removed the additional buttons as they often caused confusion. While Fig. 2 shows the positive and negative containers, a major change has been to hide them initially. These containers, along with the filters from the original interface are now accessed through the buttons on the right and left sidebar. The right sidebars "Thumbs up" and "Thumbs down" buttons open containers for the positive and negative items. The history (already seen items for this model) of the current session is accessed by the "Clock" icon on the right sidebar.

**Fig. 2.** The main Exquisitor display with positive and negative examples highlighted. (Color figure online)

Similar to the positive and negative containers, the filter panel can be opened by clicking the top button on the left sidebar. The main change to the filters, aside from the type of filters being more distinct, is to have the user apply the filters by pressing a button. Previously, whenever a filter was set in a field it was automatically applied, which was less transparent.

To alleviate the problems with search described above, the different search types, and the separation of search and relevance feedback, the new UI offers a panel that is accessed from the left sidebar and co-exists with the relevance feedback grid. Figure 3 shows the search panel, which is going to be the primary component for the conversational search.

## 3   Conversational Search

Simulating many real-world scenarios, the VBS task is not always described through a single query. Often, multiple semantic concepts, actions, or events need to be searched for before the relevant video shot is found. In addition, not all information is provided at the start when it comes to Textual Known-Item-Search tasks. Straightforward search using e.g. CLIP can form a solid base for subsequent interactive retrieval rounds. However, the user may need to rephrase or entirely reformulate the query when more information is provided or the top results are unsatisfactory. Instead of manual restarting or query reformulation, we propose a conversational search approach. The user inputs an initial query, for which a handful of the most relevant items are returned. The user is then invited to supply additional information based on the examples. They can provide textual feedback or label the items as positive or negative. We utilize the text

**Fig. 3.** The Exquisitor search panel, where conversational search will be integrated.

generation and understanding capabilities of a visual-language model (VLM) to interpret the intention of the user [3,29]. To facilitate conversational search, we use conversational query rewriting to summarize the information contained in the multi-turn retrieval context [5,20] and use the refined query.

Specifically, we use BLIP-2 [15] with the image encoder (ViT-B/32) to extract embeddings and LLM with Vicuna-7b to generate descriptions for each keyframe, which are stored in a database. The BLIP-2 text encoder transforms the textual query and searches the database for appropriate items. In the case of textual feedback, the VLM takes the text input and rewrites the original query. For the labeled feedback the generated descriptions of the keyframes are used as input for the VLM. We use two prompting methods for conversational search: few-shot prompting and chain-of-thought prompting (CoT). In few-shot prompting a set of examples or instructions are provided to help the VLM understand the semantic features of the target task to generate a better answer [7,28,31]. CoT is a step-by-step method that guides the VLM to refine the query in multi-turn conversations [30]. To make an informed choice of query refinement and retrieval approach, we deploy an automatic evaluation framework inspired by [32].

## 4    Conclusion

In this paper, we introduce a new version of Exquisitor, which is based on two pillars. First, the user relevance feedback of the Exquisitor system is used to refine a model to support a user's information need through content-based feedback. Second, conversational search is used to interactively refine or build upon queries to either directly solve an information need or to provide information to enhance the relevance feedback process. The user interface of Exquisitor has been fully redesigned based on experiences from previous participations and demonstrations to better support conversational search, and to improve the experience for novice users.

**Acknowledgments.** This work was supported by Icelandic Research Fund grant 239772-051.

# References

1. Amato, G., et al.: VISIONE at video browser showdown 2023. In: Dang-Nguyen, D.T., et al. (eds.) MMM 2023. LNCS, vol. 13833, pp. 615–621. Springer, Cham (2023). https://doi.org/10.1007/978-3-031-27077-2_48
2. Arnold, R., Sauter, L., Schuldt, H.: Free-form multi-modal multimedia retrieval (4MR). In: Dang-Nguyen, D.T., et al. (eds.) MMM 2023. LNCS, vol. 13833, pp. 678–683. Springer, Cham (2023). https://doi.org/10.1007/978-3-031-27077-2_58
3. Brown, T., et al.: Language models are few-shot learners. In: Advances in Neural Information Processing Systems 33, pp. 1877–1901 (2020)
4. Carreira, J., Noland, E., Hillier, C., Zisserman, A.: A short note on the kinetics-700 human action dataset. arXiv preprint arXiv:1907.06987 (2019)
5. Dalton, J., Xiong, C., Callan, J.: CAsT 2020: the conversational assistance track overview. In: Proceedings of TREC (2021)
6. Guðmundsson, G.Þ., Jónsson, B.Þ., Amsaleg, L.: A large-scale performance study of cluster-based high-dimensional indexing. In: Proceedings of the International Workshop on Very-Large-Scale Multimedia Corpus, Mining and Retrieval (VLS-MCM), Firenze, Italy (2010)
7. Jagerman, R., Zhuang, H., Qin, Z., Wang, X., Bendersky, M.: Query expansion by prompting large language models. arXiv preprint arXiv:2305.03653 (2023)
8. Jaided AI: EasyOCR. https://github.com/JaidedAI/EasyOCR
9. Jónsson, B.Þ, Khan, O.S., Koelma, D.C., Rudinac, S., Worring, M., Zahálka, J.: Exquisitor at the video browser showdown 2020. In: Ro, Y.M., et al. (eds.) MMM 2020. LNCS, vol. 11962, pp. 796–802. Springer, Cham (2020). https://doi.org/10.1007/978-3-030-37734-2_72
10. Khan, O.S., et al.: Exquisitor at the video browser showdown 2021: relationships between semantic classifiers. In: Lokoč, J., et al. (eds.) MMM 2021. LNCS, vol. 12573, pp. 410–416. Springer, Cham (2021). https://doi.org/10.1007/978-3-030-67835-7_37
11. Khan, O.S., Jónsson, B.Þ.: User relevance feedback and novices: anecdotes from Exquisitor's participation in interactive retrieval competitions. In: Proceedings of the Content-Based Multimedia Indexing, CBMI 2023, Orléans, France (2023)
12. Khan, O.S., et al.: Interactive learning for multimedia at large. In: Jose, J.M., et al. (eds.) ECIR 2020. LNCS, vol. 12035, pp. 495–510. Springer, Cham (2020). https://doi.org/10.1007/978-3-030-45439-5_33
13. Khan, O.S., et al.: Exquisitor at the video browser showdown 2022. In: Þór Jónsson, B., et al. (eds.) MMM 2022. LNCS, vol. 13142, pp. 511–517. Springer, Cham (2022). https://doi.org/10.1007/978-3-030-98355-0_47
14. Kratochvíl, M., Veselý, P., Mejzlík, F., Lokoč, J.: SOM-hunter: video browsing with relevance-to-SOM feedback loop. In: Ro, Y.M., et al. (eds.) MMM 2020. LNCS, vol. 11962, pp. 790–795. Springer, Cham (2020). https://doi.org/10.1007/978-3-030-37734-2_71

15. Li, J., Li, D., Savarese, S., Hoi, S.: BLIP-2: bootstrapping language-image pre-training with frozen image encoders and large language models. arXiv preprint arXiv:2301.12597 (2023)

16. Lokoč, J., et al.: Interactive video retrieval in the age of effective joint embedding deep models: lessons from the 11th VBS. Multimedia Syst. **29**, 3481–3504 (2023)

17. Lokoč, J., et al.: Interactive search or sequential browsing? A detailed analysis of the video browser showdown 2018. ACM TOMM **15**(1), 1–18 (2019)

18. Lokoč, J., Kovalčík, G., Souček, T.: VIRET at video browser showdown 2020. In: Ro, Y.M., et al. (eds.) MMM 2020. LNCS, vol. 11962, pp. 784–789. Springer, Cham (2020). https://doi.org/10.1007/978-3-030-37734-2_70

19. Lokoč, J., et al.: Is the reign of interactive search eternal? Findings from the video browser showdown 2020. ACM Trans. Multimedia Comput. Commun. Appl. (TOMM) **17**(3), 1–26 (2021)

20. Mao, K., Dou, Z., Qian, H.: Curriculum contrastive context denoising for few-shot conversational dense retrieval. In: Proceedings of the 45th International ACM SIGIR Conference, pp. 176–186 (2022)

21. Mettes, P., Koelma, D.C., Snoek, C.G.: The ImageNet shuffle: reorganized pre-training for video event detection. In: Proceedings of the 2016 ACM on International Conference on Multimedia Retrieval, ICMR 2016, New York, NY, USA, pp. 175–182. Association for Computing Machinery (2016)

22. Nguyen, T.N., et al.: VideoCLIP: an interactive CLIP-based video retrieval system at VBS2023. In: Dang-Nguyen, D.T., et al. (eds.) MMM 2023. LNCS, vol. 13833, pp. 671–677. Springer, Cham (2023). https://doi.org/10.1007/978-3-031-27077-2_57

23. Radford, A., et al.: Learning transferable visual models from natural language supervision. In: International Conference on Machine Learning, pp. 8748–8763. PMLR (2021)

24. Ragnarsdóttir, H., et al.: Exquisitor: breaking the interaction barrier for exploration of 100 million images. In: Proceedings of the ACM Multimedia, Nice, France (2019)

25. Sauter, L., Amiri Parian, M., Gasser, R., Heller, S., Rossetto, L., Schuldt, H.: Combining boolean and multimedia retrieval in vitrivr for large-scale video search. In: Ro, Y.M., et al. (eds.) MMM 2020. LNCS, vol. 11962, pp. 760–765. Springer, Cham (2020). https://doi.org/10.1007/978-3-030-37734-2_66

26. Schoeffmann, K., Stefanics, D., Leibetseder, A.: diveXplore at the video browser showdown 2023. In: Dang-Nguyen, D.T., et al. (eds.) MMM 2023. LNCS, vol. 13833, pp. 684–689. Springer, Cham (2023). https://doi.org/10.1007/978-3-031-27077-2_59

27. Song, W., He, J., Li, X., Feng, S., Liang, C.: QIVISE: a quantum-inspired interactive video search engine in VBS2023. In: Dang-Nguyen, D.T., et al. (eds.) MMM 2023. LNCS, vol. 13833, pp. 640–645. Springer, Cham (2023). https://doi.org/10.1007/978-3-031-27077-2_52

28. Wang, L., Yang, N., Wei, F.: Query2doc: query expansion with large language models. arXiv preprint arXiv:2303.07678 (2023)

29. Wei, J., et al.: Finetuned language models are zero-shot learners. arXiv preprint arXiv:2109.01652 (2021)

30. Wei, J., et al.: Chain-of-thought prompting elicits reasoning in large language models. In: Advances in Neural Information Processing Systems 35, pp. 24824–24837 (2022)

31. Yu, W., et al.: Generate rather than retrieve: large language models are strong context generators. arXiv preprint arXiv:2209.10063 (2022)

32. Zahálka, J., Rudinac, S., Worring, M.: Analytic quality: evaluation of performance and insight in multimedia collection analysis. In: Proceedings of the 23rd ACM International Conference on Multimedia, MM 2015, pp. 231–240, New York, NY, USA. Association for Computing Machinery (2015)
33. Zahálka, J., Rudinac, S., Jónsson, B.Þ, Koelma, D.C., Worring, M.: Blackthorn: large-scale interactive multimodal learning. IEEE TMM **20**(3), 687–698 (2018)

# VERGE in VBS 2024

Nick Pantelidis[1]([⊠]), Maria Pegia[1,2], Damianos Galanopoulos[1],
Konstantinos Apostolidis[1], Klearchos Stavrothanasopoulos[1],
Anastasia Moumtzidou[1], Konstantinos Gkountakos[1], Ilias Gialampoukidis[1],
Stefanos Vrochidis[1], Vasileios Mezaris[1], Ioannis Kompatsiaris[1],
and Björn Þór Jónsson[2]

[1] Information Technologies Institute/Centre for Research and Technology Hellas,
Thessaloniki, Greece
{pantelidisnikos,mpegia,dgalanop,kapost,klearchos_stav,moumtzid,
gountakos,heliasgj,stefanos,bmezaris,ikom}@iti.gr
[2] Reykjavik University, Reykjavik, Iceland
{mariap22,bjorn}@ru.is

**Abstract.** This paper presents VERGE, an interactive video content
retrieval system designed for browsing a collection of images extracted
from videos and conducting targeted content searches. VERGE incorporates a variety of retrieval methods, fusion techniques, and reranking
capabilities. It also offers a user-friendly web application for query formulation and displaying top results.

## 1 Introduction

VERGE is an interactive system for video content retrieval that offers the ability to perform queries through a web application in a set of images extracted
from videos, display the top ranked results in order to find the appropriate one
and submit it. The system has continuously participated in the Video Browser
Showdown (VBS) competition [17] since its inception. This year, we enhanced
the performance of existing modalities for better performance and introduced
new ones based on the lessons learned from the VBS 2023 [17]. Moreover, we
made some subtle yet effective enhancements to the web application, all aimed
at providing a better user experience.

The structure of the paper is as follows: Sect. 2 presents the framework of the
system, Sect. 3 describes the various retrieval modalities, Sect. 4 presents the UI
and its features, concluding with Sect. 5 that outlines the future work.

## 2 The VERGE Framework

Figure 1 visualizes the VERGE framework that is composed of four layers. The
first layer consists of the various retrieval modalities that most of them are
applied beforehand to the datasets and their results are stored in a database.
The second layer contains the Text to Video Matching service that uses the

S. Rudinac et al. (Eds.): MMM 2024, LNCS 14557, pp. 356–363, 2024.
https://doi.org/10.1007/978-3-031-53302-0_32

corresponding modality to return results. The third layer consists of the API endpoints that read data either from the database or the services and return the top results for each case. The last layer is the Web Application that sends queries to the API and displays the results.

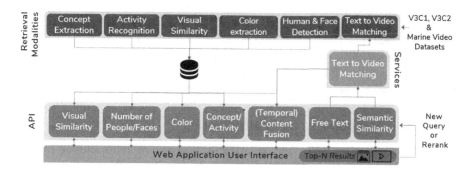

**Fig. 1.** The VERGE Framework

VERGE integrates the V3C1+V3C2 dataset [23], the marine video dataset [26] and the new dataset for this year, the LapGyn100 dataset (surgeries in laparoscopic gynecology). The shot segmentation information was provided for the initial two datasets. For the third dataset, OpenCV[1] was used to segment videos into coherent shots, with each shot lasting about 10 s. A sampling ratio of 5 was applied to reduce frame retention.

The main updates realized this year include: the addition of semantic similarity search and the late fusion using the text to video matching module which also was improved in terms of speed, the use of advanced architecture for the concept extraction module, and the application of preprocessing steps related to human/face detection for activity recognition.

## 3    Retrieval Modalities

### 3.1    Concept-Based Retrieval

This module annotates each shot's keyframe with labels from a pool of concepts consisting of 1000 ImageNet concepts, a subset of 298 concepts from the TRECVID SIN task [19], 365 scene classification concepts, 580 object labels, and 22 sports classification labels. To generate annotation scores for the ImageNet concepts, we adopted an ensemble approach, averaging the concept scores from two pre-trained models, the ViT_B_16 [13] and EfficientNet_V2_L [25]. For the subset of concepts from the TRECVID SIN task, we trained and employed a model based on the EfficientNet-B4 architecture using the official SIN task dataset. In the case of scene-related concepts, we adopted the same approach,

---

[1] https://opencv.org/.

averaging the concept scores from WideResNet [29] and ResNet50 [10] models, pre-trained on the Places365 [30] dataset. Object detection scores were extracted using models pre-trained on the MS COCO [16] and Open Images V4 [14] datasets, which include 80 and 500 detectable objects, respectively. To label sports in video frames, a custom dataset was created using web images related to sports and used to train a model based on the EfficientNetB3 architecture. For a more concise listing representation of the concept-based annotations, we applied the sentence-BERT [22] text encoding framework to measure the text similarity among all concept labels. After analyzing the results, we manually grouped very similar concepts, assigning them a common label and taking the maximum score among their members.

### 3.2   Spatio-Temporal Activity Recognition

This module processes each video shot to extract one or more human-related activities. Thus, enhances the filtering capabilities by adding activity label options to the list of concepts. A list of 400 pre-defined human-related activities was extracted for each shot using a 3D-CNN architecture deploying a pipeline similar to [8] that modifies the 3D-ResNet-152 [9] architecture. During inference, the shots were fed to the model to fit the model's input dimension, $16 \times 112 \times 112 \times 3$, and the extracted activities were sorted according to their scores, from higher to lower. Finally, a post-processing step was applied to remove false-positive activity predictions by excluding those that weren't human-related, as the human and face detection module (Sect. 3.6) did not predict human bodies or faces.

### 3.3   Visual Similarity Search

This module uses DCNNs to find similar content based on characteristics extracted from each shot. These characteristics are derived from the final pooling layer of the GoogleNet architecture [21], serving as a global image representation. To enable swift and effective indexing, we create an IVFADC index database vector with these feature vectors, following the technique proposed in [12].

### 3.4   Text to Video Matching Module

This module inputs a complex free-text query and retrieves the most relevant video shots from a large set of available video shots. We extend the $T \times V$ network [6] into the $T \times V + Objects$. The $T \times V$ network consists of two key sub-networks, one for the textual and one for the visual stream and utilizes multiple textual and visual features along with multiple textual encoders to build multiple cross-modal common latent spaces. The $T \times V + Objects$ network utilizes visual information at more than one level of granularity. Specifically, it extends $T \times V$ by introducing an extra object-based video encoder using transformers.

Regarding the training, a combination of four large-scale video caption datasets is used (e.g. MSR-VTTT [28], TGIF [15], ActivityNet [2] and Vatex

[27]), and the improved marginal ranking loss [4] is used to train the entire network. We utilize four different textual features: i) Bag-of-Words (bow), ii) Word2Vec model [20], iii) Bert [3] and iv) the open_clip ViT-G14 [11] model. These features are used as input to the textual sub-network ATT presented in [5]. As a second textual encoder, we consider a network that feedforwards the open_clip features. Similarly to the textual sub-network, we utilize three visual encoders to extract frame-level features: i) R_152, a ResNet-152 [10] network trained on the ImageNet-11k dataset, ii) Rx_101 a ResNeXt-101 network, pre-trained by weakly supervised learning on web images followed and fine-tuned on ImageNet [18] and iii) the open_clip ViT-G14 [11] model. Moreover, we utilize the object detector and the feature extractor modules of the ViGAT network [7] to obtain the initial object-based visual representations.

### 3.5 Semantic Similarity Search

The module is used to find video shots that are semantically similar, meaning they may not be visually similar but carry similar contexts. To find semantically similar shots, we utilize and adapt the $T \times V + Objects$ of Sect. 3.4 used in the text-to-video matching module. The $T \times V + Objects$ model consists of two key sub-networks (textual and visual). It associates text with video by creating new common latent spaces. We modify this model to convert a query video and the available pool of video shots into the new latent spaces where a semantic relationship is exposed. Based on these new video shot representations we retrieve the most related video shots with respect to the video shot query.

### 3.6 Human and Face Detection

This module aims to detect and report the number of humans and human faces in each keyframe of each shot to enhance user filtering by providing capabilities for distinguishing the results of single-human or multi-human activities. The detections of both human silhouettes (bodies) and human faces (heads) were extracted using YoloV4 [1]. MS COCO [16] datasets' weights were used and fine-tuned using the CrowdHuman dataset [24], where partial occlusions among humans or between humans and objects can occur in crowd-centre scenes.

### 3.7 Late Fusion Approaches for Conceptual Text Matching

This section explores two late fusion approaches for combining conceptual text and shot detection in video analysis. Thus, concepts from Sect. 3.1 or sentences from Sect. 3.4 serve as text information. These methods create a list of shots, utilizing sequential or intersection information from texts, and sorting them using

$$f(P) = \sum_{i=1}^{|P|} e^{-p_i} + \sum_{i,j=1, i \neq j}^{|P|} e^{-|p_i - p_j|}, \tag{1}$$

where $P = \{p_i\}_{i=1}^{i=n}$ are the shot probabilities computed by modules from Sect. 3.1 or Sect. 3.4.

**Conceptual Text Late Fusion.** This method focuses on assembling a list of shots that encompass specific sentences or concepts from the textual content. It identifies shots that appear in both text queries, indicating shared context and content. The process involves developing independent lists of shot probabilities for each text and subsequently calculating their intersection at shot level. The final list is then sorted using Eq. 1.

**Temporal Text Late Fusion.** The method returns a list of tuples of shots, with each element corresponding to the same video and containing all queried texts in a particular order. This approach incorporates text queries and sorts them using the same late fusion method (Eq. 1). The process includes creating a list of shot probabilities for each text, calculating their intersection at video layer, retaining the first tuple of each video while respecting the order of texts, and finally sorting the shots using the Eq. 1.

### 3.8 Semantic Segmentation in Videos

The module is designed to provide precise pixel-level annotations of surgical tools and anatomical structures within each shot frame. This capability enables the user to distinguish objects of interest against the surgical laparoscopic background easily and to locate the relevant scenes. To achieve this, we employ the state-of-the-art DCNN architecture, LWANet. The model's initial weights are acquired through pre-training on the ImageNet dataset, followed by fine-tuning on the domain-specific EndoVis2017 dataset tailored for laparoscopic surgery scenarios. During inference, the model processes every frame of a shot, generating a binary mask that precisely outlines the areas of segmented objects.

## 4    User Interface and Interaction Modes

The VERGE UI (Fig. 2) is a web-based application that allows users to create and execute queries, utilizing the retrieval modalities that were described above. The top results from the queries are displayed so as the users can browse through them, watch the corresponding videos and submit the appropriate data.

The UI is composed of three main parts. The header on the top, the menu on the left and the results panel that takes all the remaining space. The header contains, from left to right, the back button for restoring previous results, the rerank button for returning the intersection of the current results and the results of the new performed query, the logo on the center and the pagination buttons on the right for navigating through the pages. The first options on the menu are the dataset selector, which allows you to choose the dataset for conducting queries, and the query type where when the AVS option is selected, yields a single result per video. The initial search module is the free text search, allowing users to input anything in the form of free text. Subsequently, the concept/activity-based

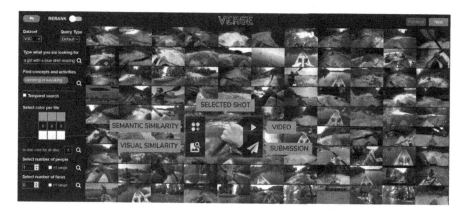

**Fig. 2.** The VERGE User Interface

search offers pre-extracted values for users to choose from. Multiple selection is also supported for late fusion as well as temporal fusion if the corresponding checkbox is checked. The color search is achieved by selecting the color for a part of the image in a 3 × 3 grid or by selecting a color for the whole image. Finally, there is possibility to search based on the count of people or faces discernible in the image, alongside the choice to conduct a flexible search by opting to include results with either an additional or one less person/face. Moreover, when a user hovers over a result/image, four buttons appear (Fig. 2). From the top-left corner and in a clockwise order, these are: the button that returns images with similar context, the one that returns visually similar images, the video play button and the button for sending this shot to the competition's server.

To showcase the VERGE functionality, we present three use cases. In the first case, we are looking for shots of canoeing and kayaking so we can select from the concepts/activities the activity with the name canoeing or kayaking. In the second use case, we want to find a girl with a blue shirt that reads a book outdoors. We can use this exact description in the free text search. The last use case involves locating a skier within a scene that encompasses both the sky and the snow. Therefore, we can color the relevant regions of the image accordingly (Fig. 2), and then reorganize the outcomes based on the visible individuals.

## 5  Future Work

This year, the newly added LapGyn100 dataset introduced unique challenges due to its nature. Our upcoming months of research will focus on exploring methods tailored to address these challenges that involve further study of images processing techniques and changes to the UI. Finally, the changes realised are expected to enhance the UI's user-friendliness for both expert and novice users while maintaining its functionality.

**Acknowledgements.** This work has been supported by the EU's Horizon 2020 research and innovation programme under grant agreements 101070250 XRECO, 883302 ISOLA, 883484 PathoCERT, 101021866 CRiTERIA, and the Horizon research and innovation programme under grant agreement 101096405 UP2030.

# References

1. Bochkovskiy, A., Wang, C.Y., Liao, H.Y.M.: YOLOv4: optimal speed and accuracy of object detection. arXiv Preprint arXiv:2004.10934 (2020)
2. Caba Heilbron, F., Escorcia, V., Ghanem, B., Carlos Niebles, J.: ActivityNet: a large-scale video benchmark for human activity understanding. In: Proceedings of the IEEE Conference on Computer Vision and Pattern Recognition, pp. 961–970 (2015)
3. Devlin, J., Chang, M., Lee, K., Toutanova, K.: BERT: pre-training of deep bidirectional transformers for language understanding. CoRR abs/1810.04805 (2018). http://arxiv.org/abs/1810.04805
4. Faghri, F., Fleet, D.J., Kiros, J.R., Fidler, S.: VSE++: improving visual-semantic embeddings with hard negatives. In: Proceedings of the British Machine Vision Conference (BMVC) (2018)
5. Galanopoulos, D., Mezaris, V.: Attention mechanisms, signal encodings and fusion strategies for improved ad-hoc video search with dual encoding networks. In: Proceedings of the ACM International Conference on Multimedia Retrieval (ICMR 2020). ACM (2020)
6. Galanopoulos, D., Mezaris, V.: Are all combinations equal? Combining textual and visual features with multiple space learning for text-based video retrieval. In: Karlinsky, L., Michaeli, T., Nishino, K. (eds.) ECCV 2022. LNCS, pp. 627–643. Springer, Cham (2023). https://doi.org/10.1007/978-3-031-25069-9_40
7. Gkalelis, N., Daskalakis, D., Mezaris, V.: ViGAT: bottom-up event recognition and explanation in video using factorized graph attention network. IEEE Access **10**, 108797–108816 (2022). https://doi.org/10.1109/ACCESS.2022.3213652
8. Gkountakos, K., Touska, D., Ioannidis, K., Tsikrika, T., Vrochidis, S., Kompatsiaris, I.: Spatio-temporal activity detection and recognition in untrimmed surveillance videos. In: Proceedings of the 2021 International Conference on Multimedia Retrieval, pp. 451–455 (2021)
9. Hara, K., et al.: Can spatiotemporal 3D CNNs retrace the history of 2D CNNs and ImageNet? In: Proceedings of IEEE CVPR 2018 (2018)
10. He, K., Zhang, X., Ren, S., Sun, J.: Deep residual learning for image recognition. In: Proceedings of the IEEE Conference on Computer Vision and Pattern Recognition, pp. 770–778 (2016)
11. Ilharco, G., et al.: OpenCLIP (2021). https://doi.org/10.5281/zenodo.5143773, if you use this software, please cite it as below
12. Jegou, H., Douze, M., Schmid, C.: Product quantization for nearest neighbor search. IEEE Trans. Pattern Anal. Mach. Intell. **33**(1), 117–128 (2010)
13. Kolesnikov, A., et al.: An image is worth $16 \times 16$ words: transformers for image recognition at scale (2021)
14. Kuznetsova, A., et al.: The open images dataset v4: unified image classification, object detection, and visual relationship detection at scale. Int. J. Comput. Vis. **128**(7), 1956–1981 (2020)

15. Li, Y., et al.: TGIF: a new dataset and benchmark on animated GIF description. In: Proceedings of the IEEE Conference on Computer Vision and Pattern Recognition, pp. 4641–4650 (2016)
16. Lin, T.-Y., et al.: Microsoft COCO: common objects in context. In: Fleet, D., Pajdla, T., Schiele, B., Tuytelaars, T. (eds.) ECCV 2014. LNCS, vol. 8693, pp. 740–755. Springer, Cham (2014). https://doi.org/10.1007/978-3-319-10602-1_48
17. Lokoč, J., et al.: Interactive video retrieval in the age of effective joint embedding deep models: lessons from the 11th VBS. Multimed. Syst. **29**, 3481–3504 (2023)
18. Mahajan, D., et al.: Exploring the limits of weakly supervised pretraining. In: Proceedings of the European Conference on Computer Vision (ECCV), pp. 181–196 (2018)
19. Markatopoulou, F., Moumtzidou, A., Galanopoulos, D., et al.: ITI-CERTH participation in TRECVID 2017. In: Proceedings of TRECVID 2017 Workshop, USA (2017)
20. Mikolov, T., Chen, K., Corrado, G., Dean, J.: Efficient estimation of word representations in vector space. In: 1st International Conference on Learning Representations, Workshop Track Proceedings, ICLR 2013 (2013)
21. Pittaras, N., Markatopoulou, F., Mezaris, V., Patras, I.: Comparison of fine-tuning and extension strategies for deep convolutional neural networks. In: Amsaleg, L., Guðmundsson, G.Þ, Gurrin, C., Jónsson, B.Þ, Satoh, S. (eds.) MMM 2017. LNCS, vol. 10132, pp. 102–114. Springer, Cham (2017). https://doi.org/10.1007/978-3-319-51811-4_9
22. Reimers, N., Gurevych, I.: Sentence-BERT: sentence embeddings using Siamese BERT-networks. arXiv preprint arXiv:1908.10084 (2019)
23. Rossetto, L., Schuldt, H., Awad, G., Butt, A.A.: V3C – a research video collection. In: Kompatsiaris, I., Huet, B., Mezaris, V., Gurrin, C., Cheng, W.-H., Vrochidis, S. (eds.) MMM 2019. LNCS, vol. 11295, pp. 349–360. Springer, Cham (2019). https://doi.org/10.1007/978-3-030-05710-7_29
24. Shao, S., Zhao, Z., Li, B., Xiao, T., Yu, G., et al.: CrowdHuman: a benchmark for detecting human in a crowd. arXiv Preprint arXiv:1805.00123 (2018)
25. Tan, M., Le, Q.: Efficientnetv2: smaller models and faster training. In: International Conference on Machine Learning, pp. 10096–10106. PMLR (2021)
26. Truong, Q.T., et al.: Marine video kit: a new marine video dataset for content-based analysis and retrieval. In: Dang-Nguyen, D.T., et al. (eds.) MMM 2023. LNCS, vol. 13833, pp. 539–550. Springer, Cham (2023). https://doi.org/10.1007/978-3-031-27077-2_42
27. Wang, X., Wu, J., Chen, J., Li, L., Wang, Y.F., Wang, W.Y.: VATEX: a large-scale, high-quality multilingual dataset for video-and-language research. In: Proceedings of the 2019 IEEE/CVF International Conference on Computer Vision (ICCV), pp. 4581–4591 (2019)
28. Xu, J., Mei, T., Yao, T., Rui, Y.: MSR-VTT: a large video description dataset for bridging video and language. In: Proceedings of the IEEE Conference on Computer Vision and Pattern Recognition, pp. 5288–5296 (2016)
29. Zagoruyko, S., Komodakis, N.: Wide residual networks. arXiv preprint arXiv:1605.07146 (2016)
30. Zhou, B., Lapedriza, A., Khosla, A., Oliva, A., Torralba, A.: Places: a 10 million image database for scene recognition. IEEE Trans. Pattern Anal. Mach. Intell. **40**(6), 1452–1464 (2017)

# Optimizing the Interactive Video Retrieval Tool Vibro for the Video Browser Showdown 2024

Konstantin Schall[✉], Nico Hezel, Kai Uwe Barthel, and Klaus Jung

Visual Computing Group, HTW Berlin, University of Applied Sciences,
Wilhelminenhofstraße 75, 12459 Berlin, Germany
konstantin.schall@htw-berlin.de
http://visual-computing.com/

**Abstract.** *Vibro* (*Video Browser*) is an interactive video retrieval system and the winner of the Video Browser Showdown 2022 and 2023. This paper gives an overview of the underlying concepts of this tool and highlights the changes that were implemented for the upcoming competition of 2024. Additionally, we propose a way to evaluate retrieval engine performance for the specific use case of known-item search, by using logged query data from previous competitions. This evaluation helps in finding an optimal embedding model candidate for image- and text-to-image retrieval and making the right decisions to improve the current system.

**Keywords:** Content-based Video Retrieval · Exploration · Visualization · Image Browsing · Visual and Textual co-embeddings

## 1 Introduction

The Video Browser Showdown (VBS) represents an annual interactive video retrieval competition with participants from all over the world. Up to this year, the competition consisted of solving several tasks from three categories: *visual Known-Item Search* (v-KIS), *textual Known-Item Search* (t-KIS) and *Ad-Hoc Video Search* (AVS). In the first case, participants are shown a short segment from a video clip and the task is solved by not only finding the required video, but also providing the exact segment's timestamp. The second category is similar, with the distinction that no video segment is shown to the users, but a textual description that is extended with details over the task's duration. The last category is different in the sense that not only one item has to be found, but as many fitting segments from different videos as possible. An example from previous competitions is 'find all shots of hot air balloons flying over mountains'. In the next competition of 2024, there will be a new type of task, namely the Question-Answering category. Instead of finding fitting shots or an known-item, questions such as 'How many distinct videos containing a bear are in the dataset?' have to be answered. Moreover, after adding the Marine Video Kit

S. Rudinac et al. (Eds.): MMM 2024, LNCS 14557, pp. 364–371, 2024.
https://doi.org/10.1007/978-3-031-53302-0_33

[17], a rather small and specific video collection of underwater diving scenes, next time, another very domain specific dataset is added: VBSLHE, a medical image dataset. This makes model selection even more challenging, as networks that generalize well and generate meaningful representations of the 17000 different videos from the V3C dataset [1], but are also able to understand thematically narrow datasets, are desired for feature extraction.

Another important change is that for the first time since the pandemic, a novice session will be held. Users with little to zero experience in solving known-item search tasks will have to solve the same type of tasks in concurrence with the other participating systems. It has been shown in the analysis of previous competitions that the performance of single power-users can strongly influence a systems final result [12]. These power-users often have many years of experience in using their software and are strongly involved in its development. However, in the novice session, this experience has no impact on the success of the system and an easy to use and elegant UI can help the untrained users to use and remember all of the systems features.

For these reasons, we decided to not modify the underlying concepts of Vibro and moreover focus on the improvement of the UI and the retrieval components. This paper first gives a brief overview of all of Vibro's functionalities, then the UI changes that where implemented to adjust to the upcoming changes are highlighted and finally, a new way to evaluate performance of retrieval models in interactive retrieval settings is proposed. This evaluation is based on the log-data from previous competitions and is meant to help making the right decision on choosing models and improving the system.

## 2    System Overview

Even though Vibro is capable of searching and browsing videos, most of the retrieval functions are performed with still images. To achieve this, we first sub-sample each video with two frames per second. Next, these frames are merged into two levels of abstraction. The first level is called 'keyframes' and is meant to represent slight changes of objects and positions inside a single scene. The second level is called 'shots' and is meant to represent the different scenes of each video. The detailed merging process can be found in [7]. In 2023 this merging led to 3.8 million keyframes and 1.6 million shots for the V3C dataset [1].

The presented video browsing system has many different features that allow to search and browse videos, or more specifically video keyframes and shots. However, the three most important and most widely used features are the following: Text-to-image search, which relies on a joint-embedding model, such as CLIP [10], to match a user's text query with previously extracted keyframe-embeddings. The second feature is image-to-image search, enabling users to select any image from Vibros UI as a query. The third feature involves presenting results in a 2D sorted map. Other, less used functions are the temporal search, in which two subsequent shots within a video can be searched for, and browsing the entire video collection with a hierarchical graph [6]. Optical character search (OCR) and text search of spoken words are also supported but where not used during the last two competitions. None of these underlying mechanics of the system have been changed in this years version and more details can be found in [15] and [7]. A screenshot of Vibro's 2023 UI can be found in [15].

## 3   User-Interface Changes

Most introduced changes to the UI aim to make Vibro less overwhelming for new users. To achieve this, we removed features that showed little to no usage during the last two competitions. This includes the removal of the sketch-based search interface, which first, did not show very good performance and second, took away to much space in Vibros search interface on the left side of the UI. Another removed feature is the arrangement of each video's keyframe on a 2D map due to low usage. Only one new button has been added. It is used for the question-answering task category. Other changes consist of the clarification of the textual descriptions in the search interface.

## 4   Improving Query Performance

Analysis of previous competitions has shown that strong retrieval models play an important role in the overall performance of the system. A high ranking of the target video requires less interaction and browsing, and can therefore be found and submitted more quickly [4,9]. Additionally, an extended post-evaluation [12] of the top three scoring teams from the VBS 22 has highlighted that a strong text-to-image retrieval model is not enough to consistently being able to solve KIS tasks and an image-to-image model is more helpful in complicated target shot scenarios. Even though Vibro already included models for both modalities, we had no appropriate metric to evaluate them for the specific task of interactive retrieval and chose models according to their performance on general-purpose retrieval benchmarks like GPR1200 [13] or the large-scale evaluation introduced in [14]. The problem with these evaluations is that all of them assess the performance of neural networks in content-based image retrieval with exactly one class per image. However, most extracted keyframes from the videos in the VBS datasets, which are used as image queries in Vibro, contain concepts that can not be represented by a single class. More specifically, if a user selects an image that shows a car driving through a forest, they are more likely to search for keyframes that combine the terms "car", "road" and "forest", as opposed to keyframes that only show cars that are visually similar to the one in the query. The following paragraph therefore introduces a way to evaluate retrieval models for the more specific task of keyframe retrieval in the context of the Video Browser Showdown and Vibro.

### 4.1   Evaluation Settings

VBS participants are usually asked to log their data during the competition, where the most important logs consist of a query, including a detailed modality description and the top 10000 results for each query. This data allows different analysis as reported in [4,9]. Since the results of the top scoring teams were quite close, three teams have agreed to perform a more extensive evaluation with the 2022 version of their systems. 57 visual-KIS tasks were solved and each system was operated by four different users, allowing a much higher volume of queries and data, compared to the usual competition settings. The results of this evaluation are reported in [12]. The four Vibro users have formulated a total of 434 textual and 308 visual queries during this extended competition. One type of analysis was the performance of the different text and image retrieval models. This led to the idea to reuse the relatively large number of queries to evaluate the performance of new models. The question to be answered is, if another model had been used to formulate this set of queries, how would this have affected the ranking of the target video or shot? To evaluate a single model, we have to extract embeddings for all of the 3.8 million keyframes and additionally embed the query-text in the case of the textual retrieval evaluation (all image-queries where performed with one of the 3.8 million keyframes). After that, each query can be rerun with the new feature vectors and since the target videos and time-frames are known, the video-rank and the shot-rank can be computed for each query. The video-rank is the rank of the target video in relation to the other videos, where all videos are represented by their keyframe closest to the query. The shot-rank is the rank of the best shot within the target time frame and closest to the query, relation to all other shots, sorted according to their relevance to the query. To make results comparable to similar previous analysis and to tackle outliers, we set the rank for both video and shot to 10000 if the actual rank is higher than this number. The final metrics for each model are the average video-rank and the average shot-rank.

## 4.2 Image-to-Image Queries

The model that was used for image-to-image queries in the 2022 version of Vibro was a Swin-L transformer [8] and was fine-tuned for content-based image retrieval similarly as described in [14]. It achieved a GPR1200 mAP of 73.4. The 2023 version used a CLIP pre-trained ViT-L [2,10] and was also fine-tuned for image retrieval, i.e. it was the best performing model from the mentioned paper [14]. The GPR1200 mAP of this model is 86.1. Both models produce binary vectors with 1024 dimensions. Even though the later network shows a much higher general-purpose image retrieval performance, the Vibro users were left with the feeling that the keyframe retrieval performance was worse in 2023 compared to 2022 after the latest VBS competition. The introduced evaluation with the V3C keyframes confirms this observation as shown in Table 1. Next, we evaluated CLIP based image encoders and confirm that those models are not particularly suited for image-to-image retrieval. OpenCLIP-ViT-H was trained with over 2 billion image/text pairs from the LAION2B dataset [16] and the EVACLIP-ViT-E model is the biggest visual transformer that was open sourced up to today. It was trained iteratively with billions of image and text pairs as described in [3].

**Table 1.** Image-to-image retrieval evaluation with the V3C dataset

| Model | Dim | GPR1200 mAP | Avg. Video-Rank | Avg. Shot-Rank |
|---|---|---|---|---|
| Vibro22 | 1024 | 73.4 | 1997 | 2576 |
| Vibro23 | 1024 | **86.1** | 2429 | 3099 |
| OpenCLIP-ViT-H | 1024 | 77.6 | 2526 | 3327 |
| EVACLIP-ViT-E | 1024 | 81.9 | 2015 | 2584 |
| MixedSwin0.0 | 1536 | 64.3 | 1723 | 2254 |
| MixedSwin0.5 | 1536 | 79.8 | **1427** | **2057** |
| MixedSwin1.0 | 1536 | 83.7 | 1870 | 2448 |

The next group of models is produced by a linear interpolation of two sets of weights from a Swin-L transformer. This technique was first introduced in [18] and is called WiseFT. The first set of weights is taken from the model that was pre-trained for classification with the ImageNet21k dataset [11] and the second weights are obtained by fine-tuning this network for image-retrieval [14]. Table 1 shows these results as MixedSwin $\alpha$, where $\alpha$ is the interpolation factor of both model weights. Figure 1 shows why this could be useful for keyframe retrieval. The query is taken from the ImageNet-Adversarial test-set [5] and pictures a stick insect sitting on a black chair. While the first model ($\alpha = 0.0$) only retrieves chairs, the fine-tuned model ($\alpha = 1.0$) solely focuses on the stick insect. Both behaviors are not optimal and an interpolation of those weights with an $\alpha = 0.5$ yields better results. Our evaluation also shows that the interpolation with $\alpha = 0.5$ achieves the best average video- and shot-ranks across all of the evaluated models.

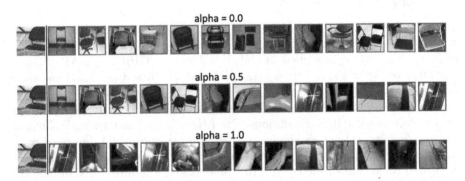

**Fig. 1.** Retrieval results with the MixedSwin model with different interpolation factors.

### 4.3   Text-to-Image Queries

The same evaluation procedure is repeated with the text-based queries from the extended post-evaluation and we compare four different CLIP based text-to-image retrieval models. The first one is the best performing model from the original CLIP paper [10], the ViT-L@336 transformer. The next two models are taken from OpenClip. The ViT-L was used in the 2023 version of Vibro and the ViT-H is the largest available visual transformer. Both were trained on the LAION2B dataset [16]. The last one is a EVA trained ViT-E, which has also been evaluated for image-to-image search. Table 2 depicts the results of this comparison. The biggest gap in performance can be observed between the original CLIP and the other networks. This can be explained by the large disparity in the used training data (400 million vs 2 billion). The other three models are closer to each other, but the EVACLIP model comes out on top with a 23% lower average shot-rank compared to the ViT-L and a 18% lower average compared to the ViT-H, and will be used as the text-to-image retrieval model in the upcoming version of Vibro.

**Table 2.** Text-to-image retrieval evaluation with the V3C dataset

| Model | Dim | Avg. Video-Rank | Avg. Shot-Rank |
|---|---|---|---|
| CLIP-ViT-L | 768 | 897 | 1730 |
| OpenCLIP-ViT-L (Vibro23) | 768 | 685 | 1343 |
| OpenCLIP-ViT-H | 1024 | **645** | 1262 |
| EVACLIP-ViT-E | 1024 | 682 | **1036** |

## 5   Conclusion

This paper presents the next iteration of Vibro for the video browser competition VBS 2024 which will be held at the MMM24 conference in Amsterdam. Due to the re-introduction of the novice session, the UI was modified and simplified. The introduced evaluation with a relatively large set of queries and know-items, allowed to pick the optimal candidates for image-to-image and text-to-image retrieval. Vibro 2024 will therefore use a MixedSwin model with an interpolation factor of 0.5 for image search and the EVACLIP-ViT-E model for text-based searches. All other Vibro functionalities have not been modified in this version.

## References

1. Berns, F., Rossetto, L., Schoeffmann, K., Beecks, C., Awad, G.: V3C1 dataset: an evaluation of content characteristics. In: Proceedings of the 2019 on International Conference on Multimedia Retrieval, ICMR 2019 (2019)
2. Dosovitskiy, A., et al.: An image is worth 16 × 16 words: transformers for image recognition at scale. CoRR (2020)

3. Fang, Y., et al.: EVA: exploring the limits of masked visual representation learning at scale (2022). https://doi.org/10.48550/ARXIV.2211.07636. https://arxiv.org/abs/2211.07636

4. Heller, S., et al.: Interactive video retrieval evaluation at a distance: comparing sixteen interactive video search systems in a remote setting at the 10th video browser showdown. Int. J. Multim. Inf. Retr. **11**(1), 1–18 (2022). https://doi.org/10.1007/s13735-021-00225-2

5. Hendrycks, D., Zhao, K., Basart, S., Steinhardt, J., Song, D.: Natural adversarial examples (2019). https://doi.org/10.48550/ARXIV.1907.07174. https://arxiv.org/abs/1907.07174

6. Hezel, N., Barthel, K.U.: Dynamic construction and manipulation of hierarchical quartic image graphs. In: Proceedings of the 2018 ACM on International Conference on Multimedia Retrieval, ICMR 2018, pp. 513–516. Association for Computing Machinery, New York (2018)

7. Hezel, N., Schall, K., Jung, K., Barthel, K.U.: Efficient search and browsing of large-scale video collections with vibro. In: Þór Jónsson, B. (ed.) MMM 2022. LNCS, vol. 13142, pp. 487–492. Springer, Cham (2022). https://doi.org/10.1007/978-3-030-98355-0_43

8. Liu, Z., et al.: Swin transformer: hierarchical vision transformer using shifted windows. arXiv preprint arXiv:2103.14030 (2021)

9. Lokoč, J., et al.: Interactive video retrieval in the age of effective joint embedding deep models: lessons from the 11th vbs. Multimedia Systems, 24 August 2023. https://doi.org/10.1007/s00530-023-01143-5

10. Radford, A., et al.: Learning transferable visual models from natural language supervision. CoRR abs/2103.00020 (2021)

11. Russakovsky, O., et al.: ImageNet large scale visual recognition challenge. Int. J. Comput. Vis. **115**(3), 211–252 (2015). https://doi.org/10.1007/s11263-015-0816-y

12. Schall, K., Bailer, W., Barthel, K.U., et al.: Interactive multimodal video search: an extended post-evaluation for the VBS 2022 competition, 11 September 2023. Preprint, currently under review. Available at Research Square https://doi.org/10.21203/rs.3.rs-3328018/v1

13. Schall, K., Barthel, K.U., Hezel, N., Jung, K.: GPR1200: a benchmark for general-purpose content-based image retrieval. In: Þór Jónsson, B., et al. (eds.) MMM 2022, Part I. LNCS, vol. 13141, pp. 205–216. Springer, Cham (2022). https://doi.org/10.1007/978-3-030-98358-1_17

14. Schall, K., Barthel, K.U., Hezel, N., Jung, K.: Improving image encoders for general-purpose nearest neighbor search and classification. In: Proceedings of the 2023 ACM International Conference on Multimedia Retrieval, ICMR 2023, pp. 57–66. Association for Computing Machinery, New York (2023). https://doi.org/10.1145/3591106.3592266

15. Schall, K., Hezel, N., Jung, K., Barthel, K.U.: Vibro: video browsing with semantic and visual image embeddings. In: Dang-Nguyen, D.T., et al. (eds.) MultiMedia Modeling, MMM 2023. LNCS, vol. 13833, pp. 665–670. Springer, Cham (2023). https://doi.org/10.1007/978-3-031-27077-2_56

16. Schuhmann, C., et al.: LAION-5B: an open large-scale dataset for training next generation image-text models (2022). https://doi.org/10.48550/ARXIV.2210.08402. https://arxiv.org/abs/2210.08402

17. Truong, Q.T., et al.: Marine video kit: a new marine video dataset for content-based analysis and retrieval. In: Dang-Nguyen, D.T., et al. (eds.) MultiMedia Modeling - 29th International Conference, MMM 2023, Bergen, Norway, 9–12 January 2023, vol. 13833, pp. 539–550. Springer, Cham (2023). https://doi.org/10.1007/978-3-031-27077-2_42

18. Wortsman, M., et al.: Robust fine-tuning of zero-shot models (2021). https://doi.org/10.48550/ARXIV.2109.01903. https://arxiv.org/abs/2109.01903

# DiveXplore at the Video Browser Showdown 2024

Klaus Schoeffmann$^{(\boxtimes)}$ and Sahar Nasirihaghighi

Klagenfurt University, Institute of Information Technology (ITEC),
Klagenfurt, Austria
{klaus.schoeffmann,sahar.nasirihaghighi}@aau.at

**Abstract.** According to our experience from VBS2023 and the feedback from the IVR4B special session at CBMI2023, we have largely revised the diveXplore system for VBS2024. It now integrates OpenCLIP trained on the LAION-2B dataset for image/text embeddings that are used for free-text and visual similarity search, a query server that is able to distribute different queries and merge the results, a user interface optimized for fast browsing, as well as an exploration view for large clusters of similar videos (e.g., weddings, paraglider events, snow and ice scenery, etc.).

**Keywords:** video retrieval · interactive video search · video analysis

## 1 Introduction

The Video Browser Showdown (VBS) [8,9,13], which was started in 2012 as a video browsing competition with a rather small data set (30 videos with an average duration of 77 min), has evolved significantly since then and became an extremely challenging international evaluation platform for interactive video retrieval tools. It not only uses several datasets with different content, as shown in Table 1, but VBS also evaluates several types of query in different sessions.

Table 1. Video datasets used for the VBS

| Dataset | Content | Files | Hours |
|---|---|---|---|
| V3C1 + V3C2 [3,15] | Diverse content uploaded by users on Vimeo | 17,235 | 2,500.00 |
| Marine videos [22] | Diving videos | 1,371 | 12.25 |
| Surgical videos | Surgery videos from laparoscopic gynecology | 75 | 104.75 |

Known item search (KIS) queries require competitors to find a specific segment in the entire video collection, which could be as small as a 2-s clip. KIS queries would be issued as *visual* KIS, where the desired clip is presented to the

S. Rudinac et al. (Eds.): MMM 2024, LNCS 14557, pp. 372–379, 2024.
https://doi.org/10.1007/978-3-031-53302-0_34

participants via the shared server and/or the on-site projector, or as a *textual* description of the clip, which is even more challenging, as it leaves room for subjective interpretation and imagination. Visual KIS simulates the typical situation where one knows of a specific video segment and knows that it is somewhere contained in the collection, but does not know where to find it. On the other hand, textual KIS simulates a situation where someone wants to find a specific video segment, but does not know how it looks like (or cannot search himself/herself and only describes it to a video retrieval expert). For KIS queries, it is crucial to be as fast and accurate as possible. The faster the clip is found, the more points a team will get. Wrong submissions will be penalized. Ad-hoch search (AVS) queries are another type of search where no specific segment needs to be found, but many examples that fit a specific description (e.g., *clips where people are holding a baloon*). For AVS tasks, it is important to find as many examples as possible, whereas diversity in terms of video files will get rated higher. Question answering (QA) queries are the final type of search, where answers to specific questions (e.g., *What is the name of the bride in the wedding on the beach in Thailand?*) need to be sent as plain text by a team (this type of query is new for VBS2024).

At the VBS, queries are performed by different users. In a first session, the teams themselves (i.e., the video search *experts*) solve various queries for the different datasets. Then, typically the next day, volunteers from the audience (the *novices*) are recruited to use the systems of the teams and solve queries (usually a little easier ones, such as visual KIS and AVS in the V3C datset). Therefore, video search systems at VBS have to be very efficient and effective at the retrieval itself, but also fast, flexible, and easy in their usage to both experts and novices. VBS systems have to be designed very carefully and with the many facets of the VBS in mind.

The Klagenfurt University diveXplore system has been participating in the VBS for several years already [7,18]. However, for VBS2024 it was significantly redesigend and optimized in several components, so that it should be much more competitive than in previous years, while still being easy and effective to use. The most important changes include (i) the integration of OpenCLIP [4] trained on the LAION-2B [19] dataset for image/text embeddings that are used during free-text and similarity search, (ii) the redesign of the query server that is now able to perform and merge parallel queries (which can be temporal queries, or metadata and embeddings combinations, for example), and (iii) user interface optimizations for quick inspection of the context of search results, quick navigation of results, and clustering of similar videos.

## 2  DiveXplore 2024

Figure 1 shows the that architecture of diveXplore consists of three major components: the backend, the middleware, and the frontend. Before the system can be used, all videos are analyzed by the backend, which stores image/text embeddings in a FAISS [5] index, and all other results in the metadata database (with

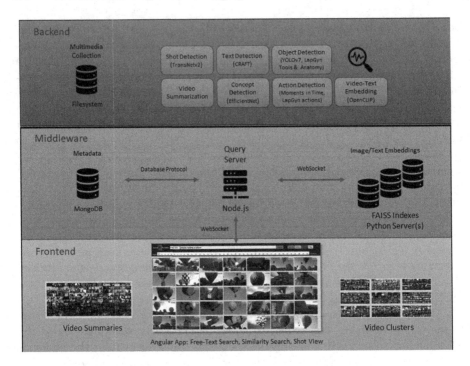

**Fig. 1.** diveXplore 2024 architecture

MongoDB[1]). The middleware and the frontend are used together and communicate via a WebSocket protocol; while the middleware is responsible for the actual search and retrieval, the frontend is used to present the results and to interact with the user.

## 2.1 Video Analysis

The first step of the analysis is shot segmentation and keyframe extraction, which is done with TransNetv2 [20] for V3C and a specific keyframe extraction algorithm for endoscopic videos [17] that is based on ORB keypoint tracking. As keyframes, the middle of the shot is used. For videos in the marine dataset, we simply use uniform temporal subsampling, with a 1-s interval. The keyframes are then used for further analysis. All keyframes are analyzed with OpenCLIP ViT-H/14 (laion2b) [4,19] and the obtained embeddings are added to a FAISS index [5], which is later used by the query server for free text and similarity search. The V3C keyframes are also analyzed for contained objects (COCO), concepts (Places365), text (CRAFT), and actions (Moments in Time), as shown in Table 2 (see also [18]). The keyframes from the surgery videos are analyzed for surgical tools, anatomical structures, and a few common surgical actions,

---

[1] https://www.mongodb.com.

such as cutting, cauterization, and suturing [11]. Furthermore, all keyframes of a video are used to create video summaries and similarity clusters of videos, as detailed in [16].

**Table 2.** Overview of analysis components used by diveXplore 2024 (V = V3C dataset, M = Marine dataset, S = Surgery dataset)

| V | M | S | Type | Model | Ref. | Description |
|---|---|---|------|-------|------|-------------|
| x | - | - | Shots | TransNetv2 | [20] | Deep model to detect shot boundaries |
| - | x | - | Segments | Uniform subsampling | | 1 s-segments |
| - | - | x | Segments | Endoscopy-Segments | [17] | Semantic segments with coherent content |
| x | - | - | Concepts | EfficientNet B2 | [21,24] | Places365 categories |
| x | - | - | Objects | YOLOv7 | [23] | 80 MS COCO categories |
| - | - | x | Med.Objects | Mask R-CNN | [6] | Tools and anatomy in gynecology |
| - | - | x | Actions | Bi-LSTM | [11] | Common actions in gynecology |
| x | - | - | Events | Moments in Time | [10] | 304 Moments in Time action events |
| x | - | x | Texts | CRAFT | [1,2] | Text region localization and OCR |
| x | x | x | Embeddings | OpenCLIP ViT-H/14 | [12] | Embeddings with 1024 dimensions |
| x | x | x | Similarity | OpenCLIP ViT-H/14 | [12] | Embeddings with 1024 dimensions |

## 2.2   Query Server

The query server is implemented with Node.js and communicates with the frontend via a websocket connection. Its main purpose is to receive queries from the frontend, analyze these queries, split and forward them, and collect and merge results. This design allows not only for scalability (e.g., to send consecutive queries to alternating index servers), but also to issue several queries in parallel. For example, the query from the frontend could contain free-text that should be forwarded to the FAISS index and an object detected by YOLO that is stored in the metadata database and should be forwarded to MongoDB. Then the query server would wait for results of both queries and merge or filter them, whereas different strategies could be selected for this, e.g., filter the results from FAISS with the video IDs returned by MongoDB, or fuse results from both sources with a particular ranking scheme. Also, the query from the frontend could contain a number of free-text queries that are sent to several FAISS indexes at the same time and the query server would merge the results in a temporal way, i.e., that only results from those videos are finally returned to the frontend that contain matches for all queries in a specific order and in temporal proximity.

## 2.3   User Interface

The main user interface of diveXplore 2024 is shown in Fig. 2. It consists of a query selection (Free-Text, Temporal, Objects, etc.), a search bar, and a paging-based result list below. Whenever the user moves the mouse over a result, buttons for additional features appear. For example, the user could inspect the

**Fig. 2.** diveXplore User Interface

corresponding video summary or the entire video with a video player, the shot list, and all meta-data (see Fig. 3a and b). The user could also perform a similarity search for the keyframe or send it directly to the VBS evaluation server (DRES) [14]. When the user moves the mouse horizontally over the top area of the result, different keyframes from the same video are presented as the mouse position moves; this should allow the user to very quickly inspect the video context of the keyframe. The result list is strongly optimized for keyboard-only use. The *space* key can used to open and close the video summary for a keyframe, the *arrow* keys are used to navigate between results and pages. The search bar supports an expert mode that can combine several different search modalities (e.g., **c**oncepts, **o**bjects, **e**vents, **t**exts, etc.) with simple prefixes (e.g., -c, -o, -e, -t etc.). This way it is easy to combine search for embeddings with metadata filters, but also to enter temporal search by simply using less < and greater > characters. In case the user rather wants to explore, he/she can also use the Xplore button, which opens another view of large video clusters with similar content (see Fig. 3c), or a consecutive list of video summaries for all videos.

(a) Video summary for quick context inspection.

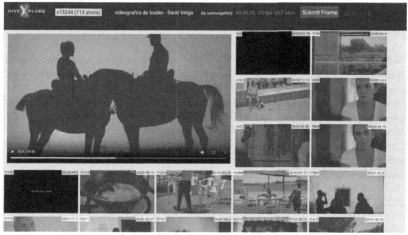

(b) Video-view with player and shot list.

(c) Exploration view of clusters with similar videos.

**Fig. 3.** Different views of diveXplore

## 3 Conclusion

We introduce diveXplore for the VBS2024 competition. The system has been significantly improved in many ways, in particular the improved index of Open-CLIP embeddings, extracted with the model trained on LAION2B, and the flexible and efficient search possibilities with the distributed query server make our system very competitive. Last but not least, we would like to mention that we integrate specific content analysis for the surgical video dataset, which is crucial to find content in these videos with highly redundant content.

**Acknowledgements.** This work was funded by the FWF Austrian Science Fund by grant P 32010-N38.

## References

1. Baek, J., et al.: What is wrong with scene text recognition model comparisons? Dataset and model analysis. In: Proceedings of the IEEE/CVF International Conference on Computer Vision, pp. 4715–4723 (2019)
2. Baek, Y., Lee, B., Han, D., Yun, S., Lee, H.: Character region awareness for text detection. In: Proceedings of the IEEE/CVF Conference on Computer Vision and Pattern Recognition, pp. 9365–9374 (2019)
3. Berns, F., Rossetto, L., Schoeffmann, K., Beecks, C., Awad, G.: V3C1 dataset: an evaluation of content characteristics. In: Proceedings of the 2019 on International Conference on Multimedia Retrieval, pp. 334–338. ACM (2019)
4. Cherti, M., et al.: Reproducible scaling laws for contrastive language-image learning (2022). https://doi.org/10.48550/ARXIV.2212.07143. https://arxiv.org/abs/2212.07143
5. Johnson, J., Douze, M., Jégou, H.: Billion-scale similarity search with GPUs. IEEE Trans. Big Data **7**(3), 535–547 (2019)
6. Kletz, S., Schoeffmann, K., Leibetseder, A., Benois-Pineau, J., Husslein, H.: Instrument recognition in laparoscopy for technical skill assessment. In: Ro, Y.M., et al. (eds.) MMM 2020, Part II. LNCS, vol. 11962, pp. 589–600. Springer, Cham (2020). https://doi.org/10.1007/978-3-030-37734-2_48
7. Leibetseder, A., Schoeffmann, K.: diveXplore 6.0: ITEC's interactive video exploration system at VBS 2022. In: Þór Jónsson, B., et al. (eds.) MMM 2022. LNCS, vol. 13142, pp. 569–574. Springer, Cham (2022). https://doi.org/10.1007/978-3-030-98355-0_56
8. Lokoč, J., et al.: Interactive video retrieval in the age of effective joint embedding deep models: lessons from the 11th vbs. Multimedia Systems, pp. 1–24 (2023)
9. Lokoč, J., et al.: Interactive search or sequential browsing? A detailed analysis of the video browser showdown 2018. ACM Trans. Multimedia Comput. Commun. Appl. **15**(1), 29:1–29:18 (2019). https://doi.org/10.1145/3295663. http://doi.acm.org/10.1145/3295663
10. Monfort, M., et al.: Moments in time dataset: one million videos for event understanding. IEEE Trans. Pattern Anal. Mach. Intell. **42**(2), 502–508 (2020). https://doi.org/10.1109/TPAMI.2019.2901464
11. Nasirihaghighi, S., Ghamsarian, N., Stefanics, D., Schoeffmann, K., Husslein, H.: Action recognition in video recordings from gynecologic laparoscopy. In: 2023 IEEE 36th International Symposium on Computer-Based Medical Systems (CBMS), pp. 29–34. IEEE (2023)

12. Radford, A., et al.: Learning transferable visual models from natural language supervision. In: International Conference on Machine Learning, pp. 8748–8763. PMLR (2021)
13. Rossetto, L., et al.: Interactive video retrieval in the age of deep learning - detailed evaluation of VBS 2019. IEEE Trans. Multimedia **23**, 243–256 (2021). https://doi.org/10.1109/TMM.2020.2980944
14. Rossetto, L., Gasser, R., Sauter, L., Bernstein, A., Schuldt, H.: A system for interactive multimedia retrieval evaluations. In: Lokoč, J., et al. (eds.) MMM 2021, Part II 27. LNCS, vol. 12573, pp. 385–390. Springer, Cham (2021). https://doi.org/10.1007/978-3-030-67835-7_33
15. Rossetto, L., Schoeffmann, K., Bernstein, A.: Insights on the V3C2 dataset. arXiv preprint arXiv:2105.01475 (2021)
16. Schoeffmann, K.: diveXB: an interactive video retrieval system for beginners. In: Proceedings of the 20th International Conference on Content-based Multimedia Indexing, CBMI 2023, pp. 1–6. IEEE (2023)
17. Schoeffmann, K., Del Fabro, M., Szkaliczki, T., Böszörmenyi, L., Keckstein, J.: Keyframe extraction in endoscopic video. Multimedia Tools Appl. **74**, 11187–11206 (2015)
18. Schoeffmann, K., Stefanics, D., Leibetseder, A.: diveXplore at the video browser showdown 2023. In: Dang-Nguyen, D.T., et al. (eds.) MultiMedia Modeling, MMM 2023. LNCS, vol. 13833. Springer, Cham (2023). https://doi.org/10.1007/978-3-031-27077-2_59
19. Schuhmann, C., et al.: LAION-400M: open dataset of clip-filtered 400 million image-text pairs. arXiv preprint arXiv:2111.02114 (2021)
20. Souček, T., Lokoč, J.: TransNet V2: an effective deep network architecture for fast shot transition detection. arXiv preprint arXiv:2008.04838 (2020)
21. Tan, M., Le, Q.: EfficientNet: rethinking model scaling for convolutional neural networks. In: International Conference on Machine Learning, pp. 6105–6114. PMLR (2019)
22. Truong, Q.T., et al.: Marine video kit: a new marine video dataset for content-based analysis and retrieval. In: Dang-Nguyen, D.T., et al. (eds.) MultiMedia Modeling, MMM 2023. LNCS, vol. 13833. Springer, Cham (2023). https://doi.org/10.1007/978-3-031-27077-2_42
23. Wang, C.Y., Bochkovskiy, A., Liao, H.Y.M.: YOLOv7: trainable bag-of-freebies sets new state-of-the-art for real-time object detectors. arXiv preprint arXiv:2207.02696 (2022)
24. Zhou, B., Lapedriza, A., Khosla, A., Oliva, A., Torralba, A.: Places: a 10 million image database for scene recognition. IEEE Trans. Pattern Anal. Mach. Intell. **40**(6), 1452–1464 (2018). https://doi.org/10.1109/TPAMI.2017.2723009

# Leveraging LLMs and Generative Models for Interactive Known-Item Video Search

Zhixin Ma[1]([✉]), Jiaxin Wu[1,2], and Chong Wah Ngo[1]

[1] School of Computing and Information Systems, Singapore Management University, Singapore, Singapore
zxma.2020@phdcs.smu.edu.sg, jiaxin.wu@my.cityu.edu.hk, cwngo@smu.edu.sg
[2] Department of Computer Science, City University of Hong Kong, Hong Kong, China

**Abstract.** While embedding techniques such as CLIP have considerably boosted search performance, user strategies in interactive video search still largely operate on a trial-and-error basis. Users are often required to manually adjust their queries and carefully inspect the search results, which greatly rely on the users' capability and proficiency. Recent advancements in large language models (LLMs) and generative models offer promising avenues for enhancing interactivity in video retrieval and reducing the personal bias in query interpretation, particularly in the known-item search. Specifically, LLMs can expand and diversify the semantics of the queries while avoiding grammar mistakes or the language barrier. In addition, generative models have the ability to imagine or visualize the verbose query as images. We integrate these new LLM capabilities into our existing system and evaluate their effectiveness on V3C1 and V3C2 datasets.

**Keywords:** Large Language Models · Generative Model · Known-Item Search · Interactive Video Retrieval

## 1 Introduction

Advancements in cross-modal embedding techniques have substantially enhanced the performance of video searches on large-scale datasets. In the annual competition of interactive video search, e.g., video browser showdown (VBS) [2,5], leveraging a pre-trained transformer-based joint embedding models (e.g., CLIP [9], CLIP4Clip [6] and BLIP [4]) has become a norm in recent years. For example, the top-performance team [12] in VBS2023 has used CLIP [9] features to measure the similarity of the input textual query and videos. Nevertheless, individual user proficiency still has a great impact on search performance. In the current VBS, the users have to inspect the return search results after initializing a search query. When the return results are not satisfying, users need to revise the query and browse the results repetitively. The effect of such trial-and-error search also depends on personal language proficiency and visual sensitivity, making search results user-dependent as also studied in [7].

© The Author(s), under exclusive license to Springer Nature Switzerland AG 2024
S. Rudinac et al. (Eds.): MMM 2024, LNCS 14557, pp. 380–386, 2024.
https://doi.org/10.1007/978-3-031-53302-0_35

In this paper, we propose using LLMs to assist users in query formulation. In recent years, LLMs such as GPT-4 [8] and LLaMA [13] have demonstrated surprising capability in natural language understanding and generation. The involvement of LLMs in interactive systems can help users improve the quality of query formulation, like correcting grammar mistakes, and diversify the query semantics by rephrasing the query in multiple ways. Table 1 shows the queries of TRECVid Ad-hoc Video Search (AVS) and VBS textual Known-Item Search (KIS) which are rephrased by GPT-4. The post-fix "-rx" indicates the x-th rephrased caption. The upper section of Table 1 compares the original and rephrased captions of query 731 from AVS 2023. The first rephrased caption keeps the keywords of the original query (i.e., a man and a baby) and expresses "is seen" in a different way. The second one details the man as a gentleman and the baby as an infant. In the example of the KIS query, the original description includes three shots. However, the current generative models cannot produce successive frames for multiple shots described in a video caption. Therefore, we designed a prompt to summarize the query and expect the LLM can find the key information to index the target query. As shown in the lower section of Table 1, the GPT-4 selects the last shot from the KIS query and rephrases it in two different ways, which helps the search engine focus on a specific scene. In the interactive system, a pop-up window is employed to display the GPT-rephrased queries and allow user to evaluate and pick a query as the search engine's input.

In addition to the query formulation, having a good understanding of the textual query in the KIS task and envisioning the target video in mind are challenging for users, especially those people who are not native English speakers. Inspired by the generative models [10] capability in understating natural language descriptions and generating high-quality images, we propose to use generative models to perceive the information need and do the imagination for users in the KIS task. Specifically, given a detailed caption describing the target video, we input the caption to a generative model (e.g., stable diffusion [10]) to generate images. The generated images are subsequently used in two ways. On one hand, the generated image can be directly employed as a visual query to optimize the rank of target shots by similarity search. On the other hand, the visualized search targets can inspire users to use them as references during the results inspection. In a search session, the user query will be fed into the generative model. The generated images will be displayed in the interface to help the user envision the target scene. Thereafter, the user can pick the generated images as a visual query to optimize the ranked list.

We conduct preliminary experiments on the V3C1 [1] and V3C2 [11] datasets to demonstrate the impact of engaging LLMs and generative models in the interactive search. Specifically, we use the visual query generated by the stable diffusion model [10] to improve the mAP performance on the TRECVid AVS task and examine the effectiveness of the LLM-rephrased query in indexing the search target on the textual KIS task. Furthermore, we enhance the ITV [14] feature in our system and achieve competitive performance compared to the state-of-the-art transformer-based methods.

**Table 1.** Comparison between the original and rephrased query by GPT-4. The postfix "-rx" indicates the x-th rephrased caption. The upper section shows the query comparison from the AVS task and the lower section presents a more detailed query from VBS textual KIS task.

| Query ID | Description |
| --- | --- |
| tv23-731 | A man is seen with a baby |
| tv23-731-r1 | A man and a baby are in the shot |
| tv23-731-r2 | A gentleman is spotted with an infant |
| vbs23-kis-t6 | A sequence of three shots: two people and a wall with posters, a balcony with laundry hanging on a rope and a train passing behind two standing tombstones. In the first shot, a person is standing around a corner of the wall on the right, the other person walks away to the left. In the last shot, there is a fence between the graveyard and the railway line, and the roof of a building is visible behind the railway line. |
| vbs23-kis-t6-r1 | A train moves past two tombstones, separated by a fence from the railroad, with a building's roof in the background. |
| vbs23-kis-t6-r2 | Two tombstones stand tall as a train goes by, a fence in the forefront and the silhouette of a building behind. |

## 2   Textual Query Enhancement

In this section, we elaborate on the engagement of the LLMs and generative models in our interactive video search system. Specifically, for a user query, we employ the LLM (i.e., GPT-4 [8]) to rephrase the textual query and generative model (i.e., stable diffusion [10]) to generate visual queries.

### 2.1   Textual Query Re-formulation with LLMs

We employ GPT-4 [8] to rephrase the query to improve the user query's quality and robustness. As shown in Table 1, the two rephrased queries, ending with "-r1" and "-r2", present the same semantics in different expressions. The prompt for query generation is "The following lines show a list of video captions. Please rephrase each video caption in 5 different ways. If there are multiple shots or scenes in a caption, please select the most distinguishable one and avoid describing multiple scenes in a caption". In our interactive system, the prompt is preset. Once the user inputs a textual query, the system will pass the query as well as the pre-defined prompt to GPT4 API and display the refined captions in a pop-up window. The users can evaluate the goodness of the candidates and select one caption for search.

VBS22-KIS-t3

#1          #2          #3

Target video          Generated images

**Fig. 1.** Examples of the generated images for two VBS Textual KIS queries (i.e., VBS22-KIS-t3), along with their target videos shown on the left.

## 2.2   Visual Query Generation with Generative Model

In the textual KIS task, given the textual description of the search target, users' imagination of the target has a great impact on the browsing. It is challenging to browse the search results and identify the target from the search rank list without a picture in mind. However, it is easy for users to select a few images imagined by the diffusion model that are close to their visual impression of the search target. The selected images are then used as a query for search. In this paper, we propose to use the generative model (e.g., stable diffusion [10]) to do the imagination for the user, i.e., generating images based on the given textual queries. Given a query, three images are generated with different seeds using the stable diffusion model. Figure 1 shows examples of the image generations for two VBS textual KIS queries, i.e., VBS22-KIS-t3[1], along with their search targets. As seen, the target video and the generated images are visually similar. In the system, we employ the generated images in two ways. Firstly, they will be displayed to the user as a visual guide to identify the search target. Moreover, the users could select a subset of the generated images as a visual query to find visually similar videos in our new interactive retrieval system.

## 3   Experiments

To evaluate the quality of the rephrased textual query and generated visual query in video search, we compare them with using original queries to search. The experiments are conducted on the V3C1 [1] and V3C2 [11] datasets. The text-video similarity is measured by three models, i.e., CLIP4Clip [6], BLIP2 [3], and enhanced ITV [14]. In the evaluation, we report mean average precision (mAP) on the TRECVid AVS tasks and the rank of the search target on VBS textual KIS tasks.

Table 2 presents the rank of the search target using different query types. When multiple shots are included in the target segment, we employ the best

---

[1] VBS23-KIS-t3: Almost static shot of a brown-white caravan and a horse on a meadow. The caravan is in the center, the horse in the back to its right, and there is a large tree on the right. The camera is slightly shaky, and there is a forested hill in the background.

**Table 2.** The rank of the search target using different models on VBS22 and VBS23 query sets. The post-fix "-t", "-r", "-u", and "-v3" denote the original query, the GPT-rephrased query, the user query, and the mean pooling of three visual queries, respectively. In the "-u" setting, a user is asked to find the target in the real KIS setting and the * indicates that the user found the search target.

| | VBS22 | | | | | | | | | | | |
|---|---|---|---|---|---|---|---|---|---|---|---|---|
| | t1 | t2 | t3 | t4 | t5 | t6 | t7 | t8 | t9 | t10 | t11 | t12 |
| CLIP4Clipt-t | 82 | 190 | 1 | 1 | 1966 | 12 | 4731 | 4 | **2** | 67747 | 261 | 6 |
| CLIP4Clipt-t-r | 2455 | 62 | 2 | 1 | 3876 | 1 | 4259 | 1 | 2 | 33742 | 582 | 150 |
| CLIP4Clipt-t-u | 61* | 1134 | 2* | 2* | **567** | 45 | 156 | 51 | 10* | **12612** | 25 | 71 |
| CLIP4Clip-v3 | **38** | 6 | 1 | 1 | 2240 | 515 | **25** | 11108 | 12 | 13638 | 8 | 17 |
| Enhanced ITV-t | 1 | 15 | 627 | 1 | 62 | 11 | 1 | 1 | 1359 | 1553 | 5 | **2** |
| Enhanced ITV-t-r | 1 | **9** | 462 | 1 | **33** | 1 | 51 | 8 | 2233 | **137** | 173 | 40 |
| Enhanced ITV-t-u | 1* | 934 | 524* | 1* | 502 | 97 | 3 | 24 | 315* | 218 | 2 | 24 |
| Enhanced ITV-v3 | 33 | 136 | 1 | 5 | 449 | 4 | 1329 | 764 | **3** | 1251 | 9 | 2 |
| BLIP-t | 2 | **30** | 1 | 1 | 26 | 229 | 1 | 3 | 13 | 955 | 24 | 3 |
| BLIP-t-r | 1 | 34 | 1 | 1 | **2** | 81 | 592 | 51 | 27 | 1067 | 26 | 27 |
| BLIP-t-u | 1* | 507 | 1* | 98* | 35 | 1513 | 12 | **2** | 14* | **168** | 1 | 1 |
| BLIP-v3 | 1 | 130 | 1 | 8 | 23 | **3** | **1** | 67 | 1 | 2380 | 39 | 1 |
| | VBS23 | | | | | | | | | | | |
| | t1 | t2 | t3 | t4 | t5 | t6 | t7 | t8 | t9 | t10 | t11 | t12 |
| CLIP4Clipt-t | 1 | 8770 | 26 | **186** | 2059 | 1927 | **145** | 6756 | **23** | 167 | **8** | 280 |
| CLIP4Clipt-t-r | 1 | 2272 | 484 | 282 | 4605 | 29 | 450 | 2574 | 127 | 169 | 10 | 1 |
| CLIP4Clipt-t-u | **1*** | 254 | 309 | 193 | **601*** | **19** | 1268 | **78** | 62* | **49*** | 58 | 25 |
| CLIP4Clip-v3 | 96 | 49 | 8 | 1135 | 3391 | 931 | 1381 | 7324 | 26 | 86 | 148 | 36 |
| Enhanced ITV-t | 1 | 1 | 170 | 394 | 225 | 88783 | 2554 | 558 | 3 | 66 | **4** | 1 |
| Enhanced ITV-t-r | 4 | 1 | **23** | 3420 | 4116 | 169 | 2608 | **323** | 3 | 15 | 41 | 2 |
| Enhanced ITV-t-u | 42* | 1 | 312 | 8938 | **44*** | **21** | 1881 | 323 | 8* | **4*** | 73 | 2 |
| Enhanced ITV-v3 | 6 | 14 | 1061 | **204** | 1208 | 19911 | 2047 | 19856 | 742 | 89 | 1996 | 5 |
| BLIP-t | 1 | 1 | 30 | **545** | 237 | 323 | 912 | 386 | 35 | 24 | 176 | 35 |
| BLIP-t-r | 2 | 1 | 53 | 5189 | 31253 | 111 | **245** | 454 | **25** | 93 | **46** | **11** |
| BLIP-t-u | **1*** | 13 | 87 | 1382 | **152*** | **16** | 2869 | **227** | 53* | **18*** | 502 | 51 |
| BLIP-v3 | 96 | 49 | 8 | 1135 | 3391 | 931 | 1381 | 7324 | 26 | 86 | 148 | 36 |

shot rank. Models ending in "-t", "-r", and "-u" denote the original, GPT-rephrased, and the user query, respectively. The suffix "-v3" refers to the mean pooling of three generated images. The user queries of "-u" are collected during the search session where we ask a novice user to locate the target in the real KIS setting using the CLIP4Clip feature. The * in the cells indicates the user finally finds the correct target. As depicted in Table 2, almost all the found queries are ranked within the top-100 results. In terms of the target rank, the user performs similarly with GPT in VBS 22 but achieves a higher rank on more tasks on VBS 23 tasks. In addition, the rephrased query enhances the search target's rank in 30% to 50% of VBS queries. Notably, the performance of queries VBS23-KIS-t6 and VBS23-KIS-t12 consistently witnessed improvement across

**Table 3.** The mAP performance comparison of different models on TRECVid AVS query sets from tv19 to tv23. The post-fix "-t" indicated the performance using the original textual query and the post-fix "-vx" the visual query using the mean pooling of "x" generated image(s). The term "Fused" refers to the fusion of "-t" and "-v3" with the weights in the parentheses.

|                  | tv19      | tv20      | tv21      | tv22      | tv23      |
|------------------|-----------|-----------|-----------|-----------|-----------|
| CLIP4Clip-t      | 0.144     | 0.153     | 0.178     | 0.157     | 0.138     |
| CLIP4Clip-v1     | 0.095     | 0.082     | 0.118     | 0.094     | 0.055     |
| CLIP4Clip-v3     | 0.106     | 0.108     | 0.142     | 0.104     | 0.077     |
| Fused (0.1, 0.9) | **0.156** | **0.167** | **0.191** | **0.170** | **0.145** |
| Enhanced ITV-t   | 0.163     | **0.359** | 0.355     | **0.282** | 0.292     |
| Enhanced ITV-v1  | 0.189     | 0.199     | 0.218     | 0.162     | 0.191     |
| Enhanced ITV-v3  | 0.219     | 0.220     | 0.240     | 0.181     | 0.212     |
| Fused (0.1, 0.9) | **0.257** | 0.347     | **0.371** | 0.243     | **0.298** |
| BLIP2-t          | 0.199     | 0.222     | **0.262** | 0.164     | 0.204     |
| BLIP2-v1         | 0.137     | 0.183     | 0.159     | 0.130     | 0.137     |
| BLIP2-v3         | 0.179     | 0.211     | 0.189     | 0.159     | 0.190     |
| Fused (0.1, 0.9) | **0.202** | **0.232** | 0.227     | **0.174** | **0.210** |

all baselines. The original captions of these two queries describe multiple scenes. For example, the query VBS23-KIS-t6 starts with "A sequence of three shots" and the three shots are almost independent. Although the query VBS23-KIS-t12 is not explicitly separated into multiple shots, the involved objects and events are scattered across diverse scenes. The query re-formulation of LLMs can help the model to concentrate on a specific scene. Oppositely, the performance on the query VBS23-KIS t4 and t5 consistently drop after revising the query. Compared to the original text, the rephrased t4 misses the key information "blue bow", "white/blue flower bouquet" and "red jacket", which are necessary to identify the target from the massive amount of wedding video shots. Although LLM can concisely summarize the query, they cannot properly identify the key information for the video search. Nevertheless, human intervention can potentially mitigate these shortcomings by manually selecting one from multiple generated queries.

Table 3 compares the mAP performance on the TRECVid AVS tasks. Similarly, the model with post-fix "-t" indicates the original textual query, while "-vx" denotes mean pooling utilizing "x" image(s) produced by stable diffusion. The term "Fused" refers to an ensemble of "-t" and "-v3" with the specified weights in the parentheses. Generally, the mAP performance with textual queries is superior to the visual ones. In addition, with the increase in visual query numbers, the mAP performance gradually gets better. A possible reason is that the mean pooling of multiple images can provide a more general and stable representation. Notably, the fusion of textual and visual queries delivers the optimal mAP performance on most of the query sets. Among the baselines, the Enhanced-ITV consistently outperforms others across all AVS query sets.

# 4  Conclusion

We have presented the major improvements to our interactive system. Specifically, our system is equipped with GPT-4 and stable diffusion, allowing users to rephrase and imagine a text query, and then select the suitable LLM-refined queries for retrieval. We also improved the embedding techniques of the VIREO search engine using the enhanced ITV feature which demonstrates its effectiveness on the TRECVid AVS tasks.

**Acknowledgments.** This research was supported by the Singapore Ministry of Education (MOE) Academic Research Fund (AcRF) Tier 1 grant.

# References

1. Berns, F., Rossetto, L., Schoeffmann, K., Beecks, C., Awad, G.: V3C1 dataset: an evaluation of content characteristics. In: Proceedings of the 2019 on International Conference on Multimedia Retrieval, ICMR 2019, pp. 334–338 (2019)
2. Heller, S., et al.: Interactive video retrieval evaluation at a distance: comparing sixteen interactive video search systems in a remote setting at the 10th video browser showdown. Int. J. Multimedia Inf. Retr. **11**, 1–18 (2022)
3. Li, J., Li, D., Savarese, S., Hoi, S.C.H.: BLIP-2: bootstrapping language-image pre-training with frozen image encoders and large language models. arXiv arXiv:abs/2301.12597 (2023)
4. Li, J., Li, D., Xiong, C., Hoi, S.C.H.: BLIP: bootstrapping language-image pre-training for unified vision-language understanding and generation. In: International Conference on Machine Learning (2022)
5. Loko, J., et al.: Is the reign of interactive search eternal? Findings from the video browser showdown 2020. ACM Trans. Multimedia Comput. Commun. Appl. (TOMM) **17**, 1–26 (2021)
6. Luo, H., et al.: CLIP4Clip: an empirical study of clip for end to end video clip retrieval. Neurocomputing **508**, 293–304 (2021)
7. Nguyen, P.A., Ngo, C.W.: Interactive search vs. automatic search. ACM Trans. Multimedia Comput. Commun. Appl. (TOMM) **17**, 1–24 (2021)
8. OpenAI: GPT-4 technical report. CoRR abs/2303.08774 (2023)
9. Radford, A., et al.: Learning transferable visual models from natural language supervision. In: International Conference on Machine Learning (2021)
10. Rombach, R., Blattmann, A., Lorenz, D., Esser, P., Ommer, B.: High-resolution image synthesis with latent diffusion models. In: 2022 IEEE/CVF Conference on Computer Vision and Pattern Recognition (CVPR), June 2022, pp. 10674–10685 (2022)
11. Rossetto, L., Schoeffmann, K., Bernstein, A.: Insights on the V3C2 dataset. arXiv preprint arXiv:2105.01475 (2021)
12. Schall, K., Hezel, N., Jung, K., Barthel, K.U.: Vibro: video browsing with semantic and visual image embeddings. In: Dang-Nguyen, D.T., et al. (eds.) MultiMedia Modeling, MMM 2023. LNCS, vol. 13833, pp. 665–670. Springer, Cham (2023). https://doi.org/10.1007/978-3-031-27077-2_56
13. Touvron, H., et al.: LLaMA: open and efficient foundation language models. arXiv arXiv:2302.13971 (2023)
14. Wu, J., Ngo, C.W., Chan, W.K., Hou, Z.: (un)likelihood training for interpretable embedding. ACM Trans. Inf. Syst. **42**, 1–26 (2023)

# TalkSee: Interactive Video Retrieval Engine Using Large Language Model

Guihe Gu[1,2,3], Zhengqian Wu[1,2,3], Jiangshan He[1,2,3],
Lin Song[1,2,3(✉)], Zhongyuan Wang[1,2,3], and Chao Liang[1,2,3(✉)]

[1] National Engineering Research Center for Multimedia Software (NERCMS),
Wuhan, China
[2] Hubei Key Laboratory of Multimedia and Network Communication Engineering,
Wuhan, China
[3] School of Computer Science, Wuhan University, Wuhan, China
{2022302191473,2023202110068,riverhill,lin_song,cliang}@whu.edu.cn

**Abstract.** The current interactive retrieval system mostly relies on collecting user's positive and negative feedback and updating the retrieval content based on this feedback. However, this method is not always sufficient to accurately express users' retrieval intent. Inspired by the powerful language understanding capability of the Large Language Model (LLM), we propose TalkSee, an interactive video retrieval engine using LLM for interaction in order to better capture users' latent retrieval intentions. We use the large language model for processing positive and negative feedback into natural language interactions. Specifically, combined with feedback, we leverage LLM to generate questions, update the queries, and conduct re-ranking. Last but not least, we design a tailored interactive user interface (UI) in conjunction with the above method for more efficient and effective video retrieval.

**Keywords:** Large Language Model · Question Generation · Query Update

## 1 Introduction

The Video Browser Showdown (VBS) [2,5] is an annual international video retrieval competition held by the International Conference on MultiMedia Modeling, which aims at evaluating the state-of-the-art video retrieval system and promoting the efficiency of large-scale video retrieval. In VBS 2024, participants are required to solve Known-Item Search (KIS), Ad-Hoc Video Search (AVS), and Video Question Answering (VQA) tasks. KIS and AVS aim to retrieve video clips with the same content or semantic description based on a follow-up video or text query, while VQA answers a given question based on video content. Although the formats of these three tasks are slightly different, they all rely on the understanding of the user's real query intent.

S. Rudinac et al. (Eds.): MMM 2024, LNCS 14557, pp. 387–393, 2024.
https://doi.org/10.1007/978-3-031-53302-0_36

Previous participants in VBS have made significant progress in the above tasks. IVIST [3] matches human-made queries (e.g., language) with various features (e.g., visual and audio) in the multimedia database effectively. Vibro [6] employs a ViT-L network pre-trained with CLIP and fine-tuned with large-scale image datasets. VISIONE [1] used an architecture called TERN (Transformer Encoder Reasoning Network) for text-to-visual retrieval. QIVISE proposed quantum-inspired interaction [7] for re-ranking. The above approaches ultimately converge on mapping text and visual features into a unified embedding space, measuring the similarity between different modalities and improving search efficiency. However, the above approach mostly conveys users' retrieval intent through positive and negative feedback. Given the presence of a semantic gap, this simple feedback form often struggles to accurately reflect the user's true retrieval intent when dealing with complex video content.

Given that the current strong emergence of LLM has led to its application to a wide range of downstream tasks [8], we propose to leverage LLM to better capture users' latent retrieval intent via active question generation and multi-modal query update. Specifically, we devise an LLM-based question generation module and an LLM-based query update module. The former produces selective questions based on user queries and natural language descriptions of user-labeled video shots. The latter updates the query and adjusts the current ranking based on feedback on the questions as well as positive and negative samples. Besides, in collaboration with LLM-based interaction, we specially design the interface for interactive video retrieval.

In summary, the features and innovations of our method are as follows: (1) We introduce LLM into the process of interactive video retrieval and utilize the powerful language understanding capability of LLM to capture more accurate user retrieval intent. (2) We craft some standard examples and prompts for LLM to guide LLM to better perform the tasks. After the feedback process, LLM will generate questions about the user's retrieval intent according to our prompt. Then LLM will summarize all information and produce a more accurate query according to our prompt. (3) We design the corresponding UI which can display the updated query in real-time, allowing the user to decide whether to use it for re-ranking. We dedicate a question-answering area for user interaction with LLM. In this area, users can see LLM-generated questions such as multiple-choice questions, and answer them as their intent.

## 2   Method

### 2.1   Overview

The block diagram of the proposed method in this paper is shown in Fig. 1: After a query, the user provides positive and negative feedback [2,9] the sorting result, which will be converted into textual descriptions and delivered to the LLM for analysis. Then LLM will generate several questions about the user's retrieval intent for the user to answer. Thereafter, LLM summarizes the answers and generates a more accurate text description for the retrieval engine to re-retrieve. The process is repeated until the user is satisfied.

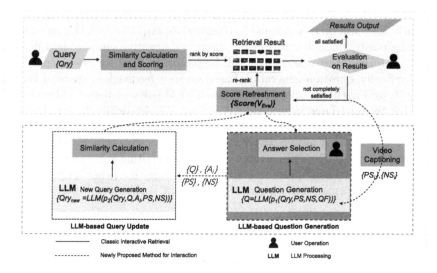

**Fig. 1.** Flowchart of Text-to-Video retrieval process.

## 2.2 General Process of Classic Interactive Retrieval

At the feedback stage, the user can label positive samples and negative samples. We call the set of positive sample feature vectors labeled by the user $\{V_{ps_i}\}_{i=0}^{p}$ and the set of negative sample feature vectors $\{V_{ns_j}\}_{j=0}^{n}$. Then the engine will re-rank the results according to the user's feedback. In this process generally, the cosine similarity formula is applied:

$$Score(V_{Eva}) = \lambda \cdot cosine(V_{Qry}, V_{Eva})$$
$$+ (1 - \lambda) \cdot \left[ \sum_{i=0}^{p} cosine(V_{ps_i}, V_{Eva}) - \sum_{j=0}^{n} cosine(V_{ns_j}, V_{Eva}) \right] \tag{1}$$

where $V_{eva}$ refers to the pre-extracted feature vector of the video clip to be evaluated in the next round, $\lambda$ is a constant less than 1 that we customize, $cosine(\cdot)$ refers to cosine similarity, $V_{Qry}$ refers to the feature vector of the query entered by the user. Besides, $p$ and $n$ represent the number of positive and negative samples, respectively.

However, We have already mentioned above that traditional interactive retrieval cannot precisely comprehend the intent of the query. We think that the solution should start from the source query and optimize the retrieval results by correcting the query.

## 2.3  LLM-Based Question Generation

To capture the precise retrieval intent of the user, we introduced LLM-based interaction. In our approach, we use mPLUG-2 [10] to pre-process the data and Blip-2 [4] as the backbone of the retrieval engine. After the retrieval engine gets feedback, they will be transformed into textual descriptions by video captioning, including positive samples' text descriptions and negative samples' text descriptions. Then these text descriptions will be passed to LLM for processing.

Specifically, question generation based on user feedback consists of the following steps: Firstly, we utilize a pre-trained VLM [10] to generate video descriptions. For the initial query, we denote it as $Qry$. We also define the set of positive samples text descriptions as $PS = \{PS_1, ..., PS_p\}$, and the set of negative samples text descriptions as $NS = \{NS_1, ..., NS_n\}$. As shown in Fig. 2, We guide LLM to generate targeted questions, e.g., for conceptual refinement, by giving LLM well-designed examples. Secondly, we pass question generating prompt $p_1$ to LLM in this format. The format of the prompt is shown in Fig. 2.

**Standard example**:

Q: The initial query is *"Find shots of someone repairing or cleaning a bicycle."*
The negative samples are *"Two persons are washing the bicycle."* and *"Two persons are riding the bicycle."*
Please provide questions about the central words of the above information to obtain the user's retrieval intent in the following format: *single-choice*.

A: Question 1: How many people do you expect to see?
        A: One      B: Two      C: More than two
    Question 2: What actions should be highlighted in your assignment?
        A: Bicycling  B: bicycle washing  C: Repairing bicycles
......

---

**Prompt**:

Here are standard examples:······
Q: *The initial query is {Qry}.*
*The positive sample text descriptions are {PS}.*
*The negative sample text description are {NS}.*
Please <u>consider the standard examples given</u> and ask questions <u>on the central words</u> of the above information to capture the user's retrieval intent in the following format: {QF}.

**Fig. 2.** Question generation prompt

The Standard example part will be inserted into the prompt and passed to the LLM along with the prompt. The contents of the curly braces indicate the information generated during user interaction, including query, textual descriptions of positive and negative samples, and so on. We denote it as $Q = LLM(p_1(Qry, PS, NS, QF))$, where QF refers to Question format, including single-choice questions, multiple-choice questions, and judgment questions.

## 2.4  LLM-Based Query Update

After the user answers the question, the selected answer $A_i$, question $Q$, the initial query $Qry$, the set of positive samples text descriptions $PS$, and the set of negative samples text descriptions $NS$ will be passed to LLM to generate a better query. Similarly, we craft targeted illustrative instances for LLM. Compared to Fig. 2, we add LLM-generated questions $Q$ and corresponding user answers $A_i$ to the prompt and update the standard example. Then we pass query update prompt $p_2$ to LLM in this format. We denote it as $Qry_{new} = LLM(p_2(Qry, Q, A_i, PS, NS))$, where $Qry_{new}$ refers to the updated query.

After getting the new query, we will recalculate the similarity score of the samples and re-rank the samples. The final similarity score is calculated using the following formula:

$$Score_{new}(V_{Eva}) = \kappa \cdot cosine(V_{Qry_{new}}, V_{Eva}) + (1 - \kappa) \cdot Score(V_{Eva}) \quad (2)$$

where $V_{Qry_{new}}$ refers to the feature vector of the new query generated by LLM, $Score(V_{Eva})$ refers to the fraction formula in the classical method mentioned above, and $\kappa$ is a constant less than 1 that we customize.

It is worth noting that positive and negative samples do not need to be available at the same time. In fact, as long as feedback exists our system can perform the tasks of question generation and query updating.

## 3 Introduction of Interface Design

Figure 3 shows the interface of our interactive retrieval system. With our designed interface, users can easily handle AVS, KIS, and VQA problems. The retrieval interface is divided into three parts. Our retrieval system is implemented using a C/S architecture. We upload the back-end to the server and the front-end

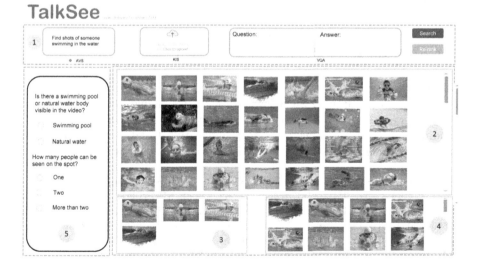

**Fig. 3.** Front-Interface Architecture of an Interactive retrieval System

page acts as a client that can be opened using a browser. In the backend, we will pre-implement the feature extraction of the video to get the feature vector of each keyframe. When the user inputs the query, it will be converted into a feature vector, which will be used to calculate the cosine score with the pre-processed video feature vector, and the result will be sorted and passed to the front-end interface and displayed. Then, after the user gives feedback on the results, the pre-designed prompt is passed to the LLM, which generates relevant questions for the user to answer. Similarly, after the end of the answer, the pre-designed prompt is passed to the LLM to guide it in generating a new query, and then the system performs a new round of score computation and re-sorting. Correspondingly, we divide the front-end interface into five parts.

The first part is our input session where the user begins their retrieval query. Depending on the task type, users can input their queries either in text form or by uploading a video. To facilitate this process, we add retrieval bars and an upload button. We design a specific field for VQA tasks to be entered, and the answer is displayed in the answer box. The updated query will be displayed in real-time, allowing the user to decide whether to use it for re-rank.

The second to the fourth part shows the retrieval results. For each query, the result of sorting all the video keyframes is displayed in the second part. Whenever the user clicks on an image, correspondingly, a few frames before and after this image are displayed in the third part, and all the keyframes of the video in which the image is located are displayed in the fourth part to give the user a quick overview of the video content. Additionally, users can double-click on any image to zoom in and give feedback on the search results, selecting positive and negative samples, skipping, adding to the submission list, and other actions.

In the fifth part where the users answer questions, users are presented with questions related to their retrieval intent and allowed to specify details. After users answer these questions, the engine iteratively refines and updates the retrieval results, ensuring they get the most relevant videos for their query.

## 4   Conclusion

This paper presents an interactive video retrieval engine based on LLM for participation in Video Browser Showdown 2024. In this engine, we will realize the model (TalkSee), which will implement video text retrieval tasks based on LLM, including video text matching, video text similarity calculation to sort retrieval results, VQA, interactive systems, iterative retrieval, and so on.

**Acknowledgement.** This work is supported by the National Natural Science Foundation of China (No. U1903214, 62372339, 62371350, 61876135), and the Ministry of Education Industry-University Cooperative Education Project (No. 202102246004, 220800006041043). The numerical calculations in this paper have been done on the supercomputing system in the Supercomputing Center of Wuhan University. We would like to express our sincere gratitude to Zhiyu Zhou for his previous contribution to this work.

# References

1. Amato, G., et al.: VISIONE at video browser showdown 2023. In: Dang-Nguyen, DT., et al. (eds.) MultiMedia Modeling, MMM 2023. LNCS, vol. 13833, pp. 615–621. Springer, Cham (2023). https://doi.org/10.1007/978-3-031-27077-2_48

2. Jónsson, B.Þ., Khan, O.S., Koelma, D.C., Rudinac, S., Worring, M., Zahálka, J.: Exquisitor at the video browser showdown 2020. In: Ro, Y.M., et al. (eds.) MMM 2020, Part II 26. LNCS, vol. 11962, pp. 796–802. Springer, Cham (2020). https://doi.org/10.1007/978-3-030-37734-2_72

3. Lee, Y., Choi, H., Park, S., Ro, Y.M.: IVIST: interactive video search tool in VBS 2021. In: Lokoč, J., et al. (eds.) MMM 2021, Part II 27. LNCS, vol. 12573, pp. 423–428. Springer, Cham (2021). https://doi.org/10.1007/978-3-030-67835-7_39

4. Li, J., Li, D., Savarese, S., Hoi, S.: BLIP-2: bootstrapping language-image pre-training with frozen image encoders and large language models. arXiv preprint arXiv:2301.12597 (2023)

5. Lokoč, J., et al.: Interactive video retrieval in the age of effective joint embedding deep models: lessons from the 11th VBS. Multimedia Syst. **29**(10), 1–24 (2023)

6. Schall, K., Hezel, N., Jung, K., Barthel, K.U.: Vibro: video browsing with semantic and visual image embeddings. In: Dang-Nguyen, D.T., et al. (eds.) MultiMedia Modeling, MMM 2023. LNCS, vol. 13833, pp. 665–670. Springer, Cham (2023). https://doi.org/10.1007/978-3-031-27077-2_56

7. Song, W., He, J., Li, X., Feng, S., Liang, C.: QIVISE: a quantum-inspired interactive video search engine in VBS2023. In: Dang-Nguyen, D.T., et al. (eds.) International Conference on Multimedia Modeling, vol. 13833, pp. 640–645. Springer, Heidelberg (2023). https://doi.org/10.1007/978-3-031-27077-2_52

8. Sun, W., Yan, L., Ma, X., Ren, P., Yin, D., Ren, Z.: Is ChatGPT good at search? Investigating large language models as re-ranking agent. arXiv preprint arXiv:2304.09542 (2023)

9. Thomee, B., Lew, M.S.: Interactive search in image retrieval: a survey. Int. J. Multimedia Inf. Retriev. **1**, 71–86 (2012)

10. Xu, H., et al.: mPLUG-2: a modularized multi-modal foundation model across text, image and video. arXiv preprint arXiv:2302.00402 (2023)

# VideoCLIP 2.0: An Interactive CLIP-Based Video Retrieval System for Novice Users at VBS2024

Thao-Nhu Nguyen[1], Le Minh Quang[1], Graham Healy[1], Binh T. Nguyen[2], and Cathal Gurrin[1]($\boxtimes$)

[1] Dublin City University, Dublin, Ireland
cathal.gurrin@dcu.ie
[2] Ho Chi Minh University of Science, Vietnam National University, Hanoi, Vietnam

**Abstract.** In this paper, we present an interactive video retrieval system named VIDEOCLIP 2.0 developed for the Video Browser Showdown in 2024. Building upon the foundation of the previous year's system, VIDEOCLIP, this upgraded version incorporates several enhancements to support novice users in solving retrieval tasks quickly and effectively. Firstly, the revised system enables search using a variety of modalities, such as rich text, dominant colour, OCR, query-by-image, and now relevance feedback. Additionally a new keyframe selection technique and a new embedding model to replace the existing CLIP model have been employed. This new model aims to obtain richer visual representations in order to improve search performance in the live interactive challenge. Lastly, the user interface has been refined to enable quicker inspection and user-friendly navigation, particularly beneficial for novice users. In this paper we describe the updates to VIDEOCLIP.

**Keywords:** Video Browser Showdown · Interactive Video Retrieval · Embedding Model · Multimodal Retrieval · Video Retrieval System

## 1 Introduction

In the multimedia content era, millions of videos are produced every day which transform how information is shared and consumed. As videos become an increasingly integral part of our digital lives and interactions, the need for efficient and precise video retrieval systems has become increasingly important, especially so for novice users. The annual Video Browser Showdown (VBS) [5] at the MMM conference aims to encourage comparative benchmarking of interactive video retrieval systems in a live, metrics-based evaluation. Similar to previous years' competitions, the participants in VBS are tasked with solving three categories of query: textual Known-Item Search (t-KIS), visual Known-Item Search (v-KIS), and Ad-hoc Video Search (AVS). While the first two tasks focus on identifying

---

T.-N. Nguyen and L.M. Quang—Contributed equally to this research.

the videos/moments matching the given textual or visual description as fast as possible, the latter's objective is to find as many target moments as possible. The data for this year's challenge consists of a combination of four datasets, including V3C1 (7,475 videos with a duration of 1,000 h) [2], V3C2 (9,760 videos with a duration of 1,300 h) [9], Marine Video Kit (MVK) (1,374 videos with a duration of 12.38 h) [13], and a small set of laparoscopic gynecology videos (LapGynLHE dataset).

Traditionally, the systems have usually relied on visual concepts extracted from the video itself, including objects, text, colors, and automatic annotations [1,6]. With the advent of large joint embedding models, these models have been applied in many research fields, particularly in computer vision applications, due to their ability to connect visual content and natural language. Video retrieval is among the applications that have been gaining increasing attention in most retrieval systems [1,6,10,12] in recent years. Recognising the potential of user-centric design to enhance the user experience when using video retrieval systems, recent research has explored user behavior, interactive feedback, and expectations when interacting successfully with state-of-the-art interactive video retrieval systems [1,10,12].

In our work, we wish to empower novice users to perform video search as effectively as expert users. This paper outlines a revised interactive video retrieval system named VIDEOCLIP 2.0, developed for the Video Browser Showdown 2024. We introduce some new functionalities aimed at better facilitating novice users in this year's system. These are the inclusion of a relevance feedback mechanism to support more-like-this search functionality, a revised and improved keyframe selection mechanism to improve the UI, a revised and updated embedding model to support more effective search, and a number of user interface enhancements to better support novice users. The VIDEOCLIP 2.0 system, introduced in this paper, represents a refinement of our previous system, VIDEOCLIP, which performed well at the VBS'23 competition [4] and was ranked overall top search engine at the related IVR4B challenge at the CBMI conference in 2023.

## 2  An Overview of VIDEOCLIP 2.0

Our system implements the same architecture as the 2023 system VIDEOCLIP. Full details can be found in [6]. In summary, VIDEOCLIP 2.0 builds upon VIDEOCLIP's underlying engine [7] with the following important modifications. Firstly, the inclusion of a relevant feedback mechanism to facilitate enhanced user engagement. Secondly, we use an updated version of the CLIP [8] model, named Open VCLIP [14], to enrich our system's capabilities in supporting multimodal search and retrieval. Thirdly, a new search modality, meta-search, is incorporated, empowering users to filter results with a list of comparative expressions. Lastly, recognising the importance of user experience, we redesign the user interface (UI) to ease the search process by modifying the result presentation, described in Sect. 2.2.

## 2.1   New Search Modalities

**Relevance Feedback Mechanism**

Relevance feedback is often used in interactive retrieval systems to facilitate users to provide feedback on the relevance of seen search results, which is then used to refine the search query and hopefully further improve the results. In VIDEOCLIP 2.0 we employ relevance feedback to facilitate the selection of any keyframe to amend to the user query. This process involves appending the metadata linked to the selected keyframe to the existing user query, ultimately generating and displaying a new ranked list. This iterative approach enables users to actively shape and fine-tune their search experience based on the perceived relevance of visual content.

**Improved Keyframe Selection**

Given the interactive use of the VIDEOCLIP 2.0 system, it becomes increasingly important for the system to select suitable keyframes for display on screen and to represent groups of sequential video shots. For this year's system, we employ the CLIP model to assist in the selection of query-relayed keyframes, rather than a pre-selected keyframe, during a user's search session. The similarity between the query and keyframe embeddings is computed for a series of query-relevant shots and the frames with the highest similarity scores are selected for display to the user, rather than all shots. In this way, we can optimise the use of screen real estate by not showing every shot keyframe from a highly ranked video or video segment, just ones likely to be relevant.

**Enhanced Embedding Models**

Due to its ability to link between visual concepts and natural language, the CLIP model [8] has been widely used in many research fields, including image classification, image similarity, and image captioning. The CLIP has proven itself to be the state-of-the-art embedding model and utilised by various teams during the challenge, including Vibro [3,11], which were the winners of the VBS challenge for two consecutive years, VBS2022 and VBS2023.

Nevertheless, the aforementioned CLIP model has some drawbacks when applied to video retrieval. For our previous work in [6], a technique called "frame-based retrieval" was applied. To be more specific, for each of the videos being considered, a frame is picked and treated as its representative. This approach has many disadvantages since it does not inherently model spatial-temporal relationships in videos, leading to reduced performance for some videos. Therefore, it is necessary to find a model that possesses the strengths of the CLIP model, but can also model spatial-temporal relationships better.

In our VIDEOCLIP 2.0 system, Open-VCLIP [14], a modified and improved version of the CLIP model introduced by Weng et al., is employed for this year's system. The latest version of this model was released by Wu et al. in October 2023 and has been shown to outperform the state-of-the-art approaches by a clear margin when applied to action recognition video datasets. In this model, a method named Interpolated Weight Optimization is introduced to leverage the weight interpolation during training and testing phrases. Thus, the Open-VCLIP

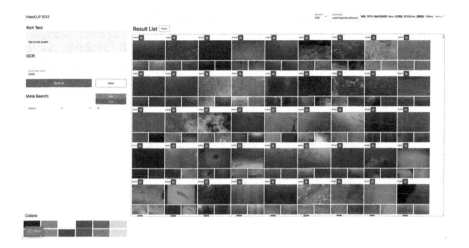

**Fig. 1.** The Prototype User Interface

is expected to be a suitable alternative option to the already used CLIP model in our newly proposed system.

## 2.2    User Interface Revisions

After participating in VBS2023, we recognised that the time taken and number of user interactions required for search, browsing and submission are critical to the system's overall performance. While our interface was good, there were opportunities to improve. To enhance the user experience, we have updated the UI with minor revisions intending to enhance both the speed and user-friendliness of inspecting ranked lists.

While maintaining the foundational layout and look & feel from the previous system (see Fig. 1), our focus is on simplifying the UI to provide users, particularly novices, with not only a better experience but also an improvement in search performance. Specifically, users will have a simplified search interface on the left side, while the results are displayed on the right side is per Fig. 1. By prioritizing and keeping only the most crucial buttons, we aim to enhance user clarity and expedite their interaction with the system.

## 3    Conclusion

In this paper, we present the latest version of our video interactive retrieval system, VIDEOCLIP 2.0, a CLIP-based system. We have updated the user interface, and the search result visualisation, along with upgrading the back end with the latest embedding models and other modifications to incrementally improve the system.

**Acknowledgments.** This research was conducted with the financial support of Science Foundation Ireland under Grant Agreement No. 18/CRT/6223, and 13/RC/2106_P2 at the ADAPT SFI Research Centre at DCU. ADAPT, the SFI Research Centre for AI-Driven Digital Content Technology is funded by Science Foundation Ireland through the SFI Research Centres Programme.

# References

1. Amato, G., et al.: VISIONE at video browser showdown 2023. In: Dang-Nguyen, D.T., et al. (eds.) MultiMedia Modeling, MMM 2023. LNCS, vol. 13833, pp. 615–621. Springer, Cham (2023). https://doi.org/10.1007/978-3-031-27077-2_48

2. Berns, F., Rossetto, L., Schoeffmann, K., Beecks, C., Awad, G.: V3C1 dataset: an evaluation of content characteristics. In: Proceedings of the 2019 on International Conference on Multimedia Retrieval, ICMR 2019, pp. 334–338, New York, NY, USA. Association for Computing Machinery (2019)

3. Hezel, N., Schall, K., Jung, K., Barthel, K.U.: Efficient search and browsing of large-scale video collections with vibro. In: Þór Jónsson, B., et al. (eds.) MMM 2022. LNCS, vol. 13142, pp. 487–492. Springer, Cham (2022). https://doi.org/10.1007/978-3-030-98355-0_43

4. Lokoč, J., et al.: Interactive video retrieval in the age of effective joint embedding deep models: lessons from the 11th VBS. Multimedia Syst. **29**(6), 3481–3504 (2023)

5. Lokoč, J., et al.: Is the reign of interactive search eternal? Findings from the video browser showdown 2020. ACM Trans. Multimedia Comput. Commun. Appl. **17**(3), 1–26 (2021)

6. Nguyen, T.N., et al.: VideoCLIP: an interactive CLIP-based video retrieval system at VBS2023. In: Dang-Nguyen, D.T., et al. (eds.) MultiMedia Modeling, MMM 2023. LNCS, vol. 13833, pp. 671–677. Springer, Cham (2023). https://doi.org/10.1007/978-3-031-27077-2_57

7. Nguyen, T.-N., Puangthamawathanakun, B., Healy, G., Nguyen, B.T., Gurrin, C., Caputo, A.: Videofall - a hierarchical search engine for VBS2022. In: Þór Jónsson, B., et al. (eds.) MMM 2022, Part II. LNCS, vol. 13142, pp. 518–523. Springer, Cham (2022). https://doi.org/10.1007/978-3-030-98355-0_48

8. Radford, A., et al.: Learning transferable visual models from natural language supervision. In: Meila, M., Zhang, T. (eds.) Proceedings of the 38th International Conference on Machine Learning, 18–24 July 2021, vol. 139 of Proceedings of Machine Learning Research, pp. 8748–8763. PMLR (2021)

9. Rossetto, L., Schoeffmann, K., Bernstein, A.: Insights on the V3C2 Dataset. CoRR, abs/2105.01475 (2021)

10. Sauter, L., et al.: Exploring effective interactive text-based video search in vitrivr. In: Dang-Nguyen, D.T., et al. (eds.) Proceedings of the 29th International Conference on MultiMedia Modeling, MMM 2023, Part I, Bergen, Norway, 9–12 January 2023, vol. 13833, pp. 646–651. Springer, Heidelberg (2023). https://doi.org/10.1007/978-3-031-27077-2_53

11. Schall, K., Hezel, N., Jung, K., Barthel, K.U.: Vibro: video browsing with semantic and visual image embeddings. In: Dang-Nguyen, D.T., et al. (eds.) Proceedings of the 29th International Conference on MultiMedia Modeling, MMM 2023, Bergen, Norway, 9–12 January 2023, Part I, vol. 13833, pp. 665–670. Springer Heidelberg (2023). https://doi.org/10.1007/978-3-031-27077-2_56

12. Schoeffmann, K., Stefanics, D., Leibetseder, A.: divexplore at the video browser showdown 2023. In: Dang-Nguyen, D.T., et al. (eds.) Proceedings of the 29th International Conference on MultiMedia Modeling, MMM 2023, Part I, Bergen, Norway, 9–12 January 2023, vol. 13833, pp. 684–689. Springer, Heidelberg (2023). https://doi.org/10.1007/978-3-031-27077-2_59
13. Truong, Q.-T., et al.: Marine video kit: a new marine video dataset for content-based analysis and retrieval. In: Dang-Nguyen, D.T., et al. (eds.) Proceedings of the 29th International Conference on MultiMedia Modeling, MMM 2023, Bergen, Norway, 9–12 January 2023, vol. 13833, pp. 539–550. Springer, Heidelberg (2023). https://doi.org/10.1007/978-3-031-27077-2_42
14. Weng, Z., Yang, X., Li, A., Wu, Z., Jiang, Y.-G.: Open-VCLIP: transforming clip to an open-vocabulary video model via interpolated weight optimization. In: Proceedings of the 40th International Conference on Machine Learning, ICML 2023. JMLR.org (2023)

# ViewsInsight: Enhancing Video Retrieval for VBS 2024 with a User-Friendly Interaction Mechanism

Gia-Huy Vuong[1,2], Van-Son Ho[1,2], Tien-Thanh Nguyen-Dang[1,2],
Xuan-Dang Thai[1,2], Tu-Khiem Le[3], Minh-Khoi Pham[3],
Van-Tu Ninh[3], Cathal Gurrin[3], and Minh-Triet Tran[1,2(✉)]

[1] University of Science, VNU-HCM, Ho Chi Minh City, Vietnam
tmtriet@hcmus.edu.vn
[2] Vietnam National University, Ho Chi Minh City, Vietnam
[3] Dublin City University, Dublin, Ireland

**Abstract.** ViewsInsight revolutionizes video content retrieval with its comprehensive suite of AI-powered features, enabling users to locate relevant videos using a variety of query types effortlessly. Its intelligent query description rewriting capability ensures precise video matching, while the visual example generation feature provides a powerful tool for refining search results. Additionally, the temporal query mechanism allows users to easily pinpoint specific video segments. The system's intuitive chat-based interface seamlessly integrates these advanced features.

**Keywords:** Video Retrieval System · Interactive Retrieval System · Spatial Insights · AI-based Assistance

## 1 Introduction

Known-item search (KIS) in large multimedia collections is a challenging task, and the Video Browser Showdown (VBS) [9] is a research benchmarking challenge that was proposed to encourage the research community to develop optimal solutions for interactive video retrieval in large-scale multimedia databases, in an open and shared environment. This paper introduces ViewsInsight an interactive, chat-based video content retrieval system tailored for the ViewsInsight uses the BLIP model [3] for zero-shot image-text retrieval, which has been shown to achieve higher recall scores than the CLIP model on the Flickr30K dataset [7]. In summary, this paper makes the following contributions:

1. Inspired by the user-interaction mechanism of ChatGPT of OpenAI and Bing AI from Microsoft, we propose ViewsInsight, which is a chat-based video retrieval system allowing the user to either augment the query or refine the ranked list via multiple rounds of chat-based context-based querying and filtering between the user and the system.

© The Author(s), under exclusive license to Springer Nature Switzerland AG 2024
S. Rudinac et al. (Eds.): MMM 2024, LNCS 14557, pp. 400–406, 2024.
https://doi.org/10.1007/978-3-031-53302-0_38

2. ViewsInsight supports a user-friendly temporal query mechanism that allows users to search for results based on a query, including temporal information (explained in detail in Sect. 3.4).

This paper is structured as follows: Sect. 2 reviews existing retrieval systems. Section 3 provides an overview of our system and discusses the extended features of the predecessor system. Finally, Sect. 4 summarizes our findings and discusses future work.

## 2    Related Work

Recent advances in video retrieval have led to the development of interactive retrieval systems that can efficiently find target images from a large pool of video content frames within a specified time limit. To address this challenging problem, researchers have employed various techniques, such as concept-based or semantic search.

Concept-based search is a methodology for improving the retrieval of images by annotating them with visual markers or concepts, such as detected objects, image captions, and text extracted from images using Optical Character Recognition (OCR). These annotations are then indexed along with other metadata and used to match the images with input keywords. Concept-based search can also leverage Natural Language Processing (NLP) to improve the retrieval experience by extracting relevant concepts and keywords from the text associated with the visual content, such as captions, tags, and annotations (system as V-FIRST 2.0 [2]).

Semantic search is a promising approach for improving the retrieval of images and videos. By incorporating the meaning of the images into the retrieval process, semantic search can lead to more precise and relevant results. Additionally, semantic search can turn traditional visual retrieval systems into semantic-driven systems, which can provide a deeper understanding of the context of the images. Therefore, numerous systems have embraced the use of vision-language pre-trained models, with a particular emphasis on the CLIP model [8], such as VideoCLIP [5], diveXplore [10], and Lokoč et al. [4]. These systems significantly improved their performance from the previous version in zero-shot image-text retrieval. These systems have exhibited a significant enhancement in their performance from their previous iterations in the realm of zero-shot image-text retrieval. Therefore, we incorporate the state-of-the-art vision-language pre-training model, BLIP [3], to construct semantic-driven video content retrieval systems. These systems use multiple AI-based features to support the retrieval process in conjunction with the semantic search function.

## 3    ViewsInsight – A Novel Video Content Retrieval System

ViewsInsight is developed based on our current lifelog retrieval system, which is LifeInsight [6], by directly extending the system to support video retrieval. The main differences between the two retrieval systems are discussed in Sects. 3.2 and 3.3.

### 3.1   An Overview of ViewsInsight

We propose a video content retrieval system called ViewsInsight, illustrated in Fig. 1. ViewsInsight accepts input queries through visual images, semantic texts, or filters (Optical Character Recognition, Audio Speech Recognition, concepts). The query semantic texts or visual images are encoded into vector embeddings by the embedding engine and then searched in the Milvus[1] – a vector database built for scalable similarity search, to retrieve the most relevant images. The preprocessed information from Optical Character Recognition (OCR) is then applied to filter the results through Elasticsearch[2]. Finally, the system caches the results to collect relevant feedback requests from users.

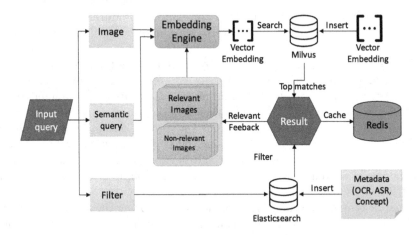

**Fig. 1.** Overview of ViewsInsight system.

Our video content retrieval system, ViewsInsight, features an interactive and user-friendly interface, as illustrated in Fig. 2. Setting it apart from traditional retrieval systems, ViewsInsight offers a diverse range of cluster view modes to elevate the user experience. When a user searches for "dogs", ViewsInsight promptly presents the results ranked by similarity scores, as shown in Fig. 2. Moreover, users can cluster images in the same video into a group using the video clustering mode (as depicted in Fig. 3). This mode allows them to see which video each image belongs to. Additionally, ViewsInsight incorporates a chatbox feature that not only stores query history but also accommodates both text and image inputs. Users can effortlessly execute advanced actions using commands. At the same time, a vertical navigation panel with handy buttons offers easy access to key features like toggling the chatbox visibility, providing feedback, and switching gallery view modes. width = 0.85.

---

[1] https://milvus.io/.

[2] https://www.elastic.co/guide/en/elasticsearch/reference/current/query-dsl-query-string-query.html.

**Fig. 2.** User-friendly interface of ViewsInsight system.

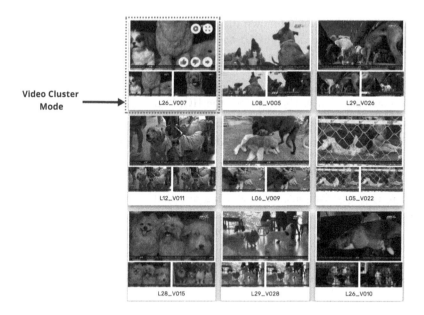

**Fig. 3.** Video Cluster Mode view of ViewsInsight system.

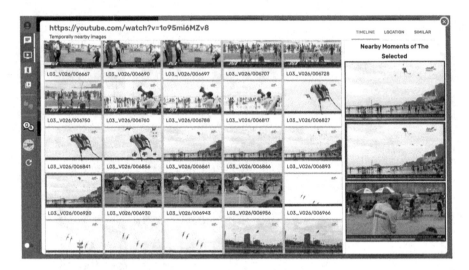

**Fig. 4.** Temporally nearby images - A collection of close frames captured in a duration.

## 3.2 Adapting Features for VBS2024

To participate in VBS2024, we revised the LifeInsight system [6] to develop ViewsInsight. While LifeInsight [6] was designed for the Lifelog Search Challenge [1], which focuses on image retrieval, ViewsInsight can also tackle video content retrieval problems. Our system allows users to group images by video ID and view related images nearby (shown in Fig. 4). Additionally, users can click on a single frame in the ViewsInsight UI to view the video, which will start playing from that frame. This feature is convenient for quickly validating the results. In other words, these features were designed specifically to tackle the task of video content retrieval. Their convenience and effectiveness have been demonstrated by securing a position in the Top-5 of the HCM AI Challenge [11].

## 3.3 ViewsInsight's Extended Features from LifeInsight

Video retrieval differs from image retrieval in that videos can contain sequential visual and audio data. This additional data dimension allows for more sophisticated retrieval techniques, such as using automatic speech recognition (ASR) to extract audio data into text for specific durations. This textual information is crucial for locating videos containing specific information like keywords or names. To facilitate this process, ViewsInsight incorporates a user-friendly temporal query mechanism that enables users to search for queries involving at least two pieces of information about past, present, or future events. This feature is detailed in Sect. 3.4.

## 3.4 Temporal Query Mechanism

The system would require three inputs from the user to use the temporal query mechanism. These include the main query and two temporal descriptions that provide details about events occurring before and after the event in the query. The algorithm then extracts relevant information from these queries within the system. Following this, it updates the scores of the "now" query, taking into account the highest score of the "before" and "after" queries associated with the same video ID. Finally, the algorithm generates a list of search results organized according to their newly updated scores.

The pseudo-code of this process is detailed in Algorithm 1. This powerful feature is seamlessly integrated into the system, enabling it to retrieve relevant results effectively. This makes the system more efficient and user-friendly.

---

**Algorithm 1.** Temporal Search Algorithm

---

**Require:** $a$, $b$, $c$
1: // The variables $a$, $b$, and $c$ represent a before, now, and after query, respectively.
2: $A \leftarrow$ Dictionary contains information retrieved from the system when $a$ is searched for.
3: $B \leftarrow$ Dictionary contains information retrieved from the system when $b$ is searched for.
4: $C \leftarrow$ Dictionary contains information retrieved from the system when $c$ is searched for.
5: **for all** $b_i$ in $B$ **do**
6:   $A_1 \leftarrow$ Subset of $A$ with video id from $b_i$
7:   $C_1 \leftarrow$ Subset of $C$ with video id from $b_i$
8:   **if** $A_1$ is not empty **then**
9:     $a_i \leftarrow$ Element in $A_1$ with the highest score
10:   **end if**
11:   **if** $C_1$ is not empty **then**
12:     $c_i \leftarrow$ Element in $C_1$ with the highest score
13:   **end if**
14:   Update score of $b_i$: $b_i.score \leftarrow b_i.score + (a_i.score \ if \ score \ in \ a_i \ else \ 0) + (c_i.score \ if \ score \ in \ c_i \ else \ 0)$
15: **end for**
16: Sort $B$ in descending order based on updated scores
17: **return** Sorted $B$ as relevant image results

---

## 4  Conclusion

In conclusion, ViewsInsight is a comprehensive video content retrieval system that leverages image-text retrieval, visual similarity search, and relevance feedback to enhance search accuracy. It further empowers users with AI-powered query rewriting and visual example generation. As a chat-based retrieval system, ViewsInsight facilitates iterative search refinement through context-based querying and filtering, enabling users to refine their search queries or the ranked list over multiple rounds. ViewsInsight also boasts a user-friendly temporal query mechanism that allows users to search for past, present, and future events effortlessly. Additionally, ViewsInsight accommodates search using transcriptions of audio data if the video contains English or Vietnamese speech generated through Automatic Speech Recognition (ASR).

**Acknowledgment.** This research was funded by Vingroup and supported by Vingroup Innovation Foundation (VINIF) under project code VINIF.2019.DA19.

# References

1. Gurrin, C., et al.: Introduction to the sixth annual lifelog search challenge, LSC23. In: Proceedings of the 2023 ACM International Conference on Multimedia Retrieval, ICMR 2023, pp. 678–679. Association for Computing Machinery, New York (2023). https://doi.org/10.1145/3591106.3592304
2. Hoang-Xuan, N., et al.: V-first 2.0: video event retrieval with flexible textual-visual intermediary for VBS 2023. In: Dang-Nguyen, D.T., et al. (eds.) MMM 2023, Part I. LNCS, vol. 13833, pp. 652–657. Springer, Cham (2023). https://doi.org/10.1007/978-3-031-27077-254
3. Li, J., Li, D., Xiong, C., Hoi, S.: BLIP: bootstrapping language-image pre-training for unified vision-language understanding and generation. In: International Conference on Machine Learning, pp. 12888–12900. PMLR (2022)
4. Lokoč, J., Vopálková, Z., Dokoupil, P., Peška, L.: Video search with clip and interactive text query reformulation. In: Dang-Nguyen, D.T., et al. (eds.) MMM 2023, Part I. LNCS, vol. 13833, pp. 628–633. Springer, Heidelberg (2023). https://doi.org/10.1007/978-3-031-27077-2_50
5. Nguyen, T.N., et al.: Videoclip: an interactive clip-based video retrieval system at VBS 2023. In: Dang-Nguyen, D.T., et al. (eds.) MMM 2023, Part I. LNCS, vol. 13833, pp. 671–677. Springer, Heidelberg (2023). https://doi.org/10.1007/978-3-031-27077-2_57
6. Nguyen-Dang, T.T., et al.: LifeInsight: an interactive lifelog retrieval system with comprehensive spatial insights and query assistance. In: Proceedings of the 6th Annual ACM Lifelog Search Challenge, LSC 2023, pp. 59–64. Association for Computing Machinery, New York (2023). https://doi.org/10.1145/3592573.3593106
7. Plummer, B.A., Wang, L., Cervantes, C.M., Caicedo, J.C., Hockenmaier, J., Lazebnik, S.: Flickr30k entities: collecting region-to-phrase correspondences for richer image-to-sentence models. In: Proceedings of the IEEE International Conference on Computer Vision, pp. 2641–2649 (2015)
8. Radford, A., et al.: Learning Transferable Visual Models From Natural Language Supervision (2021)
9. Schoeffmann, K., Lokoc, J., Bailer, W.: 10 years of video browser showdown. In: Chua, T., et al. (eds.) MMAsia 2020: ACM Multimedia Asia, Virtual Event, Singapore, 7–9 March 2021, pp. 73:1–73:3. ACM (2020). https://doi.org/10.1145/3444685.3450215
10. Schoeffmann, K., Stefanics, D., Leibetseder, A.: DiveXplore at the video browser showdown 2023. In: Dang-Nguyen, D.T., et al. (eds.) MMM 2023, Part I. LNCS, vol. 13833, pp. 684–689. Springer, Heidelberg (2023). https://doi.org/10.1007/978-3-031-27077-2_59
11. Trong-Le, D., et al.: News event retrieval from large video collection in Ho Chi Minh City AI challenge 2023. In: The 12th International Symposium on Information and Communication Technology (SOICT 2023), Ho Chi Minh, Vietnam, 7–8 December 2023 (2023). https://doi.org/10.1145/3628797.3628940

# Author Index

Printed in the United States
by Baker & Taylor Publisher Services